OXFORD MATHEMATICAL MONOGRAPHS

Series Editors

OXFORD MATHEMATICAL MONOGRAPHS

A. Belleni-Morante: *Applied semigroups and evolution equations*

I. G. Macdonald: *Symmetric functions and Hall polynomials*

J. W. P. Hirschfeld: *Projective geometries over finite fields*

N. Woodhouse: *Geometric quantization*

A. M. Arthurs: *Complementary variational principles* Second edition

P. L. Bhatnagar: *Nonlinear waves in one-dimensional dispersive systems*

N. Aronszajn, T. M. Creese, and L. J. Lipkin: *Polyharmonic functions*

J. A. Goldstein: *Semigroups of linear operators*

M. Rosenblum and J. Rovnyak: *Hardy classes and operator theory*

J. W. P. Hirschfeld: *Finite projective spaces of three dimensions*

K. Iwasawa: *Local class field theory*

A. Pressley and G. Segal: *Loop groups*

J. C. Lennox and S. E. Stonehewer: *Subnormal subgroups of groups*

Subnormal Subgroups of Groups

JOHN C. LENNOX
Reader in Pure Mathematics
University College, Cardiff

STEWART E. STONEHEWER
Reader in Mathematics
University of Warwick, Coventry

CLARENDON PRESS · OXFORD
1987

Oxford University Press, Walton Street, Oxford OX2 6DP
Oxford New York Toronto
Delhi Bombay Calcutta Madras Karachi
Petaling Jaya Singapore Hong Kong Tokyo
Nairobi Dar es Salaam Cape Town
Melbourne Auckland
and associated companies in
Beirut Berlin Ibadan Nicosia

Oxford is a trade mark of Oxford University Press

Published in the United States
by Oxford University Press, New York

British Library Cataloguing in Publication Data
Lennox, John C.
Subnormal subgroups of groups.—(Oxford mathematical monographs)
1. Groups, Theory of
I. Title II. Stonehewer, Stewart E.
512'.2 QA171

ISBN 0-19-853552-X

Library of Congress Cataloging-in-Publication Data
Lennox, John C. (John Carson)
Subnormal subgroups of groups.
(Oxford mathematical monographs)
Bibliography: p.
Includes indexes.
1. Groups, Theory of. 2. Group rings.
II. Title. III. Series.
QA171.L547 1986 512'.22¸ 86–8697

ISBN 0-19-853552-X

Set by Macmillan India Ltd, Bangalore 25
Printed in Great Britain by
St Edmundsbury Press Ltd
Bury St Edmunds, Suffolk

PREFACE

The concept of a subnormal subgroup of a group is derived from that of a normal subgroup due to the fact that normality is not a transitive relation. Thus subnormality is the transitive closure of the relation of normality. According to Karl Gruenberg, Philip Hall considered the subnormal subgroups to be the bare bones or 'skeleton' of a group, providing the framework for all the other structures.

In finite groups the significance of the subnormal subgroups is apparent since they are precisely those subgroups which occur as terms of composition series, the factors of which are of paramount importance in describing a group's structure. Thus it was with finite groups that Helmut Wielandt began his pioneering work on the subject in 1939 with a paper which has become a classic. Among many striking results, this paper contains the celebrated 'join' theorem, namely that *the subgroup generated by two subnormal subgroups of a finite group is itself subnormal.* Almost twenty years elapsed before H. Zassenhaus showed that this result can fail to hold in infinite groups. Meanwhile Wielandt continued his development of the finite theory and R. Baer, K. Hirsch and B. I. Plotkin made significant contributions more generally. Then about 1960 Philip Hall in his Cambridge lectures began to inspire certain of his students, notably D. J. S. Robinson and J. E. Roseblade, with an enthusiasm for the theory, while Baer in Frankfurt was similarly motivating H. Heineken and O. H. Kegel. Since that time the subject has proliferated, which amply justifies Hall's analogy.

It is our objective to present the theory of subnormal subgroups from its beginnings to the present day. Previously most of this material existed only in dispersed form throughout the journals and unpublished manuscripts. In this connection two points should be made. First, with the development of the theory it has been possible for us to improve and shorten many original proofs. Secondly, the nature of the subject is such that some contributions have completely superseded much previously published material. For example, on the join problem (that is discovering when the join of two subnormal subgroups is subnormal) we could have presented simply the work of John Williams (§§3.3, 3.4, and Chapter 5), omitting much of Chapter 1. We have chosen not to do this because many of the early arguments have considerable interest in their own right and are worthy of a permanent place in the record. Thus Wielandt's original proof of his join theorem (§1.3) and his theory of operators (§§4.1, 4.2) are exactly right when working within the class of finite groups, though both have very far-reaching generalizations.

The join problem has been the main research topic relating to subnormal subgroups and therefore receives fullest consideration in our book. It is the theme of Chapters 1, 3, and 5. Chapter 1 contains purely group-theoretic arguments, while Chapter 3 introduces techniques from the theory of modules over group rings and these are further exploited in Chapter 5. It is a rather long story and so Chapters 2 and 4 serve both as interludes and as a means of adding variety along the way. Nevertheless this arrangement is natural rather than artificial, each of Chapters 2 and 4 making use of what has immediately preceded. The former is mainly concerned with how the internal structure of groups generated by subnormal subgroups is restricted by those subgroups. Then in Chapter 4 we discuss the remarkable propensity which subnormal subgroups display for permuting with one another. Chapter 6 considers groups which are rich in subnormal subgroups and finally Chapter 7 describes criteria for determining when subgroups are subnormal.

For reasons of space we have been unable to include all the contributions to the theory, but following Derek Robinson's example in his books *Finiteness conditions and generalized soluble groups* [119] we have tried to compensate for omissions by giving at least statements and references wherever possible. We also decided not to give comprehensive accounts of generalizations of subnormal subgroups, for example ascendant, descendant, and serial subgroups. Apart from lengthening the book considerably, the theory of those generalizations is by no means so well developed. On the other hand they have not been ignored completely, for instance where they impinge directly on our main subject matter. A little of our material will be already familiar to some readers through certain available texts, in particular Robinson's books already mentioned. However, there are always reasons for these duplications, for example to make sections self-contained, to have important fundamental and far-reaching arguments readily available, to present familiar results in a way adapted to our needs, or to offer alternative proofs.

Our notation follows closely that of Robinson and includes Philip Hall's algebra of group classes and closure operations. This has the advantage of reducing the statement of some results to a mere half line and the disadvantage of being completely unintelligible to the uninitiated. Therefore we have made efforts not to become slaves of the symbols and for this reason readers may detect a few inconsistencies in our presentation. They are designed to help rather than hinder and in any case consistency is the hobgoblin (according to Emerson) and the vice (according to Russell) of little minds.

Some knowledge of group theory, ring theory, and linear algebra is assumed, but nothing beyond undergraduate courses. Indeed though the book has been designed as a postgraduate text, the prerequisites are such as should not deter the able and enthusiastic undergraduate student who has discovered the fascination of groups. For the postgraduate we offer much food for thought and ideas for further research. It will be seen that many last words have been written on some aspects of the theory, while elsewhere much remains to be done. Thus after more than four decades the join problem is now fairly well understood, whereas the theory of subnormalizers is possibly still in its infancy.

We are particularly grateful to many colleagues who have supplied information and commented on the manuscript during its preparation. Thus we record with pleasure our thanks to R. S. Dark, K. W. Gruenberg, B. Hartley, H. Heineken, O. H. Kegel, F. Napolitani, R. E. Phillips, D. J. S. Robinson, J. E. Roseblade, H. Wielandt, and J. S. Wilson. From this distinguished list we would like to single out Dr Roseblade who undertook the task of reading and criticizing in detail the greater part of a draft manuscript. We have endeavoured to incorporate all of his suggestions with the exception of adopting a module-theoretic approach from the beginning, and this for the reasons already given above. Accordingly we ask his forgiveness.

Finally we wish to pay tribute to the genius and inspiration of Professor Wielandt and the late Professor Hall, both of whom either directly or indirectly had the greatest influence in the development of our interest in this area. It is our sincere wish that their work will be continued by others and that our book will be of some assistance in this endeavour.

Cardiff J. C. L.
Warwick S. E. S.
October 1985

CONTENTS

INDEX OF NOTATION

1 Classes and operations

$\mathfrak{X} \leqq \mathfrak{Y}$	$\mathfrak{X}$ is a subclass of $\mathfrak{Y}$
$\mathfrak{X}\mathfrak{Y}$	$\mathfrak{X}$-by-$\mathfrak{Y}$ groups
$\mathfrak{X}_1 \mathfrak{X}_2 \cdots \mathfrak{X}_n$	inductively $(\mathfrak{X}_1 \cdots \mathfrak{X}_{n-1})\mathfrak{X}_n$
$\{\mathrm{A}_\lambda \mid \lambda \in \Lambda\}\mathfrak{X}$	smallest A_λ-closed class (all $\lambda \in \Lambda$) containing $\mathfrak{X}$
$\mathfrak{X}^{\mathrm{A}}$	largest A-closed class contained in $\mathfrak{X}$
(G)	groups isomorphic to G and groups of order 1

Special classes

$\mathfrak{A}$	abelian groups
$\mathfrak{B}\,(\mathfrak{B}_n)$	groups with every subnormal subgroup having bounded defect (defect $\leqq n$)
$\mathfrak{C}$	groups with a finite composition series
$\mathfrak{F}\,(\mathfrak{F}_p, \mathfrak{F}_\pi)$	finite (p-, π-) groups
f.r.p.d.	(155)
$\mathfrak{G}$	finitely generated groups
$\mathfrak{L}$	(127)
Max (-*n*, -*sn*)	maximal condition for (normal, subnormal) subgroups
$\mathfrak{Max}$ (-*n*, -*sn*)	groups satisfying Max (-*n*, -*sn*)
$\mathfrak{Max}$-$\lhd^2$	(33)
Min (-*n*, -*sn*)	minimal condition for (normal, subnormal) subgroups
$\mathfrak{Min}$ (-*n*, -*sn*)	groups satisfying Min (-*n*, -*sn*)
$\mathfrak{M}$	P($\mathfrak{Max}$-*sn* $\cup$ $\mathfrak{Min}$-*sn*)
$\mathfrak{N}\,(\mathfrak{N}_c)$	nilpotent groups (of class $\leqq c$)
$\mathfrak{S}$	soluble groups
$\mathfrak{S}, \mathfrak{S}_1, \mathfrak{S}_2, \mathfrak{S}^\infty, \mathfrak{S}_\infty$	(10, 13, 13, 71, 209)
$\mathfrak{T}$	groups in which normality is transitive
$\overline{\mathfrak{T}}$	$\mathfrak{T}^{\mathrm{S}}$
$\mathfrak{U}_d, \mathfrak{U}_{d,n}$	(172)

Special operations

L (39), L_n (52), M and $\overline{\mathrm{M}}$ (52), N (32), N_0 (21), N_1 (46), N_2 (23), P (13), Q (39), R (45), R_0 (130), S (13), S_n (10), U (116)

$\mathrm{A} \leqq \mathrm{B}$	$\mathrm{A}\mathfrak{X} \leqq \mathrm{B}\mathfrak{X}$ for all classes $\mathfrak{X}$

2 Elements and groups

h^g	$g^{-1}hg$
$[x, y]$	$x^{-1}y^{-1}xy$
$[x_1, \ldots, x_n]$	inductively $[[x_1, \ldots, x_{n-1}], x_n]$
$[x, {}_n y]$	$[x, \overset{\longleftarrow n \longrightarrow}{y, \ldots, y}]$
$\langle X \mid \ldots \rangle$	group or subgroup generated by X such that . . .
$H \leqq X$	H is a subgroup contained in X
$N \lhd G$	N is a normal subgroup of G
$H\ sn\ G$, $H\ asc\ G$	H is a subnormal, ascendant subgroup of G
$H\ per\ G$	H is a permutable (quasinormal) subgroup of G
$H \lhd^m G$	H is subnormal of defect at most m in G
$A\ psn\ G$	A preserves subnormality in G
$\mathrm{sn}\, G$, $\mathrm{psn}\, G$	(230)
H^g	$g^{-1}Hg$
X^Y	$\langle X^y \mid y \in Y \rangle$
$X \sigma Y$	X and Y are conjugate in $\langle X, Y \rangle$
$X \overset{G}{\cong} Y$	X and Y are G-isomorphic
$[X]$	$\langle X \rangle$
$[X, Y]$	$\langle [x, y] \mid x \in X, y \in Y \rangle$
$[X_1, \ldots, X_n]$	inductively $[[X_1, \ldots, X_{n-1}], X_n]$ for $n \geqq 3$
$[X, {}_n Y]$	$[X, \overset{\longleftarrow n \longrightarrow}{Y, \ldots, Y}]$
$\lvert S \rvert$	cardinality of S
$\lvert G : H \rvert$	index of subgroup H in group G
$C_G(H)$, $N_G(H)$	centralizer, normalizer of H in G
$P_H(K)$	permutizer of K in H
$H \upsilon K$	$HK = KH$
$\mathrm{core}_G(H)$, H_G	core of H in G
$\mathrm{Aut}\, G$, $\mathrm{Out}\, G$	automorphism, outer automorphism group of G
$\mathrm{Paut}\, G$	power automorphisms of G
$\mathrm{Hom}(X, Y)$	homomorphisms from X to Y
$H_1 \times \cdots \times H_n$	direct product
$H_1 \oplus \cdots \oplus H_n$,	direct sum
$\underset{\lambda}{Dr}\, H_\lambda$	direct product or sum
$H * K$	free product
$N] H$	semidirect product of normal subgroup N and H
$H \wr K$, $Wr\, G^\Lambda$	wreath product, power
$H \bar{\wr} K$	complete wreath product
$A \otimes B$	tensor product
$H \perp K$	$H/H' \otimes K/K' = 0$
$r(G)$, $r_\infty(G)$	rank, torsion-free rank of G
$h(G)$	Hirsch length of G

$\pi(G)$	primes dividing finite element orders in G
$\tau(G)$	primes p for which G has a quotient of order p
$\mathscr{C}(G)$, $\mathscr{C}(G:L)$	subnormal composition factors of G, between G and L
$\mathscr{L}$	subnormality lattice
A_n	alternating group of degree n
C_n	cyclic group of order n
C_{p^∞}	quasicyclic p-group
RG	group ring of G over R
$\mathfrak{g}$	augmentation ideal of RG
$\mathfrak{g}^c$	ideal generated by all products of c elements of $\mathfrak{g}$

3 Special subgroups

G'	derived subgroup $[G, G]$
$G^{(\lambda)}$	λth term of derived series of G
G^n	$\langle g^n \mid g \in G \rangle$
G_∞	(nilpotent) group G modulo its periodic subgroup
$G^{\mathfrak{X}}$	$\mathfrak{X}$-residual of G
$G_{\mathfrak{X}}$	join of all subnormal $\mathfrak{X}$-subgroups of G
$\beta_r(G)$	rth Baer radical of G
$\gamma_\lambda(G)$	λth term of lower central series of G
$\gamma_i^j(G)$	inductively $\gamma_i(\gamma_i^{j-1}(G))$
$\zeta_\lambda(G)$	λth term of upper central series of G
$\rho_{\mathfrak{X}}(G)$	$\mathfrak{X}$-radical of G
$\Phi(G)$	Frattini subgroup of G
$w(G)$	Wielandt subgroup of G
$w_n(G)$	nth term of upper Wielandt series of G

4 Miscellaneous

$<$	strict inclusion (with reference to groups and classes)
$\mathbb{Z}$, $\mathbb{Q}$	set of all integers, rationals
$\mathbb{Z}_p$	$\mathbb{Z}/p\mathbb{Z}$
$GF(p)$	field of p elements
$GL(n, p)$	general linear group of invertible $n \times n$ matrices over $GF(p)$
$GL(n, R)$	general linear group of invertible $n \times n$ matrices over ring R
p'	all primes different from p
π'	all primes not in π
ω	first limit ordinal

1

THE JOIN OF TWO SUBNORMAL SUBGROUPS

§1.1 Definitions and first properties

A subgroup N of a group G is said to be *normal* in G if N occurs as the kernel of a homomorphism of G. Then, following Wielandt, we write $N \lhd G$. Thus $N \lhd G$ if and only if $g^{-1}Ng = N$ for all $g \in G$.

A subgroup H of a group G is said to be *subnormal* in G if there are a non-negative integer m and a series

$$H = H_m \lhd H_{m-1} \lhd \cdots \lhd H_0 = G \tag{1}$$

of subgroups of G. In this situation we write

$$H \text{ sn } G \qquad \text{and} \qquad H \lhd^m G.$$

Of course, for fixed H and G, there will exist several series of type (1) of various lengths m; for example we have not required $H_i \neq H_{i-1}$. The smallest such m is called the *defect* of the subnormal subgroup H of G. Thus G has just one subnormal subgroup of defect 0, namely G itself, while the other normal subgroups of G are the subnormals of defect 1. The subgroups of order 2 in the alternating group of degree 4 and the non-central subgroups of order 2 in the dihedral group of order 8 are all examples of subnormal subgroups of defect 2. In fact the non-central subgroups of order 2 in the dihedral group of order 2^m are subnormal of defect $m-1$, for all $m \geqq 3$.

One series of type (1) is canonical. To describe it we proceed as follows. Let H be any subgroup of a group G. The *normal closure* of H in G, denoted by H^G, is the smallest normal subgroup of G containing H. So

$$H^G = \langle h^g \mid h \in H,\ g \in G \rangle.$$

(Here $h^g = g^{-1}hg$; and for any subsets X, Y of G we shall write

$$X^Y = \langle x^y \mid x \in X,\ y \in Y \rangle.)$$

If we put $H_0 = G$, then H^G becomes the first of an inductively defined descending series of subgroups, namely H_1, where for all finite $i \geqq 0$

$$H_{i+1} = H^{H_i}.$$

Therefore

$$H \leqq \cdots \lhd H_{i+1} \lhd H_i \lhd \cdots \lhd H_1 \lhd H_0 = G, \tag{2}$$

and this is the most rapidly descending series of subgroups, containing H, with each normal in the one above. For, if

$$H \leqq \cdots \lhd K_{i+1} \lhd K_i \lhd \cdots \lhd K_1 \lhd K_0 = G \tag{3}$$

is such a series and if $H_i \leqq K_i$ for some i, then

$$H_{i+1} = H^{H_i} \leqq H^{K_i} \leqq K_{i+1}.$$

Hence our claim follows by induction on i. We call (2) the *normal closure series* of H in G and we call H_i the *ith normal closure* of H in G.

For elements x, y in G, $[x, y] = x^{-1}y^{-1}xy$ is the *commutator* of x and y. Then if X, Y are subsets of G, $[X, Y] = \langle [x, y] \mid x \in X, y \in Y \rangle$. Since $[x, y] = [y, x]^{-1}$, we have $[X, Y] = [Y, X]$. For subsets $X_0, X_1, \ldots, X_n$ of G, we define $[X_0] = \langle X_0 \rangle$ and inductively

$$[X_0, X_1, \ldots, X_n] = [[X_0, X_1, \ldots, X_{n-1}], X_n], \qquad \text{for } n \geqq 2;$$

and when $X_1 = X_2 = \cdots = X_n = X$ (say), we also write this as $[X_{0,n} X]$.

Proposition 1.1.1. *Let H be a subgroup of G. Then*
(i) *the ith normal closure of H in G is $H[G, {}_iH]$;*
(ii) *$H \lhd^m G$ if and only if H coincides with its mth normal closure in G.*

Proof. (i) The statement is clear when $i = 0$ and 1. (We point out that H normalizes $[G, H]$, i.e. $[G, H]$ is normal in the subgroup which it generates with H. In fact X normalizes $[X, Y]$ provided only that X is a subgroup; see Robinson [119], Lemma 2.11(ii).) Thus if we denote the ith normal closure of H in G by H_i and suppose that

$$H_i = H[G, {}_iH]$$

for some $i \geqq 1$, then

$$H_{i+1} = H^{H_i} = H^{[G, {}_iH]} = H[G, {}_{i+1}H],$$

and the result follows by induction.

(ii) If $H \lhd^m G$, then there is a series of type (3) with $H = K_m$. Thus $H_m \leqq K_m$ implies $H = H_m$. The converse is clear. □

Remark. The argument of (i) also proves the following:
If G is generated by subgroups H and K, then, for $i \geqq 1$, the ith normal closure of H in G is $H[K, {}_iH]$.

Our first important investigation, which provides the theme of Chapter 1, is concerned with the subgroup generated by *two* subnormal subgroups. Before embarking on this major topic, it is convenient to record some elementary results.

Proposition 1.1.2
(i) *Let* $H \lhd^m G$ *and* K *be a subgroup of* G. *Then* $H \cap K \lhd^m K$. *In particular* $H \lhd^m L$ *whenever* L *is a subgroup of* G *containing* H.
(ii) *If* $H_\lambda \lhd^m G$, *for all* $\lambda \in \Lambda$ *(an indexing set), where* m *is independent of* λ, *then* $\bigcap_\lambda H_\lambda \lhd^m G$.

Proof. (i) There is a series $H = H_m \lhd H_{m-1} \lhd \cdots \lhd H_1 \lhd H_0 = G$. Thus

$$H \cap K = H_m \cap K \lhd H_{m-1} \cap K \lhd \cdots \lhd H_1 \cap K \lhd H_0 \cap K = K.$$

(ii) Again there are series $H_\lambda = H_{\lambda,m} \lhd \cdots \lhd H_{\lambda,1} \lhd H_{\lambda,0} = G$, for all $\lambda \in \Lambda$. Then the result follows since

$$\bigcap_\lambda H_{\lambda,i+1} \lhd \bigcap_\lambda H_{\lambda,i}. \quad \square$$

Some facts are so obvious that their statement is almost an embarrassment. The following is one such.

Proposition 1.1.3. *If* $H \lhd^m K \lhd^n G$, *then* $H \lhd^{m+n} G$.

Indeed subnormality is the transitivization of normality. We also record

Proposition 1.1.4. *If* $H \lhd^m G$ *and* θ *is a homomorphism of* G, *then* $H^\theta \lhd^m G^\theta$. *Thus if* $N \lhd G$, *then* $HN/N \lhd^m G/N$ *and* $HN \lhd^m G$. *In fact the normal closure series of any subgroup* K *of* G *is mapped by* θ *onto the normal closure series of* K^θ *in* G^θ.

§1.2 First results on joins

Wielandt [157] tells us that Robert Remak asked, in a seminar in the mid-1930s, whether, in a finite group, the subgroup generated by two subnormal subgroups (that is their *join*) is itself always subnormal. As we shall shortly see, this was answered in the affirmative by Wielandt in his 1939 paper [152], which provided much of the impetus for the subsequent development of the subject. Then Zassenhaus constructed an example in 1958 which shows that a join of two subnormal subgroups can fail to be subnormal in an infinite group. The determination of interesting necessary and sufficient conditions for a join to be subnormal is probably the most important unsolved problem in this area of group theory today. Many sufficient conditions are known and it is the object of the present chapter (and §§3.2, 3.4 and Chapter 5) to give a comprehensive account of them. The starting point is

Theorem 1.2.1. *Let* H, K *be subnormal subgroups of* G *and set* $J = \langle H, K \rangle$. *If* $H^K = H$, *then* J *is subnormal in* G. *More precisely, if* $H \lhd^m G$ *and* $K \lhd^n G$, *then* $J \lhd^{mn} G$. *(Note that* $H^K = H^J$.*)*

Proof. Denote the ith normal closure of H in G by H_i. It is clear, by induction on i, that K normalizes H_i for all i. Thus

$$H_{i+1} \lhd H_i K$$

and $K \lhd^n H_i K$ (by Proposition 1.1.2(i)). Hence $H_{i+1} K \lhd^n H_i K$, by Proposition 1.1.4. Then

$$J = HK = H_m K \lhd^{mn} H_0 K = G,$$

by Proposition 1.1.3. □

Corollary 1.2.2. *If $H \lhd^2 G$ and K is subnormal in G, then $J = \langle H, K \rangle$ is subnormal in G. In fact if $K \lhd^n G$, then $J \lhd^{2n} G$.*

Proof. By Proposition 1.1.1(ii), $H \lhd H^G$ and therefore $H^g = g^{-1} Hg \lhd H^G$, for all $g \in G$. It follows that $H^K \lhd H^G \lhd G$ and, since K normalizes H^K, Theorem 1.2.1 applies. □

Situations when the hypothesis $H^K = H$ is satisfied will be considered in Chapter 4 (see also Theorem 1.3.11). However, as we shall see in §1.5, $H \lhd^2 J$ is not sufficient to imply that J is subnormal in G.

It is now possible to establish two elementary conditions, each equivalent to the fact that a join of two subnormal subgroups is itself subnormal.

Theorem 1.2.3. *Suppose that H and K are subnormal subgroups of G and $J = \langle H, K \rangle$. Then each of the following conditions implies the other two.*

(i) *J is subnormal in G;*
(ii) *H^K is subnormal in G;*
(iii) *$[H, K]$ is subnormal in G.*

Proof. (i) ⇒ (ii): Let J be subnormal in G. Since H^K is normal in J, it follows, from Proposition 1.1.3, that H^K *sn* G.

(ii) ⇒ (iii): Suppose that H^K *sn* G. Since $[H, K] \lhd J$, we see that $[H, K] \lhd H^K$ and then Proposition 1.1.3 shows that $[H, K]$ *sn* G.

(iii) ⇒ (i): Finally let $[H, K]$ be subnormal in G. Since K normalizes $[H, K]$, Theorem 1.2.1 gives $[H, K]K$ *sn* G. However, H normalizes $[H, K]K$ and so a further application of Theorem 1.2.1 shows that

$$J = H[H, K]K \text{ sn } G. \quad \square$$

Corollary 1.2.4. *If H, K are subnormal subgroups of G and G' (the derived subgroup of G) is nilpotent, then $\langle H, K \rangle$ is subnormal in G.*

Proof. Since $[H, K] \leq G'$ and all subgroups of a nilpotent group (of class c) are subnormal (of defect $\leq c$), we have $[H, K]$ *sn* $G' \lhd G$. Thus the corollary follows from Theorem 1.2.3. □

The key result so far has been Theorem 1.2.1. It is also the essential ingredient in the proof of the next very useful property, which may be viewed as a natural extension of Theorem 1.2.1.

Theorem 1.2.5. *If $H \lhd^m G$, $K \lhd^n G$ and $J = HK = KH$, then $J \lhd^s G$, where $s \leqq mn(n+1)(n+2) \cdots (n+m-1)$.*

Proof. By Proposition 1.1.2(i), $H \lhd^m J$. Thus let $H \lhd^r J$, for some $r \leqq m$, and proceed by induction on r to show that $J \lhd^s G$, where

$$s \leqq mn(n+1)(n+2) \cdots (n+r-1).$$

When $r = 0$ we interpret this product as m and the result is clear. The case $r = 1$ is covered by Theorem 1.2.1. Therefore suppose that $r \geqq 2$ and assume the usual induction hypothesis. We have a series

$$H = H_r \lhd H_{r-1} \lhd \cdots \lhd H_0 = J.$$

Let $K_1 = H_1 \cap K$. Then $K_1 \lhd K$ and so K_1 *sn* G. Now $H_1 = H_1 \cap (HK) = HK_1$ with $H \lhd^{r-1} H_1$ and $K_1 \lhd^{n+1} G$. Hence by induction $H_1 \lhd^{s_1} G$, where $s_1 \leqq m(n+1)(n+2) \cdots (n+r-1)$. Since K normalizes H_1, a further application of Theorem 1.2.1 shows that $J \lhd^{s_1 n} G$ as required. □

Robinson [111] (Lemma 2.4) obtained the above result under the apparently weaker hypothesis $J = HKH$. However, as J. S. Wilson has pointed out, the two hypotheses are equivalent.

Proposition 1.2.6. *If H, K are subnormal in G and $J = \langle H, K \rangle = HKH$, then $J = HK$.*

Proof. Let $H \lhd^r J$ and proceed by induction on r. If $r \leqq 1$, the result is clear. Assume $r \geqq 2$ and the usual induction hypothesis. Then with $H^J = H_1$,

$$H_1 = H(H_1 \cap KH) = H(H_1 \cap K)H = H(H_1 \cap K),$$

by induction, since $H \lhd^{r-1} H_1$. Therefore

$$J = H_1 K = H(H_1 \cap K)K = HK. \quad \square$$

In Theorem 4.3.1 we shall see that if H and K are *orthogonal* (that is if $H/H' \otimes K/K' = 0$), then $HK = KH$ whenever H and K are subnormal subgroups of a group. (Also if H and K are *not* orthogonal, then there is a group G with subnormal subgroups $H_0 \cong H$ and $K_0 \cong K$ such that $H_0 K_0 \neq K_0 H_0$ (Theorem 4.3.8).) Thus in particular whenever H and K are subnormal subgroups of G and H is perfect (that is $H = H'$), then $\langle H, K \rangle = HK$ is subnormal in G. (See also Theorem 1.6.6.)

If J is the join of subnormal subgroups H and K of G and if $J = HKHK$, then we shall see in Theorem 1.5.7 that J can fail to be subnormal in G.

However, it appears to be unknown whether the apparently stronger hypothesis $J = HKH \cup KHK$ always implies that J is subnormal in G.

§1.3 Wielandt's Join Theorem

In 1939 Wielandt [152] (Satz 7) proved that the join of two subnormal subgroups of a finite group is always subnormal. This result, one of the most important in the theory, will be extended to many classes of infinite groups in later sections, using arguments of ever increasing complexity. Therefore it is important historically and mathematically to give Wielandt's original proof at the outset. In fact we shall give two further proofs, both of interest because of their simplicity.

Theorem 1.3.1. *Let H, K be subnormal subgroups of a finite group G. Then $\langle H, K \rangle$ is subnormal in G.*

Wielandt's proof used a double induction argument. However, in a footnote, he stated that his proof could be applied to cases where G is not necessarily finite, provided that G satisfies Max-*sn*, that is *the maximal condition for subnormal subgroups.* (This means that every non-empty set of subnormal subgroups of G contains at least one maximal member; or equivalently every strictly ascending chain of subnormal subgroups of G has finite length.) Thus we have

Theorem 1.3.2. *Let G satisfy* Max-*sn and let H, K be subnormal subgroups of G. Then $\langle H, K \rangle$ is subnormal in G.*

Proof. We argue by induction on the defect, m say, of H in G. When $H \lhd G$, the result is clear from Proposition 1.1.4. Therefore suppose that $m \geqq 2$ and assume the usual induction hypothesis. Now $H \lhd^{m-1} H^G$ and so, by a second induction on r, the join of any finite number r of conjugates of H in G is subnormal in G. Since G satisfies Max-*sn*, it follows that H^K is generated by finitely many conjugates of H by elements of K and thus $H^K \operatorname{sn} G$. So $\langle H, K\rangle \operatorname{sn} G$, by Theorem 1.2.3. □

The same method suffices to prove

Theorem 1.3.3. *If H, K are finite subnormal subgroups of G, then $\langle H, K\rangle$ is finite and subnormal in G.*

Theorem 1.3.1 is a corollary of each of the above results. Also Theorem 1.3.2 is a corollary of the next result which may be thought of as a second proof of Theorem 1.3.1.

Theorem 1.3.4. *Let H, $K \operatorname{sn} G$ and suppose that the set of subnormal*

subgroups of G lying between H and $J = \langle H, K \rangle$ *contains a maximal member. Then J sn G.*

It is convenient to isolate the key step in the argument:

Lemma 1.3.5. *Let S be a subgroup of G and suppose that H is a maximal member of the set of subnormal subgroups of G lying in S. Then* $H \lhd S$.

Proof. Denote the subnormal defect of H in S by m and suppose, for a contradiction, that $m \geqq 2$. If H_i is the ith normal closure of H in S, then there is an element x in H_{m-2} (and so H^x normalizes H) such that

$$H < HH^x = H^*,$$

say. Then $H^* \leq S$ and, by Theorem 1.2.1, H^* *sn* G, giving the desired contradiction. □

Proof of Theorem 1.3.4. Without loss of generality we may assume that H is a maximal member of the set of subnormal subgroups of G lying in $J = \langle H, K \rangle$. Thus $H^K = H$, by Lemma 1.3.5, and the result follows from Theorem 1.2.1. □

Corollary 1.3.6. *Let H, K sn G,* $J = \langle H, K \rangle$ *and suppose that* $|J:H|$ *is finite. Then J sn G.*

Readers familiar with the method of establishing isomorphic refinements of series (see Lemma 2.3.5) will have no difficulty deducing

Corollary 1.3.7. *The set of subnormal subgroups H of a group G, for which there exists a composition series of finite length between H and G, forms a lattice with respect to the operations of join and intersection.*

Variations on the theme of the proof of Theorem 1.3.4 are possible and we include two such. The first was given by Philip Hall in his Cambridge lectures.

Theorem 1.3.8. *Let H, K sn G,* $J = \langle H, K \rangle$ *and suppose that the set*

$$\{\langle H, H^{x_1}, H^{x_2}, \ldots, H^{x_s} \rangle | x_1, \ldots, x_s \in J, s \geqq 0\} \tag{1}$$

satisfies the maximal condition. Then J sn G.

Proof. We may assume that H is a maximal member of the set of subnormal subgroups of G which belong to (1). Then, as in the proof of Lemma 1.3.5, we see that $H \lhd J$ and Theorem 1.2.1 establishes the result. □

Suppose that $N \lhd G$ and that $X < Y$ are subgroups of G with $NX = NY$. Then $Y = (NX) \cap Y = (N \cap Y)X$ and therefore

$$N \cap X < N \cap Y. \tag{2}$$

This fact shows that the class of *noetherian* groups, that is groups satisfying Max (the maximal condition for subgroups), is closed under forming extensions. The same is true for Max-*sn*, also Min and Min-*sn*, the corresponding minimal conditions. We can use (2) in connection with Theorem 1.3.8 to give

Corollary 1.3.9 [111] (Corollary 4.2). *If H, K sn G and $[H, K]$ satisfies* Max, *then $\langle H, K\rangle$ sn G.*

Proof. Denote the ith normal closure of H in $J = \langle H, K\rangle$ by H_i. Since $H_1 = H[H, K]$ (by the remark after Proposition 1.1.1), it follows that

$$H_i = H(H_i \cap [H, K]) = H_{i+1}(H_i \cap [H, K]),$$

for all $i \geqq 1$, and hence H_i/H_{i+1} satisfies Max. Now (2) and induction on i decreasing show that the set of subgroups between H_1 and H satisfies Max. Thus the set (1) in Theorem 1.3.8 satisfies Max and so that theorem applies. □

Lemma 1.3.5 is also the key step in the following.

Theorem 1.3.10 [158] (10.5). *Let $\{H_\lambda | \lambda \in \Lambda\}$ be a set of subnormal subgroups of G and let J be their join. Then J is subnormal in G if and only if the set of subnormal subgroups of G lying in J contains a maximal member.*

Proof. Let M be a subnormal subgroup of G maximal with respect to lying in J. By Lemma 1.3.5, $M \lhd J$ and so $H_\lambda M$ *sn* G, by Theorem 1.2.1. Therefore $H_\lambda \leqq M$ for all $\lambda \in \Lambda$ and hence $J = M$ *sn* G. □

Of course Theorem 1.3.4 is a special case of this result.

Yet another proof of Theorem 1.3.1 was suggested by Kegel and uses

Theorem 1.3.11 [159]. *Let N be a minimal normal subgroup of the finite group G. Then N normalizes every subnormal subgroup of G.*

Proof. We proceed by induction on $|G|$. Let H *sn* G, $H \neq G$ and put $H_1 = H^G$. So $H_1 < G$. If $N \nleqq H_1$, then $N \cap H_1 = 1$ and thus $[N, H_1] = 1$. Hence $[N, H] = 1$. On the other hand, if $N \leqq H_1$, then $N = N_1^G$ for some minimal normal subgroup N_1 of H_1. Indeed each conjugate N_1^g, for $g \in G$, will be a minimal normal subgroup of H_1 and so will normalize H, by induction. Therefore N normalizes H. □

In Theorem 4.6.1 we shall extend this result to certain infinite groups. At this point we are interested only in its application to Theorem 1.3.1.

Third proof of Theorem 1.3.1. We recall that H, K are subnormal subgroups of the finite group G and we must prove that $J = \langle H, K \rangle \ sn\, G$. Proceed by induction on $|G|$ and let N be a minimal normal subgroup of G. By induction $JN/N \ sn\, G/N$ and so, by Theorem 1.3.11, $J \lhd JN \ sn\, G$. □

Robinson weakens the hypothesis of Theorem 1.3.2 in [111] (Theorem 4.3(i)):

Theorem 1.3.12. *Let $H, K \ sn\, G$ and suppose that G' satisfies* Max-*sn*. *Then $\langle H, K \rangle \ sn\, G$.*

Proof. Let m denote the defect of H in G and proceed by induction on m. Let $H_1 = H^G$ so that $H'_1 \lhd G'$ and hence H'_1 satisfies Max-*sn*. Therefore, by induction on m and a second induction on s,

$$\mathscr{X} = \{\langle H^{x_1}, H^{x_2}, \ldots, H^{x_s} \rangle | x_1, \ldots, x_s \in J, s \geqq 1\}$$

is a set of subnormal subgroups of G. We show that $H^K \in \mathscr{X}$ and then Theorem 1.2.3 applies. Thus suppose, for a contradiction, that $H^K \notin \mathscr{X}$. Then there is an infinite strictly ascending chain of subgroups

$$X_1 < X_2 < \cdots \tag{3}$$

with $X_i = \langle H, H^{x_1}, H^{x_2}, \ldots, H^{x_i} \rangle$, say, where $x_1, \ldots, x_i \in K$. Let $L_i = \langle [H, x_1], [H, x_2], \ldots, [H, x_i] \rangle$. (Here $[H, x_i] = [H, \{x_i\}]$.) Then

$$L_i \lhd \langle H, L_i \rangle = X_i \ sn\, G$$

and so $L_i \ sn\, G'$. It follows from the hypothesis that there exists $r \geqq 1$ such that $L_r = L_{r+1} = L_{r+2} = \cdots$ and then $X_r = X_{r+1} = X_{r+2} = \cdots$, contradicting (3). □

Another condition, sufficient for the subnormality of a join J of subnormal subgroups H and K, is that J satisfies Max-*sn*. However, this is superseded by

Corollary 1.3.13 [111]. *Let H, $K \ sn\, G$ and $J = \langle H, K \rangle$ and suppose that J' satisfies* Max-*sn*. *Then $J \ sn\, G$.*

Proof. By Theorem 1.3.12 and induction on s,

$$\{\langle H^{x_1}, H^{x_2}, \ldots, H^{x_s} \rangle | x_1, \ldots, x_s \in J, \ s \geqq 1\}$$

is a set of subnormal subgroups of J. Let $X_1 \leqq X_2 \leqq \cdots$ be an ascending series of subgroups, where $X_i = \langle H, H^{x_1}, \ldots, H^{x_{s_i}} \rangle$, $s_1 \leqq s_2 \leqq \cdots$ and all $x_i \in J$. Define $L_i = \langle [H, x_1], \ldots, [H, x_{s_i}] \rangle$. Then

$$\cdot L_i \lhd \langle H, L_i \rangle = X_i \ sn\, J$$

and so $L_i \ sn\, J'$. Hence, by hypothesis, there is an integer $r \geqq 1$ such that $L_r = L_{r+1} = \cdots$ and then $X_r = X_{r+1} = \cdots$. Thus Theorem 1.3.8 applies to show that J is subnormal in G. □

§1.4 The class $\mathfrak{S}$ of groups with the subnormal join property

After Wielandt's theorem one of the major steps forward in the join problem was made by Roseblade in 1964. He showed that if subnormal subgroups H, K satisfy Min-*sn* (the minimal condition for subnormal subgroups), then their join J is also subnormal [123]. The following year Roseblade obtained the same conclusion when H and K satisfy Max-*sn* [125]. However, these and other results established rather more than the subnormality of J by showing how the internal structure of H and K restricts that of J. This gave rise to the concept of subnormal coalescence which we consider in §1.6 and §1.7.

For the present we shall continue to investigate the class, which we denote following Robinson by $\mathfrak{S}$, of all those groups in which the join of two subnormal subgroups is always subnormal. Our results so far show that $\mathfrak{S}$ contains all groups G such that

(i) G' is nilpotent (Corollary 1.2.4) or
(ii) G' satisfies Max-*sn* (Theorem 1.3.12).

We shall see in §1.5 that $\mathfrak{S}$ is *not* the class of all groups. Since any group can be embedded in a simple group [111], it follows that the class $\mathfrak{S}$ is not closed with respect to forming subgroups. On the other hand there is the elementary

Proposition 1.4.1. *Subnormal subgroups of groups in $\mathfrak{S}$ also belong to $\mathfrak{S}$.*

Philip Hall defined a *closure operation* $\mathbf{A}$ to be a map which assigns to any class $\mathfrak{X}$ of groups a class $\mathbf{A}\mathfrak{X}$ such that (i) $\mathfrak{X} \leqq \mathbf{A}\mathfrak{X} = \mathbf{A}^2\mathfrak{X}$, (ii) $\mathbf{A}\mathfrak{X} \leqq \mathbf{A}\mathfrak{Y}$ whenever $\mathfrak{X} \leq \mathfrak{Y}$ and (iii) $\mathbf{A}(1) = (1)$. Here, for any group G, (G) denotes the class of all groups isomorphic to G and all groups of order 1. A class $\mathfrak{X}$ is said to be $\mathbf{A}$-*closed* if $\mathfrak{X} = \mathbf{A}\mathfrak{X}$. For any class $\mathfrak{X}$ of groups, $\mathbf{s}_n\mathfrak{X}$ denotes the class of all subnormal subgroups of $\mathfrak{X}$-groups (that is groups in $\mathfrak{X}$). Then $\mathbf{s}_n$ is a closure operation and Proposition 1.4.1 says that $\mathbf{s}_n\mathfrak{S} = \mathfrak{S}$. The *subnormal subgroup interior* of a class $\mathfrak{X}$, denoted by $\mathfrak{X}^{\mathbf{S}_n}$, is the class of those groups in which all subnormal subgroups belong to $\mathfrak{X}$. Thus writing $\mathfrak{G}$ for the class of finitely generated groups, $G \in \mathfrak{G}^{\mathbf{S}_n}$ if and only if every subnormal subgroup of G is finitely generated. It does not appear to be known whether $G \in \mathfrak{G}^{\mathbf{S}_n}$ always implies that G satisfies Max-*sn*, though the converse fails because of the existence of non-finitely generated simple groups.

If $\mathfrak{Y}$ is another class of groups, then $\mathfrak{X}\mathfrak{Y}$ is defined to be the class of all groups G which have a normal subgroup N in $\mathfrak{X}$ with $G/N \in \mathfrak{Y}$. Such a group G is said to be an $\mathfrak{X}$-by-$\mathfrak{Y}$ group. One of the major results concerning the class $\mathfrak{S}$ was proved by Robinson [111] (Theorem 5.2*):

Theorem 1.4.2. $\mathfrak{S}\mathfrak{G}^{\mathbf{S}_n} = \mathfrak{S}$.

Thus if a group G has a normal subgroup N in $\mathfrak{S}$ and every subnormal subgroup of G/N is finitely generated, then $G \in \mathfrak{S}$. For the proof (and some later applications) we need an important technical result from [129]. If X is a subset of a group and $j \geqq 1$, we write

$$X(j) = \{x_1 x_2 \cdots x_j \mid x_i \in X, 1 \leqq i \leqq j\}.$$

Lemma 1.4.3. *Let H, K be subgroups of G and suppose that K is generated by a subset X. Then, for any $n \geqq 0$,*

$$H^K = \langle H, H^X, H^{X(2)}, \ldots, H^{X(n)} \rangle [H,_{n+1} X]^K.$$

Proof. Since $\langle [H,X], K \rangle = K^H$, we see that

$$[H,X]^K = [H,X]^{\langle [H,X],K \rangle} = [H,X]^{K^H} = [H,X]^{HK^H} = [H,X]^{\langle H,K \rangle}$$

and so $[H,X]^K \lhd \langle H,K \rangle$. Thus

$$H^K = H[H,X]^K. \tag{1}$$

Define $H_0 = H$ and $H_{i+1} = [H_i, X]$, for $i \geqq 0$. So (1) becomes

$$H_0^K = H_0 H_1^K. \tag{2}$$

Then, by induction on n,

$$H^K = H_0 H_1 \cdots H_n H_{n+1}^K, \tag{3}$$

for all $n \geqq 1$. (Replace H_0 in (2) by H_n for the induction step.)

Clearly

$$H_0 H_1 = \langle H_0, H_0^X \rangle. \tag{4}$$

Suppose we have shown that

$$H_0 H_1 \cdots H_{n-1} = \langle H_0, H_0^X, \ldots, H_0^{X(n-1)} \rangle, \tag{5}$$

for some $n \geqq 2$. Since $[H_i H_{i+1}, X] = H_{i+1} H_{i+2}$, for all $i \geqq 0$ (using the elementary commutator identity $[h_1 h_2, x] = [h_1, x]^{h_2}[h_2, x]$), we may replace H_i in (5) by $H_i H_{i+1}$. Then from (4) we see that

$$H_0 H_1 \cdots H_n = \langle H_0, H_0^X, H_0^{X(2)}, \ldots, H_0^{X(n)} \rangle. \tag{6}$$

Thus (6) holds for all $n \geqq 0$, by induction. The lemma now follows from (3). □

Corollary 1.4.4. *Let H, K be subgroups of G with $N \lhd K$ and suppose that K/N can be generated by r elements, where r is finite. Then, for any $n \geqq 0$,*

$$H^K = L^N [H,_{n+1} K],$$

where L is generated by at most $1 + r + r^2 + \cdots + r^n$ conjugates of H by elements of K.

Proof. There is an r-element set X such that $K = \langle X, N \rangle$. Replace X in Lemma 1.4.3 by the set XN. Then there is a subgroup L generated by at most

$1+r+r^2+\cdots+r^n$ conjugates of H under K such that

$$H^K = L^N[H,_{n+1}XN]^K \leqq L^N[H,_{n+1}K] \leqq H^K,$$

giving the required result. □

We can now give a somewhat simpler proof of Theorem 1.4.2 than that in [111].

Proof of Theorem 1.4.2. Let $G \in \mathfrak{S}\mathfrak{G}^{S_n}$ so that there is an $N \lhd G$ with $N \in \mathfrak{S}$ and $G/N \in \mathfrak{G}^{S_n}$ and suppose that H, K are subnormal subgroups of G. We prove that $\langle H, K\rangle$ *sn* G by induction on the defect m of H in G. Suppose that K has defect n ($\geqq 1$) in G. If $A = K \cap N$, then

$$K/A \cong KN/N \text{ sn } G/N$$

and so K/A is finitely generated. Therefore, by Corollary 1.4.4,

$$H^K = L^A[H,_nK]$$

where L is generated by finitely many (say r) conjugates of H under K. It is sufficient, by Theorem 1.2.3, to show that H^K *sn* G. However, $[H,_nK] \lhd K$ *sn* G and therefore, by Theorem 1.2.5, it is sufficient to show that

$$L^A \text{ sn } G. \qquad (7)$$

Clearly $H^G \in \mathfrak{S}\mathfrak{G}^{S_n}$. Thus, by induction on m and a second induction on r, L is subnormal in H^G and hence in G. Let s ($\geqq 1$) denote the defect of L in G. Since A *sn* G, (7) will follow, according to Theorem 1.2.3, provided we show that A^L *sn* G. Now a second application of Corollary 1.4.4 gives

$$A^L = \langle A^{l_1}, \ldots, A^{l_t}\rangle^{L\cap N}[A,_sL]$$

for finitely many elements $l_1, \ldots, l_t$ in L. Therefore

$$A^L \lhd \langle A^{l_1}, \ldots, A^{l_t}, L\cap N, [A,_sL]\rangle \text{ sn } N \lhd G,$$

since the $t+2$ subgroups in the join are all subnormal in N and $N \in \mathfrak{S}$. Thus A^L *sn* G, as required. □

The product of group classes defined before Theorem 1.4.2 is not associative and so to avoid ambiguity we define

$$\mathfrak{X}_1\mathfrak{X}_2\cdots\mathfrak{X}_n = (\mathfrak{X}_1\cdots\mathfrak{X}_{n-1})\mathfrak{X}_n$$

inductively for $n \geqq 3$ and group classes $\mathfrak{X}_1, \mathfrak{X}_2, \ldots, \mathfrak{X}_n$. We also introduce the notation $\mathfrak{A}$ and $\mathfrak{N}$ for the classes of abelian and nilpotent groups, respectively, and $\mathfrak{Max}$ for the class of groups satisfying Max.

Corollary 1.4.5. *Any extension of an $\mathfrak{S}$-group by a group satisfying* Max *is an $\mathfrak{S}$-group, that is $\mathfrak{S}(\mathfrak{Max}) = \mathfrak{S}$. In particular $\mathfrak{N}\mathfrak{A}(\mathfrak{G}\cap\mathfrak{N}) \leqq \mathfrak{S}$.*

The first part of the corollary is clear since a group G satisfies Max if and only if every subgroup of G is finitely generated. The second part follows from Corollary 1.2.4 and the fact that finitely generated nilpotent groups satisfy Max [119] (p. 55). A special case yields

$$\mathfrak{G} \cap \mathfrak{A}^3 \leqq \mathfrak{S},$$

that is finitely generated soluble groups of derived length at most 3 satisfy the subnormal join property. This is of interest since we shall see in §1.5 that $\mathfrak{G} \cap \mathfrak{A}^4 \nleqq \mathfrak{S}$ and $\mathfrak{A}^3 \nleqq \mathfrak{S}$.

A further result concerning $\mathfrak{S}$ is a substantial improvement on Theorem 1.3.12, due to Robinson [113] (Theorem 6.2). First, if $\mathfrak{X}$ is a class of groups, then $\mathrm{P}\mathfrak{X}$ is the class of all groups G for which there is a series of finite length,

$$1 = G_0 \lhd G_1 \lhd \cdots \lhd G_n = G,$$

with $G_{i+1}/G_i \in \mathfrak{X}$, $0 \leqq i < n$. Thus P (for poly) is a closure operation. Denote by $\mathfrak{Max}$-*sn* (respectively $\mathfrak{Min}$-*sn*) the class of groups satisfying the maximal (respectively minimal) condition for subnormal subgroups. Using this notation, Robinson's result can be stated in the following concise form.

Theorem 1.4.6. $\{\mathrm{P}(\mathfrak{Max}\text{-}sn \cup \mathfrak{Min}\text{-}sn)\}\mathfrak{S} = \mathfrak{S}$.

It will be convenient to postpone the proof until §2.6.

We have considered the class of groups in which a join of two subnormal subgroups is always subnormal. Robinson [111] also considered the class of joins themselves. For any class $\mathfrak{X}$ of groups, $\mathrm{s}\mathfrak{X}$ is the class of all subgroups of $\mathfrak{X}$-groups. Then s is another closure operation. The *subgroup interior* of a class $\mathfrak{X}$, denoted by $\mathfrak{X}^{\mathrm{S}}$, is the class of all groups in which all subgroups belong to $\mathfrak{X}$. Now let $\mathfrak{S}_1$ denote the subgroup interior of the class of all groups J such that whenever $J = \langle H, K \rangle$ is a subgroup of G with H, K *sn* G, then J *sn* G. Clearly $\mathfrak{S}_1 \leqq \mathfrak{S}$ and $\mathfrak{S}_1 \neq \mathfrak{S}$, since, as we have already remarked (before Proposition 1.4.1), $\mathfrak{S}$ is not s-closed (that is $\mathrm{s}\mathfrak{S} \neq \mathfrak{S}$). By Theorem 1.2.1, the class $\mathfrak{A}$ of abelian groups is contained in $\mathfrak{S}_1$. However, Corollary 1.3.13 implies more than this, namely

$$(\mathfrak{Max})\,\mathfrak{A} \leqq \mathfrak{S}_1.$$

By analogy with Corollary 1.4.5, Robinson also obtained

$$\mathfrak{S}_1\,(\mathfrak{Max}) = \mathfrak{S}_1$$

[111] (Theorem 5.2).

A third class is the $\mathrm{s_n}$-interior $\mathfrak{S}_2$ of the class of all $\mathfrak{S}$-groups J such that whenever $J = \langle H, K \rangle$ is a subgroup of G with H, K *sn* G, then J *sn* G. Thus

$$\mathfrak{S}_1 = \mathfrak{S}_2^{\mathrm{S}}.$$

We shall see in Theorem 1.5.1 that there are nilpotent groups of class 2 which are not in $\mathfrak{S}_2$. Also simple groups lie in $\mathfrak{S}_2$ and any group embeds in a simple group. Therefore $\mathfrak{S}_2$ is *not* s-closed and

$$\mathfrak{S}_1 < \mathfrak{S}_2 < \mathfrak{S}$$

(< denotes strict inclusion).

By Theorem 1.3.12 and its corollary,

$$(\mathfrak{Max}\text{-}sn)\,\mathfrak{A} \leqq \mathfrak{S}_2.$$

Robinson's main result concerning $\mathfrak{S}_2$ corresponds to Theorem 1.4.2:

$$\mathfrak{S}_2\mathfrak{G}^{S_n} = \mathfrak{S}_2$$

[111] (Theorem 5.2*).
There is also the following:

$$\mathfrak{N}\mathfrak{S}_2 \leqq \mathfrak{S}$$

[111] (Theorem 5.3(ii)).

§1.5 Counterexamples

We turn now to examples of a join of subnormal subgroups which is not itself subnormal. The first of these is due to Zassenhaus and appeared in 1958 as exercise 23 on page 235 of [171]. This led to some variations by Philip Hall which he included in his Cambridge lectures during the early 1960s. One of these is described by Robinson in [111]. Finally, a more general construction, offering infinitely many examples, was given by Roseblade and Stonehewer in [129].

We begin with the Zassenhaus result, which is certainly worth stating as a theorem. We shall indicate the construction, but leave the routine verifications, as was originally intended, as an exercise for the reader. For any integer $c \geqq 0$, $\mathfrak{N}_c$ denotes the nilpotent groups of class $\leqq c$.

Theorem 1.5.1. *There is a countable group G with subnormal subgroups H and K such that $\langle H, K \rangle$ is not subnormal in G. More precisely*

(i) $H, K \lhd^3 G$,
(ii) *H and K are free abelian groups of infinite rank,*
(iii) $G \in \mathfrak{A}\mathfrak{N}_2$,
(iv) $J = \langle H, K \rangle \in \mathfrak{N}_2$.

By Corollary 1.2.2, the subnormal defects of H and K in G must be precisely 3. Also, by Corollary 1.2.4, G cannot be metabelian and, by Theorem 1.4.2, G cannot be finitely generated. By Theorem 1.2.1, J cannot be abelian. In Theorem 4.3.1, we shall see that if H and K are groups such that the tensor product $H/H' \otimes K/K'$ is trivial, then $HK = KH$ whenever H and K are subnormally embedded in any group G. (The definition of tensor product can be found for example in [34].) So $J = HK$ will be subnormal in G. In fact in Chapter 5 we shall see that if $H/H' \otimes K/K'$, modulo its maximal periodic divisible subgroup, has finite rank (see p. 39), then $J = \langle H, K \rangle$ will always be subnormal in G. Therefore in many ways Theorem 1.5.1 gives the best possible counterexample in relation to the join problem. The construction is as follows.

For each integer $r \geqq 0$, let A_r denote the set of all integer sequences

$$a = (a_0, a_1, \ldots, a_r)$$

with $0 \leqq a_0 < a_1 < \cdots < a_r$. Let V be a free $\mathbb{Z}$-module with basis

$$\{u(a, b) | (a, b) \in A_r \times A_r, r \geqq 0\}.$$

Suppose that $b = (b_0, b_1, \ldots, b_r)$. If π, π' are any permutations of $a_0, a_1, \ldots, a_r$ and $b_0, b_1, \ldots, b_r$, respectively, such that $a_0\pi = a_0$ if $a_0 = 0$ and $b_0\pi' = b_0$ if $b_0 = 0$, then define

$$u(a\pi, b\pi') = (\text{sign}\, \pi)\, (\text{sign}\, \pi')u(a, b).$$

Now let $r \geqq 1$ and let a, b be *any* sequences of non-negative integers of length $r+1$ for which $u(a, b)$ has not already been defined. Then set $u(a, b) = 0$. Thus with $a = (a_0, a_1, \ldots, a_r)$ and $b = (b_0, b_1, \ldots, b_r)$, $u(a, b) = 0$ in each of the following cases:

(i) $a_i = 0$ for some $i \neq 0$,
(ii) $b_i = 0$ for some $i \neq 0$,
(iii) $a_i = a_j$ for some $i \neq j$,
(iv) $b_i = b_j$ for some $i \neq j$.

For all $i \geqq 1$, define endomorphisms ξ_i and η_i of V by

$$\xi_i\colon u(a, b) \mapsto \delta_{0,a_0} u((i, a_1, a_2, \ldots, a_r), b),$$

$$\eta_i\colon u(a, b) \mapsto \delta_{0,b_0} u(a, (i, b_1, b_2, \ldots, b_r)) - u((0, a_1, \ldots, a_r, a_0), (b_0, b_1, \ldots, b_r, i)),$$

where $\delta_{j,k}$ is the Kronecker symbol. The Zassenhaus exercise proceeds as follows. (We write the maps ξ_i, η_i on the right.)

(1) $\xi_i\xi_k = \eta_i\eta_k = 0$ for all $i, k \geqq 1$.

Let $\gamma_{ik} = \xi_i\eta_k - \eta_k\xi_i$. Then

(2) $\xi_j\gamma_{ik} = \gamma_{ik}\xi_j$ and $\eta_j\gamma_{ik} = \gamma_{ik}\eta_j$ for all $i, j, k \geqq 1$.

(3) $\gamma_{11}\gamma_{22} \cdots \gamma_{rr} \neq 0$ for any $r \geqq 1$.

The maps $x_i = 1 + \xi_i$, $y_i = 1 + \eta_i$ are automorphisms of V (since, by (1), $1 - \xi_i$, $1 - \eta_i$, respectively, are their inverses). Let

$$J = \langle x_i, y_i \,|\, i \geqq 1 \rangle$$

and form the natural semidirect product $G = V\,]\,J$.

(4) J is not subnormal in G. (Use (3).)

Now put $H = \langle x_i \,|\, i \geqq 1 \rangle$, $K = \langle y_i \,|\, i \geqq 1 \rangle$, so that $J = \langle H, K \rangle$. Then

(5) $H, K \lhd^3 G$. (Show that J, VH and VK belong to $\mathfrak{N}_2$.)

Finally the verification of part (ii) of the theorem follows without difficulty using (1). □

Hall's first example replaces the free $\mathbb{Z}$-modules V, H, K by elementary abelian 2-groups, with some simplification of the computations involved.

Theorem 1.5.2 [111]. *There is a countable group G with subnormal subgroups H, K such that*

(i) $H, K \lhd^3 G$,
(ii) *H and K are infinite elementary abelian 2-groups,*
(iii) $G \in \mathfrak{A}\mathfrak{N}_2$,
(iv) $J = \langle H, K \rangle \in \mathfrak{N}_2$,
(v) *J is not subnormal in G, indeed J is self-normalizing in G and $J^G = G$.*

Proof. Let

$\mathscr{S} = \{S \mid S \subseteq \mathbb{Z}, \exists$ integers $m < n$ such that S contains all $i \leqq m$ and no $i \geqq n\}$.

Then let A, B be elementary abelian 2-groups with bases

$$\{a_S \mid S \in \mathscr{S}\}, \qquad \{b_S \mid S \in \mathscr{S}\},$$

respectively, and put $V = A \times B$. For all $n \in \mathbb{Z}$, define automorphisms x_n, y_n of V:

$$x_n\colon a_S \mapsto a_S, \qquad y_n\colon a_S \mapsto a_S b_{S*n},$$
$$b_S \mapsto b_S a_{S*n}, \qquad b_S \mapsto b_S$$

for all $S \in \mathscr{S}$. Here we are adopting the following notation: if the integers $n_1, n_2, \ldots, n_r$ are distinct and none belongs to S, then

$$a_{S*n_1*\cdots*n_r} = a_T,$$

where $T = S \cup \{n_1, n_2, \ldots, n_r\}$; otherwise $a_{S*n_1*\cdots*n_r} = 1$. The same applies with b in place of a.

Thus $x_n^2 = y_n^2 = 1$; and

$$x_m x_n\colon b_S \mapsto b_S a_{S*m} a_{S*n},$$

showing that $x_m x_n = x_n x_m$. Similarly $y_m y_n = y_n y_m$. Therefore

$$H = \langle x_n \mid n \in \mathbb{Z} \rangle \quad \text{and} \quad K = \langle y_n \mid n \in \mathbb{Z} \rangle$$

are elementary abelian 2-groups. Let $J = \langle H, K \rangle$ and $G = V] J$, the natural semidirect product of V and J. Then

$$[B, H] = A, \quad [A, K] = B, \quad [A, H] = [B, K] = 1.$$

Direct calculations show that, with $z_{mn} = [x_m, y_n]$,

$$z_{mn}\colon a_S \mapsto a_S a_{S*m*n}$$
$$b_S \mapsto b_S b_{S*m*n}$$

and $[z_{mn}, x_l] = [z_{mn}, y_l] = 1$. Hence $J \in \mathfrak{N}_2$.

Now $H_1 = H^G = VH[H, K]$ and

$$H^{H_1} = H^V = H^{AB} = H^B = AH.$$

Since $H \lhd AH$, it follows that $H \lhd^3 G$. Similarly $K \lhd^3 G$. However,

$$J^G = H^G K^G = VJ = G$$

and so J is not subnormal in G. In fact it is not hard to see that $J = N_G(J)$, the *normalizer* of J in G. □

The above group G has two natural automorphisms. First there is the involution t_1 defined by

$$a_S \mapsto b_S, \quad b_S \mapsto a_S, \quad x_n \mapsto y_n, \quad y_n \mapsto x_n$$

for all S in $\mathscr{S}$ and n in $\mathbb{Z}$. Let $G_1 = G] \langle t_1 \rangle$ and observe that $z_{mn} = z_{nm}$. Also one checks easily that $z_{mn}^2 = 1$. Therefore t_1 fixes z_{mn}, for all m, n. Moreover t_1 fixes $x_m y_m$ and $a_S b_S$, for all m and S. Hence

$$\langle t_1 \rangle \lhd^4 G_1.$$

Now $J \lhd J\langle t_1 \rangle = \langle H, t_1 \rangle = J_1$, say, and so J_1 cannot be subnormal in G_1. On the other hand $H \lhd^4 G_1$. Thus we have proved

Theorem 1.5.3 (P. Hall). *The join of a subnormal, countably infinite, elementary abelian 2-subgroup and a subnormal subgroup of order 2 need not be subnormal.*

The second automorphism t_2 of G in Theorem 1.5.2 is induced by the shift map $n \mapsto n+1$ of $\mathbb{Z}$. So t_2 is defined by

$$x_n \mapsto x_{n+1}, \quad y_n \mapsto y_{n+1}, \quad a_S \mapsto a_{S'}, \quad b_S \mapsto b_{S'},$$

where $m \in S'$ if and only if $m-1 \in S$. Let $G_2 = G] \langle t_2 \rangle$. Then it is easy to see that $G_2 = \langle a_{S_0}, b_{S_0}, x_0, y_0, t_2 \rangle$, where S_0 is the set of all integers ≤ 0. Thus, $H, K \lhd^4 G_2$, but J is not subnormal in G_2. Therefore we can state

Theorem 1.5.4 (P. Hall). *There is a finitely generated soluble group of derived length* 4 *in which the join of two subnormal subgroups is not subnormal.* (Cf. p. 13.)

So far we have exhibited four groups not in $\mathfrak{S}$. Our final construction yields infinitely many such examples.

Theorem 1.5.5 (Roseblade and Stonehewer [129]). *Let $S(\neq 0)$ be a commutative ring with* 1, *let A be a free S-module of infinite dimension and let C be the additive subgroup of S generated by* 1. *Then there exists a group G with subnormal subgroups $H \cong A$ and $K \cong C$ such that $J = \langle H, K \rangle \neq G$ and J is self-normalizing in G.*

Proof. Let R be the exterior algebra on A, that is $R = T/I$, where

$$T = S + \sum_{n \geq 1} \overleftrightarrow{A \otimes_S \cdots \otimes_S A}^{\,n}$$

is the tensor algebra on A and I is the ideal generated by $\{a^2 \mid a \in A\}$. Then A embeds in R and

$$a^2 = 0 \tag{1}$$

for all a in A. Also, for x in R,

$$xA = 0 \qquad \text{if and only if } x = 0. \tag{2}$$

This follows without difficulty from the facts that $\dim_S A$ is infinite and I is a homogeneous ideal of T, that is contains its homogeneous components.

For all a in A, define 2×2 matrices over R by

$$\lambda_0(a) = \begin{pmatrix} 1 & a \\ 0 & 1 \end{pmatrix}.$$

Then, under multiplication, $H = \{\lambda_0(a) \mid a \in A\} \cong A$ (under addition). Let

$$\xi = \begin{pmatrix} 1 & 0 \\ 1 & 1 \end{pmatrix}, K = \langle \xi \rangle \text{ (under multiplication)}$$

and $J = \langle H, K \rangle$. Note that $K \cong C$. Also $H \otimes K \cong A \otimes_C C \cong A$ does not have finite rank modulo its maximal periodic divisible subgroup, a necessary condition for J to fail to be subnormal, as we mentioned on p. 14. We claim that

$$H^K, K^H \in \mathfrak{N}_2. \tag{3}$$

In order to see this, write

$$\lambda_n(a) = \begin{pmatrix} 1+na & a \\ -n^2a & 1-na \end{pmatrix} \quad \text{and} \quad \delta(x) = \begin{pmatrix} 1+x & 0 \\ 0 & 1+x \end{pmatrix}$$

for $a \in A$, $x \in R$ and $n \in \mathbb{Z}$. Then

$$[\delta(x), \xi] = 1 \tag{4}$$

for all x in R. Also, if $a, b \in A$, then $(a+b)^2 - a^2 - b^2 = 0$, by (1), and so $ab = -ba$. Thus $abc = cab$, for c in A, and hence

$$[\delta(ab), \lambda_0(c)] = 1. \tag{5}$$

It is routine to check that $\lambda_n(a)^\xi = \lambda_{n+1}(a)$ and so

$$H^K = \langle \lambda_n(a) \mid n \in \mathbb{Z}, a \in A \rangle.$$

Then

$$[\lambda_n(b), \lambda_m(a)] = [\lambda_0(b), \lambda_{m-n}(a)]^{\xi^n} = \delta((m-n)^2ab)^{\xi^n} = \delta((m-n)^2ab)$$

which lies in the centre of J, by (4) and (5). Thus $H^K \in \mathfrak{N}_2$.

For all a in A, write

$$\beta(a) = \xi^{\lambda_0(a)}, \qquad \gamma(a) = [\beta(a), \xi].$$

Then

$$\beta(a) = \begin{pmatrix} 1-a & 0 \\ 1 & 1+a \end{pmatrix}, \quad \gamma(a) = \begin{pmatrix} 1 & 0 \\ 2a & 1 \end{pmatrix} \quad \text{and} \quad [\beta(a), \gamma(b)] = 1,$$

for all b in A. Since $K^H = \langle \beta(a) \mid a \in A \rangle$, it follows that $\gamma(a) \in \zeta_1(K^H)$, the centre of K^H. Therefore, *modulo* $\zeta_1(K^H)$, $\beta(a)$ and ξ commute. Thus $[\beta(a-b), \xi] = 1$ and hence (conjugating by $\lambda_0(b)$), $[\beta(a), \beta(b)] = 1$. Hence $K^H \in \mathfrak{N}_2$ and (3) is established.

The group J acts in a natural way on the direct sum V of two copies of R. Thus if $(x, y) \in V$, then

$$(x, y)^{\lambda_0(a)} = (x, xa + y), \tag{6}$$

$$(x, y)^{\xi} = (x + y, y). \tag{7}$$

Let $G = V]J$ and $V_1 = \{(x, 0) \mid x \in R\}$, $V_2 = \{(0, y) \mid y \in R\}$. By (2) and (6), $C_V(H) = V_2$. Similarly, from (7), $C_V(K) = V_1$. Again by (6), $[V, H] \leqq V_2$ and hence $[V, H, H] = 1$. Therefore $VH \in \mathfrak{N}_2$. In the same way, (7) gives $VK \in \mathfrak{N}_2$. Since $G/V \cong J$, it follows from (3) that

$$H \lhd^2 VH \lhd^3 G, \qquad K \lhd^2 VK \lhd^3 G.$$

Hence $H, K \lhd^5 G$. However, if $N_G(J) > J$, then $C_V(J) \neq 1$. But

$$C_V(J) = C_V(H) \cap C_V(K) = V_2 \cap V_1 = 1,$$

and so J must be self-normalizing in G. □

To complete our counterexamples, we describe a variation of the previous construction, due to Rex Dark, which shows that J can fail to be subnormal in G even when $J = HKHK$ (cf. Proposition 1.2.6). In place of the exterior algebra in Theorem 1.5.5, we consider *commutative* rings R such that

(i) $2r = r^2 = 0$ for all r in R and

(ii) $rR \neq 0$ for all $r (\neq 0)$ in R.

For convenience we shall call any *countable*, commutative ring R satisfying conditions (i) and (ii) a *Dark* ring. (It is easy to see that there are no finite Dark rings $\neq 0$.) The key result is then

Lemma 1.5.6. *There is a Dark ring $R (\neq 0)$ such that $R = \{rs \mid r, s \in R\}$.*

Proof. Let R be any Dark ring and let $a \in R$. We show first that

there is a Dark ring $S \geqq R$ with elements x, y such that $xy = a$. (8)

Clearly we may suppose that $a \neq 0$. Let F be a field of 2 elements and define

$R' = F \oplus R$ as additive groups. Then define multiplication on R' by

$$(e+r)(f+s) = ef + (fr + es + rs), \tag{9}$$

for all $e, f \in F$ and $r, s \in R$, where $0r = 0$ and $1r = r$. So R' is a commutative, associative and countable ring with 1, containing R as an ideal. Now let $A = \{\rho \in R' \mid a\rho = 0\}$, an ideal of R'.

Let P be the ring $R'[x, y]$ with relations $x^2 = y^2 = Ax = Ay = 0$. Then if $T = R'/A$, we can write the elements of P uniquely in the form

$$e + r + \alpha x + \beta y + \gamma xy$$

with $e \in F$, $r \in R$ and $\alpha, \beta, \gamma \in T$. For $r \in R'$, put $\bar{r} = A + r (\in T)$ and let

$$I = \{ra - \bar{r}xy \mid r \in R'\}.$$

Since $a \in A$, we have $(ra - \bar{r}xy)x = (ra - \bar{r}xy)y = 0$ and hence I is an ideal of P. Moreover if $S = \{r + \alpha x + \beta y \mid r \in R \text{ and } \alpha, \beta \in T\}$, then $S \cap I = 0$ and so S embeds in $Q = P/I$. Therefore S is a countable, commutative ring, containing R, in which $xy = a$. It is easy to see that S satisfies (i). In order to see that S satisfies (ii), consider $\sigma = r + \alpha x + \beta y \in S$, $\sigma \neq 0$. If $r \neq 0$, then there is an element s in R with $rs \neq 0$ and thus $\sigma s \neq 0$, as required. If $\alpha = A + s \neq 0$ $(s \in R')$, then

$$\sigma y = as + ry \neq 0$$

since $s \notin A$. Similarly if $\beta = A + t \neq 0$ $(t \in R')$, then $\sigma x = at + rx \neq 0$. Therefore S is a Dark ring and (8) follows.

Now let R be a Dark ring. Then we can use (8) to show that

$$\textit{there is a Dark ring } S \textit{ with } R \leqq S \textit{ and } R \subseteq \{xy \mid x, y \in S\}. \tag{10}$$

For, suppose that $R = \{r_1, r_2, \ldots\}$ and take $S_0 = R$. Define S_i $(i \geqq 1)$ inductively as follows. Using (8), choose a Dark ring S_i with $x_i, y_i \in S_i$ such that $R \leqq S_{i-1} \leqq S_i$ and $x_i y_i = r_i$. Then $S = \bigcup_{i \geqq 0} S_i$ satisfies (10).

Finally let M be a countably infinite vector space over F and let R_0 be the subring generated by M of the exterior algebra on M. (R_0 is commutative since $|F| = 2$.) Then R_0 is a Dark ring. Define R_i $(i \geqq 1)$ inductively as follows. Using (10), choose a Dark ring R_i with $R_{i-1} \subseteq \{xy \mid x, y \in R_i\}$. Then $R = \bigcup_{i \geqq 0} R_i$ satisfies the requirements of the lemma. □

We can now prove

Theorem 1.5.7. *There is a group G with subnormal subgroups H, K such that $J = \langle H, K \rangle = HKHK$, but J is not subnormal in G.*

Proof. According to Lemma 1.5.6, there is a Dark ring $R \neq 0$ such that

$$R = \{rs \mid r, s \in R\}. \tag{11}$$

As in the lemma, we take $R' = F \oplus R$ with multiplication given by (9). (F is a field of 2 elements.) For all $r \in R$, define invertible 2×2 matrices over R' by

$$\lambda(r) = \begin{pmatrix} 1 & r \\ 0 & 1 \end{pmatrix}, \qquad \mu(r) = \begin{pmatrix} 1 & 0 \\ r & 1 \end{pmatrix}, \qquad \delta(r) = \begin{pmatrix} 1+r & 0 \\ 0 & 1+r \end{pmatrix}.$$

Then the multiplicative groups $H = \{\lambda(r) \mid r \in R\}$, $K = \{\mu(r) \mid r \in R\}$ and $L = \{\delta(r) \mid r \in R\}$ are all elementary abelian 2-groups. Also

$$[\lambda(r), \mu(s)] = \lambda(r)\mu(s)\lambda(r)\mu(s) = \begin{pmatrix} 1+rs & r \\ s & 1 \end{pmatrix}^2 = \delta(rs),$$

for all $r, s \in R$. Therefore, by (11), $[H, K] = L \leqq HKHK$. Hence, with $J = \langle H, K \rangle$, we have $J = H[H, K]K = HKHK$. Since $L \leqq \zeta_1(J)$, it follows that J is nilpotent of class 2.

Let V_1, V_2 be copies of R and take $V = V_1 \oplus V_2$ as additive groups. Then J acts as a group of automorphisms of V according to

$$(x, y)^{\lambda(r)} = (x, xr + y),$$
$$(x, y)^{\mu(r)} = (x + yr, y),$$

for all $x, y, r \in R$. By property (ii) of Dark rings, $C_V(H) = V_2$ and $C_V(K) = V_1$. Let $G = V \rtimes J$. Then

$$N_G(J) = C_V(J)J = (C_V(H) \cap C_V(K))\, J = J.$$

However,

$$H \lhd V_2 H \lhd VHL \lhd G$$

and similarly $K \lhd^3 G$. □

§1.6 Subnormal coalescence and the permutizer

A class $\mathfrak{X}$ of groups is said to be a *normal coalition* class if the product of two normal $\mathfrak{X}$-subgroups of a group is always an $\mathfrak{X}$-group. In this case we also say that the class is N_0-*closed*. Then N_0 becomes a closure operation by defining $\mathrm{N}_0\mathfrak{Y}$, for any class $\mathfrak{Y}$, to be the intersection of all N_0-closed classes $\mathfrak{X}$ such that $\mathfrak{X} \geqq \mathfrak{Y}$. One of the most famous results about normal coalition classes is due to Fitting [33]:

Theorem 1.6.1. *The nilpotent groups form a normal coalition class. More precisely, if $H, K \lhd G$ and if H, K are nilpotent of classes c, d, respectively, then HK is nilpotent of class at most $c + d$.*

Proof. The commutator identity $[hk, l] = [h, l]^k[k, l]$ shows that if L is also a normal subgroup of G, then $[HK, L] = [H, L]\,[K, L]$. For any group X and all $i \geqq 1$, write

$$\gamma_i(X) = [\overleftarrow{X, X, \ldots, X}^{i}],$$

the ith term of the lower central series of X ($\gamma_1(X) = X$). By induction on j, we obtain

$$\gamma_{j+1}(HK) = \Pi[X_1, X_2, \ldots, X_{j+1}],$$

where $X_i = H$ or K and all 2^{j+1} possible factors in the product are included. Taking $j = c + d$, each factor must contain at least $c + 1$ Hs or $d + 1$ Ks among the X_i. Since $H, K \lhd G$, it follows that all $\gamma_i(H)$, $\gamma_i(K)$ are normal in G and so each factor lies in $\gamma_{c+1}(H)$ or $\gamma_{d+1}(K)$, both of which are 1. Thus $HK \in \mathfrak{N}_{c+d}$. □

The finite case of the above theorem is due to Wendt [149].

The concept of a normal coalition class has been extended in a way which is relevant to our discussion of the join problem. A class $\mathfrak{X}$ of groups is called a *subnormal coalition* (or *coalescence*) class if the join of two (and hence finitely many) subnormal $\mathfrak{X}$-subgroups of any group G is always a subnormal $\mathfrak{X}$-subgroup of G. We also say that $\mathfrak{X}$ is *subnormally coalescent*. Then the examples of §1.5 show that the class of *all* groups is *not* a subnormal coalition class. In fact they show considerably more, namely that Fitting's theorem (above) is no longer valid if *normal* is replaced by *subnormal* in the statement. Thus, Hall's group G of Theorem 1.5.2 is generated by two elementary abelian 2-subgroups, namely AH and BK, both subnormal of defect 2 in G. However, G cannot be nilpotent as it possesses the self-normalizing subgroup J. Hall's second example G_1 (in Theorem 1.5.3) is generated by AH and $\langle t_1 \rangle$. Here AH is an elementary abelian 2-subgroup, subnormal of defect 3 in G_1, and $\langle t_1 \rangle$ has order 2 and is subnormal of defect at most 5. Again G_1 cannot be nilpotent.

In contrast to the above examples, Baer [5] (§3, Satz 3) had previously shown that the finitely generated nilpotent groups form a subnormal coalition class. This and all the other major theorems on coalescence can now be obtained as corollaries of one main result (Theorem 1.6.11, though see also Theorem 3.4.1). However, Lemma 1.4.3 allows us to prove Baer's theorem very quickly as a special case of the following.

Theorem 1.6.2 (Roseblade and Stonehewer [129]). *If* $\mathfrak{X} = \mathrm{s_n}\mathfrak{X} = \mathrm{N_0}\mathfrak{X}$, *then the finitely generated* $\mathfrak{X}$*-groups form a subnormal coalition class.*

(The hypothesis $\mathfrak{X} = \mathrm{s_n}\mathfrak{X}$ cannot be removed here; see Theorem 2.1.8. However, the result will be significantly improved in Theorem 3.2.1.)

Two subgroups H, K of a group are said to *permute* (or be *permutable* with each other) if $HK = KH$, and this is the case precisely when $HK = \langle H, K \rangle$. In order to prove that the join of two subnormal $\mathfrak{X}$-subgroups is an $\mathfrak{X}$-group in Theorem 1.6.2, we need to consider first the special case in which H and K permute. Also, with later applications in mind, we choose to give rather more information about joins by introducing an operation N_2 on group classes. Thus for any class $\mathfrak{X}$, $\mathrm{N}_2\mathfrak{X}$ will denote *all those groups which are the product of two normal $\mathfrak{X}$-subgroups.* Since $\mathrm{N}_0\mathfrak{X}$ is always N_2-closed, $\mathfrak{X} \leqq \mathrm{N}_0\mathfrak{X}$ implies $\mathrm{N}_2\mathfrak{X} \leqq \mathrm{N}_0\mathfrak{X}$. We shall call the class

$$(\mathrm{N}_2\mathrm{S}_\mathrm{n})^t\mathfrak{X}$$

the tth *subjunction* of $\mathfrak{X}$, for each $t \geqq 0$. (The 0th subjunction is $\mathfrak{X}$ itself.)

Proposition 1.6.3. *Let H, K be subnormal subgroups of J of defects m and n respectively and suppose that $J = HK$. Then there is an integer t, depending only on s, such that J belongs to the tth subjunction of $(H) \cup (K)$, whenever $m+n \leqq s$. In particular, $J \in \mathfrak{X}$ whenever H, $K \in \mathfrak{X} = \mathrm{S}_\mathrm{n}\mathfrak{X} = \mathrm{N}_0\mathfrak{X}$.*

Proof. We proceed by induction on s. Then we may assume that $s \geqq 3$ and, without loss of generality, $m \geqq 2$. Let $H_1 = H^K$, so that $H_1 = H(H_1 \cap K)$. Now H has defect $m-1$ in H_1 and $H_1 \cap K$ has defect at most n. Since $H_1 \cap K \lhd K$, we see that $H_1 \cap K \in \mathrm{S}_\mathrm{n}(K)$. Therefore, by induction, H_1 belongs to the t_1th subjunction of $(H) \cup (K)$, for some $t_1 = t_1(s)$.

Finally, $J = H_1K$ and $1+n < m+n \leqq s$. A second application of the induction hypothesis now gives the result. □

Proof of Theorem 1.6.2. Let H, K be finitely generated subnormal $\mathfrak{X}$-subgroups of G and suppose that H, K have defects m, n, respectively, in $J = \langle H, K \rangle$. We prove that J is a subnormal $\mathfrak{X}$-subgroup of G by induction on m. The case $m \leqq 1$ is covered by Theorem 1.2.1 and Proposition 1.6.3. Assume, therefore, that $m \geqq 2$.

By Lemma 1.4.3,

$$H^K = L[H, {}_nK],$$

where L is generated by finitely many conjugates of H under K. Thus, by induction on m and a second induction on the number of these conjugates, L is a (finitely generated) subnormal $\mathfrak{X}$-subgroup of G. Also

$$[H, {}_nK] \lhd K,$$

and hence $[H, {}_nK]$ is a subnormal $\mathfrak{X}$-subgroup of G. Then H^K is subnormal in G, by Theorem 1.2.5, and $H^K \in \mathfrak{X}$, by Proposition 1.6.3. Thus J is subnormal in G according to Theorem 1.2.3. Finally $J = H^KK \in \mathfrak{X}$ on applying Proposition 1.6.3 once more. □

By taking the class of nilpotent groups for $\mathfrak{X}$, we obtain Baer's theorem. Similarly we see that the finite groups form a subnormal coalition class, though this was proved in Theorem 1.3.3. Other subnormal coalition classes (from Theorem 1.6.2) are

(i) the finitely generated groups ($\mathfrak{G}$),

(ii) the finitely generated soluble groups ($\mathfrak{G} \cap \mathbf{P}\mathfrak{A}$),
(iii) the groups satisfying the maximal condition for subgroups ($\mathfrak{Max}$).

Remark. Using the more precise information of Proposition 1.6.3 and keeping track of defects, the argument of Theorem 1.6.2 shows that when H and K have subnormal defects m and n in G and are generated by r and s elements, then $J = \langle H, K \rangle$ is subnormal of defect at most d in G and J lies in the tth subjunction of $(H) \cup (K)$, where d and t depend only on m, n, r and s.

The main goal throughout the remainder of this section is to prove a result (Theorem 1.6.11) which has *all* the important results on subnormal coalescence as corollaries. Our starting point is the following interesting and important result due to Roseblade.

Theorem 1.6.4. *Let $J = \langle H, K \rangle$ with H and K subnormal of defects m and n, respectively, in J. Then there is an integer λ, depending only on s, such that*

$$J^{(\lambda)} \leqq HK$$

whenever $m + n \leqq s$.

Here $J^{(\lambda)}$ is the λth term of the derived series of J. We also point out that HK will not be a subgroup in general. In order to prove this theorem we shall need

Lemma 1.6.5. *Suppose that $H_1, H_2, \ldots, H_r$ are subgroups of G and $X \lhd G$. If $J = \langle H_1, H_2, \ldots, H_r \rangle$, then*

$$[X, J] = [X, H_1]\,[X, H_2] \cdots [X, H_r]. \tag{1}$$

Proof. Denote the right-hand side of (1) by Y. Since H_i centralizes $X/[X, H_i]$, H_i normalizes Y and centralizes X/Y, for all i. Therefore $[X, J] \leqq Y$. The reverse inclusion is clear. □

Let $H \lhd^m G$ and $N \lhd G$. For any $a \geqq 0$, $N^{(a)} \lhd G$. Modulo $N^{(a)}$, N has derived length $\leqq a$ and therefore this is also true of the factors of the normal closure series of H in HN. Thus

$$(HN)^{(am)} \leqq HN^{(a)}. \tag{2}$$

Proof of Theorem 1.6.4. Put $\lambda(0) = \lambda(1) = 0$ and

$$\lambda(s) = (s-2)!\, 4^{s-2}$$

for $s \geqq 2$. We show, by induction on s, that $J^{(\lambda)} \leqq HK$ whenever $m + n \leqq s$.

If H or K is normal in J, then the result is trivial. So we may assume that $m, n \geqq 2$. Denote the ith terms of the normal closure series of H, K in J by H_i,

K_i, respectively. Then, by induction on s, we have

$$J^{(l)} \leqq K_{n-1}H \cap H_{m-1}K,$$

where $l = \lambda(s-1)$. Let $X = [H, K]^{(l)} \lhd J$. Thus

$$X \leqq J^{(l)} \cap H_1 \leqq H_{m-1}K \cap H_1 = H_{m-1}(K \cap H_1)$$

and

$$X \leqq J^{(l)} \cap K_1 \leqq K_{n-1}H \cap K_1 = K_{n-1}(H \cap K_1).$$

From $H \lhd H_{m-1}$ and $K \lhd K_{n-1}$ we obtain

$$[X, H] \leqq H^X \leqq H^{H_{m-1}(K \cap H_1)} = H^{K \cap H_1} \leqq \langle H, K \cap H_1 \rangle$$

and similarly $[X, K] \leqq \langle K, H \cap K_1 \rangle$.

Now applying induction to $H \lhd^{m-1} H_1$ and $K \cap H_1 \lhd^n H_1$, we have

$$[X, H]^{(l)} \leqq \langle H, K \cap H_1 \rangle^{(l)} \leqq H(K \cap H_1) \leqq HK$$

and similarly $[X, K]^{(l)} \leqq HK$. Since H normalizes $[X, H]^{(l)}$, this gives

$$[X, H]^{(l)}[X, K]^{(l)} \leqq [X, H]^{(l)}HK = H[X, H]^{(l)}K = HK.$$

By Lemma 1.6.5, $[X, J] = [X, H][X, K]$ with both factors normal in X. Also

$$[X, J]/[X, H]^{(l)}[X, K]^{(l)}$$

is soluble of derived length $\leqq 2l$ and so $[X, J]^{(2l)} \leqq HK$. Write $U = [H, K]$. So $U \lhd J$ and

$$U^{(l+1+2l)} = X^{(1+2l)} \leqq [X, J]^{(2l)} \leqq HK.$$

Finally $H_1 = HU$ and $K_1 = UK$. Putting $l_1 = (1+3l)(m-1)$ and $l_2 = (1+3l)(n-1)$, it follows from (2) that $H_1^{(l_1)} \leqq HU^{(1+3l)} \leqq HK$ and similarly $K_1^{(l_2)} \leqq HK$. Therefore, since $J/H_1^{(l_1)}K_1^{(l_2)}$ has derived length $\leqq l_1 + l_2$, we have

$$J^{(l_1+l_2)} \leqq H_1^{(l_1)}K_1^{(l_2)} \leqq H_1^{(l_1)}HK = HH_1^{(l_1)}K = HK.$$

Moreover $l_1 + l_2 = (1+3l)\,(m+n-2) \leqq 4l(s-2) = (s-2)!\,4^{s-2}$. □

Remark. We shall not compute explicit bounds like $\lambda(s)$ in general. On the other hand there is no harm in exhibiting an occasional one when it is easy to compute and not unattractive.

Theorem 1.6.4 will be extended in §3.1. As a bonus at this point, however, we can use it to establish the remarkably good behaviour of perfect subnormal subgroups in relation to joins. This was shown for groups with a composition series by Wielandt [152] (Satz 23) and later generalized to arbitrary groups as a special case of Roseblade's work on orthogonality ([124] and §4.3). The original arguments were quite different from the ones given here.

Theorem 1.6.6. *Let H and K be subnormal subgroups of a group and let* $J = \langle H, K \rangle$. *If H is perfect, then* $J = HK$.

Proof. By Theorem 1.6.4, there is an integer $\lambda \geqq 0$ such that $J^{(\lambda)} \leqq HK$ and so $J^{(\lambda)}K \leqq HK$. However $H = H' = H^{(\lambda)} \leqq J^{(\lambda)}$ and thus $HK \leqq J^{(l)}K$. Therefore

$$HK = J^{(\lambda)}K,$$

a subgroup, and hence $HK = J$. □

From Theorem 1.2.5 we deduce

Corollary 1.6.7. *The join of a perfect subnormal subgroup and an arbitrary subnormal subgroup is always subnormal.*

Remarks

1. We see that the perfect groups form a subnormal coalition class.
2. A perfect subnormal subgroup is not normal in general, as can be seen in the group $A_5 \wr C_2$, the wreath product of the alternating group A_5 and the cyclic group C_2 of order 2. However, see Theorem 2.5.10.

Arbitrary subnormal subgroups do not in general permute (see, for example, the dihedral group of order 8). Nevertheless, given *any* subgroups H and K of a group, there is always at least one subgroup of H (namely 1) which permutes with K. In fact there is a unique largest subgroup $P_H(K)$ of H whose product with K is a subgroup. It is the join of all the subgroups L of H for which $LK = KL$. This subgroup $P_H(K)$ is called the *permutizer* of K in H. (The term is not etymologically sound, but conceptually belongs to the centralizer–normalizer family.) The permutizer of one subnormal subgroup in another will be of fundamental importance. To begin with, it enables us to see Theorem 1.6.6 as a special case of the following general result.

Theorem 1.6.8. *Let H and K be subnormal subgroups of a group. Then there is an integer* $\lambda \geqq 0$ *such that* $H^{(\lambda)} \leqq P_H(K)$; *furthermore* λ *depends only on the sum of the subnormal defects m and n of H and K in their join J.*

Proof. From Theorem 1.6.4 we have $J^{(\lambda)} \leqq HK$ for some λ depending only on $m+n$. Therefore $J^{(\lambda)}K \leqq HK$ and $J^{(\lambda)}K = (H \cap J^{(\lambda)}K)K$. Thus $H^{(\lambda)} \leqq H \cap J^{(\lambda)}K \leqq P_H(K)$ as required. □

The main reason for introducing and studying the permutizer $P_H(K)$ is that it is subnormal in G whenever H and K are subnormal in G. (This contrasts with the fact that $N_H(K)$, although contained in $P_H(K)$, is *not* in general subnormal in G; see for example the symmetric group of degree 4.)

Theorem 1.6.9 (Roseblade and Stonehewer [129]). *There is a function* $d(m, n)$ *such that if* $H \lhd^m G$ *and* $K \lhd^n G$, *then* $P_H(K)$ *is subnormal in G of defect at most* $d(m, n)$.

Proof. We argue by induction on m. Let $P = P_H(K)$. If $H \lhd G$, then $P = H$ and so cases $m = 0, 1$ are clear. Suppose that $m \geqq 2$ and assume the usual induction hypothesis. Let

$$H_1 = H^G,\ K_1 = P^K \cap K.$$

So

$$P^K = PK_1. \tag{3}$$

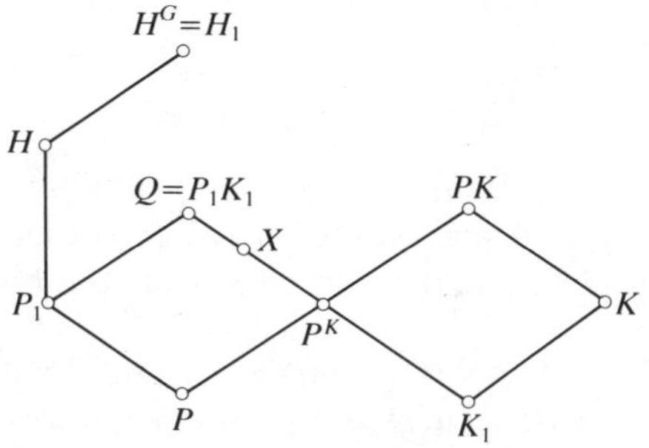

Now $K_1 \lhd^{n+1} G$ and therefore $K_1 \lhd^{n+1} H_1$.
Also $H \lhd^{m-1} H_1$. Thus, by induction,

$$P_1 = P_H(K_1) \lhd^{d(m-1,n+1)} H_1 \lhd G.$$

Then, by Theorem 1.2.5, $Q = P_1K_1 \lhd^f G$, for some f depending only on m and n.

By (3), $P \leqq P_1$ and $P^K = PK_1 \leqq P_1K_1 = Q$. Therefore the fth term X of the normal closure series of P^K in G lies in Q. Moreover K normalizes X, since K normalizes every term of this normal closure series. Thus $X = (P_1 \cap X)K_1$ and so $XK = (P_1 \cap X)K$, that is $P_1 \cap X \leqq P$. Therefore

$$P_1 \cap X = P \lhd^{d(m,n)} G,$$

where $d(m, n)$ is the larger of $d(m-1, n+1)$ and f. □

We can now prove a result linking permutable subgroups and joins of subnormal subgroups.

Theorem 1.6.10 [129]. *Suppose that L, M, H, K are subgroups of G with $L \leqq H \lhd^m G$, $M \leqq K \lhd^n G$ and $LM = ML$. Then there is a subnormal subgroup X of G such that*

(i) *$LM \leqq X \leqq HK$,*

(ii) *the subnormal defect of X in G is bounded by a function of m and n, and*

(iii) *X belongs to the tth subjunction of $(H) \cup (K)$, for some t depending only on* *m and n; in particular*

$$X \in \{\mathrm{N}_0, \mathrm{S}_n\}((H) \cup (K)),$$

the smallest N_0*- and* S_n*-closed class containing H and K.*

Proof. Let $\mathfrak{X}_i$ be the ith subjunction of $(H) \cup (K)$, $J = \langle H, K\rangle$ and $H \lhd^{m_0} J$. If $H \lhd J$, then $J = HK \lhd^{mn} G$, by Theorem 1.2.1, and $J \in \mathfrak{X}_{t_1}$, for some $t_1 = t_1(n)$, by Proposition 1.6.3. Thus we can take $X = J$ in this case.

Suppose that $m_0 \geqq 2$ and argue by induction on m_0. Let $H_1 = H^K$ and $K_1 = H_1 \cap K$. Then $L \leqq H \lhd^{m_0 - 1} H_1$,

$$H_1 \cap M \leqq K_1 \lhd K \lhd^n G$$

and $H_1 \cap LM = L(H_1 \cap M)$ is a subgroup. Therefore, by induction, there is an integer $t_2 = t_2(m, n)$ and a subnormal $\mathfrak{X}_{t_2}$-subgroup X_1 of G with $L(H_1 \cap M) \leqq X_1 \leqq HK_1$; and the defect d of X_1 in G is bounded by a function of m and n.

Now $L^M = L(L^M \cap M) \leqq L(H_1 \cap M) \leqq X_1$ and L^M is normalized by M. Thus M normalizes the dth term Q of the normal closure series of L^M in G and $Q \leqq X_1$. Hence

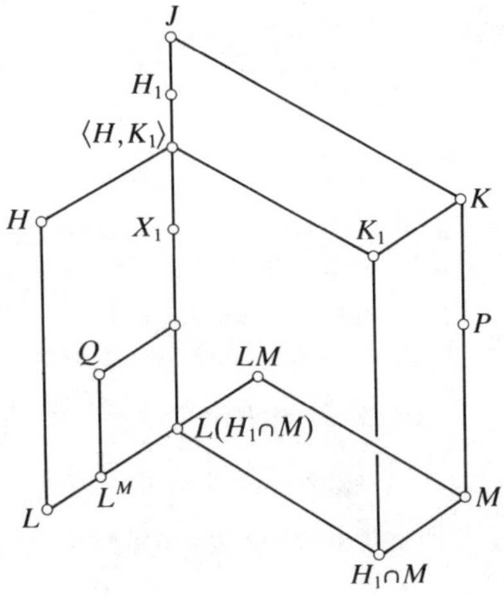

$$M \leqq P_K(Q) = P,$$

say. By Theorem 1.6.9, P is subnormal in G with defect bounded by a function of m and n. Hence, by Theorem 1.2.5,

$$X = QP$$

is subnormal in G with defect similarly bounded. Also

$$LM \leqq QP = X \leqq X_1 K \leqq HK.$$

Finally, $Q\ sn\ X_1 \in \mathfrak{X}_{t_2}$ and $P\ sn\ K \in \mathfrak{X}_0$. Therefore, by Proposition 1.6.3, there is an integer $t = t(m, n)$ such that $X \in \mathfrak{X}_t$. □

A precursor of this result appeared in Wielandt [152] (Satz 31) where he proved that if L, M are subgroups of a group G with a composition series and if H, K are the minimal subnormal subgroups of G containing L, M respectively, then $LM = ML$ implies $HK = KH$. In fact it is clear from Theorem 1.6.10 that this result extends to the case where G satisfies Min-*sn*.

The main result relating to subnormal coalescence which has been proved by purely group-theoretic means is due to J. E. Roseblade (unpublished). Later in §3.4 we shall apply module theory over group rings to obtain even deeper results.

Theorem 1.6.11 (Roseblade). *Suppose that L, M, H, K are subgroups of G with $L \leqq H \lhd^m G$, $M \leqq K \lhd^n G$ and L/L', M/M' each generated by at most r elements, where r is finite. Then there is a subgroup X of G such that*

(i) $\langle L, M \rangle \leqq X \leqq \langle H, K \rangle$;

(ii) *$X \lhd^d G$, where d depends only on m, n and r; and*

(iii) *X belongs to the tth subjunction of $(H) \cup (K)$, where t depends only on m, n and r.*

Proof. Let $J = \langle H, K \rangle$ and $H \lhd^{m_0} J$. For $i \geqq 0$, denote the ith subjunction of $(H) \cup (K)$ by $\mathfrak{X}_i$. We argue by induction on m_0. If $H \lhd J$, then $J = HK \lhd^{mn} G$, by Theorem 1.2.1; and $J \in \mathfrak{X}_{t_0}$, for some $t_0 = t_0(m, n)$, by Proposition 1.6.3. Thus we suppose that $m_0 \geqq 2$ and assume the usual induction hypothesis. Clearly we may also assume that $n \geqq 2$.

By hypothesis there is a subgroup Y of M, generated by at most r elements, such that $M = YM'$. By Lemma 1.4.3, there is a subset Y_0 of Y with $|Y_0| \leqq 1 + r + r^2 + \cdots + r^{n-1}$ such that

$$L^Y = L^{Y_0}[L, {}_n Y].$$

Thus $\langle L, Y \rangle = L^{Y_0} Y_1$ where $Y_1 = Y[L, {}_n Y] \leqq K$. The join of two subgroups with derived quotients generated by at most r elements has derived quotient generated by at most $2r$ elements. Therefore, by induction on m_0 and a second induction on $|Y_0|$, we know that there is a subgroup S of G with

$$L^{Y_0} \leqq S \leqq H^{Y_0} \leqq J \tag{4}$$

such that $S \lhd^{d_1} G$ and $S \in \mathfrak{X}_{t_1}$, where d_1 and t_1 depend only on m, n and r. Now, by Theorem 1.6.10, there is a subgroup T of G with

$$\langle L, Y \rangle = L^{Y_0} Y_1 \leqq T \leqq SK, \tag{5}$$

$T \lhd^{d_2} G$ and $T \in \mathfrak{X}_{t_2}$, where d_2 and t_2 depend only on m, n and r.

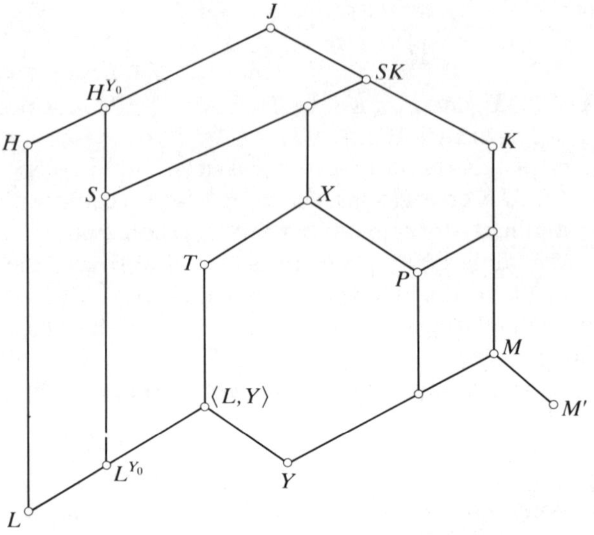

Let $P = P_K(T)$. By Theorem 1.6.9, P is subnormal in G with defect δ bounded by a function of only m, n and r. Therefore $P \lhd^{\delta} K$ and so $P \in \mathfrak{X}_1$. Thus, by Proposition 1.6.3, there is an integer $t = t(m, n, r)$ such that

$$X = TP \in \mathfrak{X}_t.$$

By Theorem 1.2.5, $X \lhd^{d} G$, where $d = d(m, n, r)$. Now,

$$P \geqq T \cap K \geqq Y$$

and so $M = \langle Y, M' \rangle = \langle P \cap M, M' \rangle$. In a soluble group it is clear that the only subnormal subgroup supplementing the derived subgroup is the group itself. (A more general result appears as Lemma 3.2.4.) Since $P \cap M \ sn\ M$, it follows that

$$M = \langle P \cap M, M^{(\lambda)} \rangle$$

for each $\lambda \geqq 0$. By Theorem 1.6.8, we can find λ (depending only on d_2 and n, that is on m, n and r) such that

$$K^{(\lambda)} \leqq P.$$

Clearly $M^{(\lambda)} \leqq K^{(\lambda)}$ and so $M = P \cap M$, that is $M \leqq P$. Therefore

$$\langle L, M \rangle \leqq X \leqq \langle H, K \rangle,$$

completing the proof of the theorem. □

When H/H' and K/K' are themselves finitely generated, we can take $L = H$ and $M = K$ to obtain

Theorem 1.6.12. *Let H, K be subnormal subgroups of G with H/H', K/K' finitely generated and let $J = \langle H, K\rangle$. Then*
(i) *$J = H^*K$, where H^* is generated by k conjugates of H under K;*
(ii) *$J \lhd^d G$;*
(iii) *J belongs to the tth subjunction of $(H) \cup (K)$; and k, d, and t depend only on the defects of H and K in G and the number of generators of H/H' and K/K'.*

Proof. In the notation of Theorem 1.6.11, $K = M \leqq P \leqq K$. Therefore $P = K$ and $J = X = TK$. However, $S = H^{Y_0}$, according to (4), and so, by (5), $H^{Y_0}Y_1 \leqq T \leqq H^{Y_0}K$. Thus

$$H^{Y_0}K = TK = J$$

and we can take $H^* = H^{Y_0}$. □

Remarks
1. The theorem itself shows that H^* *sn* G.
2. The theorem will be shown to hold under even weaker hypotheses in §3.4.

Clearly Theorem 1.6.2 is a special case of the above result. So also is the main theorem of Roseblade and Stonehewer [129] (Theorem A). In order to state the latter, we make two definitions. First a class $\mathfrak{X}$ is said to be *subjunctive* if $\mathfrak{X} = \mathrm{N}_0\mathfrak{X} = \mathrm{s}_n\mathfrak{X}$. (Finite group theorists call such an $\mathfrak{X}$ a *Fitting* class.) Secondly, we say that a class $\mathfrak{X}$ is *locally coalescent* if, whenever H and K are subnormal $\mathfrak{X}$-subgroups of a group G and F is a finitely generated subgroup of $J = \langle H, K\rangle$, there is a subnormal $\mathfrak{X}$-subgroup X of G with $F \leqq X \leqq J$. Of course, every subnormal coalescence class is locally coalescent.

We now have

Theorem 1.6.13 [129]. *Subjunctive classes are locally coalescent.*

This follows from Theorem 1.6.11, since if H, K *sn* G and $J = \langle H, K\rangle$, then every finitely generated subgroup of J is contained in a subgroup of the form $\langle L, M\rangle$, where L, M are finitely generated subgroups of H, K.

Suppose that $\mathfrak{X}$ is a locally coalescent class and $H_1, H_2, \ldots, H_s$ are subnormal $\mathfrak{X}$-subgroups of a group G. Let $J = \langle H_1, H_2, \ldots, H_s\rangle$. Induction on s shows that if F is a finitely generated subgroup of J, then there is a subnormal $\mathfrak{X}$-subgroup X of G with $F \leqq X \leqq J$. Thus, as a corollary of the above theorem, we have

Theorem 1.6.14. *Let $\mathfrak{X}$ be a subjunctive class and suppose that J is generated by subnormal $\mathfrak{X}$-subgroups of G. Then every finitely generated subgroup F of J lies in a subnormal $\mathfrak{X}$-subgroup X of G with $F \leqq X \leqq J$.*

Taking for $\mathfrak{X}$ the class of all groups, we see that a join of subnormal subgroups is always the set-theoretic union of subnormal subgroups.

For any class $\mathfrak{X}$ we denote by $\mathrm{N}\mathfrak{X}$ the class of groups G which can be generated by (possibly infinitely many) subnormal $\mathfrak{X}$-subgroups. Then

Corollary 1.6.15. *Let $\mathfrak{X}$ be a subjunctive class. Then $G \in \mathrm{N}\mathfrak{X}$ if and only if every finitely generated subgroup of G lies in a subnormal $\mathfrak{X}$-subgroup.*

In particular

a join of subnormal soluble subgroups is locally soluble. (6)

(In §3.1 (Theorem 3.1.1) we shall prove an even better result, namely *a join of finitely many subnormal soluble subgroups is actually soluble.*) Of course, a join of subnormal nilpotent subgroups is locally nilpotent and the easiest way to prove this is to use the Hirsch–Plotkin theorem, which says that the class of locally nilpotent groups is N_0-closed (Theorem 2.1.1 (iii)). Thus (6) is especially interesting, bearing in mind that the class of locally soluble groups is *not* N_0-closed (Theorem 2.1.2 (iii)).

If a subjunctive class $\mathfrak{X}$ is s-closed, then Corollary 1.6.15 shows that $\mathrm{N}\mathfrak{X}$ is also s-closed. Therefore we have

Theorem 1.6.16. *If a group is generated by subnormal soluble subgroups, then so is every subgroup.*

§1.7 Some subnormal coalition classes

Suppose that subnormal subgroups H, K of a group G satisfy Max-*sn*. Then H/H', K/K' are finitely generated and so Theorem 1.6.12 applies to show that any subjunctive subclass of $\mathfrak{Max}$-*sn* is subnormally coalescent. Roseblade proved the following slightly stronger result in 1965.

Theorem 1.7.1 [125]. *If $\mathfrak{X} = \mathrm{N}_0 \mathfrak{X} \leqq \mathfrak{Max}$-sn, then $\mathfrak{X}$ is a subnormal coalition class.*

Proof. Let H, K be subnormal $\mathfrak{X}$-subgroups of G and let $J = \langle H, K \rangle$. By the above remarks, we know that J is subnormal in G and J satisfies Max-*sn*. Thus if

$$H_{i+1} \lhd H_i \lhd H_{i-1}$$

is part of the normal closure series of H in J, then H_i satisfies Max-*sn* and so H_i is the join of finitely many conjugates of H_{i+1} (under H_{i-1}), all normal in H_i. By induction on i decreasing, we see that

$$H^K \in \mathrm{N}_0(H) \leqq \mathfrak{X}.$$

Similarly $K^H \in \mathfrak{X}$ and hence $J \in \mathfrak{X}$. □

We recall from §1.4 that $\mathfrak{G}^{\mathrm{S_n}}$ is the class of all groups in which the subnormal subgroups are finitely generated. Then there is

Theorem 1.7.2 [111]. *If* $\mathfrak{X} = \mathrm{N}_0 \mathfrak{X} \leqq \mathfrak{G}^{\mathrm{S_n}}$, *then* $\mathfrak{X}$ *is a subnormal coalition class.*

Proof. Let H, K be subnormal $\mathfrak{X}$-subgroups of G and $J = \langle H, K \rangle$. Certainly H and K, and therefore H/H' and K/K', are finitely generated. So Theorem 1.6.12 applies and J is subnormal in G. Also, because $\mathfrak{G}^{\mathrm{S_n}}$ is subjunctive (even P-closed), $J \in \mathfrak{G}^{\mathrm{S_n}}$. Thus if H_i is the ith normal closure of H in J, then H_i is finitely generated and hence is the product of finitely many conjugates of H_{i+1} under H_{i-1}. Therefore, by induction on i decreasing, we see that

$$H_1 = H^K \in \mathrm{N}_0 \mathfrak{X} = \mathfrak{X}.$$

Similarly $K^H \in \mathfrak{X}$ and then $J = H^K K^H \in \mathrm{N}_0 \mathfrak{X} = \mathfrak{X}$. □

Roseblade obtained an (unpublished) improvement on the coalescence of $\mathfrak{Max}$-*sn*. We denote the class of groups satisfying the maximal condition for subnormal subgroups of defect at most 2 by $\mathfrak{Max}$-$\lhd^2$. In fact R. S. Dark has shown that the class $\mathfrak{Max}$-$\lhd^2$ is strictly larger than the class $\mathfrak{Max}$-*sn*. However, superseding Theorem 1.7.1 we have

Theorem 1.7.3. *If* $\mathfrak{X} = \mathrm{N}_0 \mathfrak{X} \leqq \mathfrak{Max}\text{-}\lhd^2$, *then* $\mathfrak{X}$ *is a subnormal coalition class.*

Proof. We suppose that H, K are subnormal $\mathfrak{X}$-subgroups of G and $J = \langle H, K \rangle$. Then H/H' and K/K' satisfy Max and so, by Theorem 1.6.12,

$$J = LK \text{ sn } G$$

where L is subnormal in G and is generated by finitely many conjugates of H under K.

We show, by induction on the defect of H in J, that $J \in \mathfrak{Max}$-$\lhd^2$. Let L_i be the ith normal closure of L in J and let X be any join of conjugates of L_2 in J such that $X \geqq L_2$. So $X \lhd^2 J$ and therefore $X \cap K \lhd^2 K$. Also

$$X = X \cap LK = L(X \cap K).$$

Thus, since $K \in \mathfrak{Max}$-$\lhd^2$, L_1 is generated by finitely many conjugates of L_2 under K, that is

$$L_1 = \langle L_2^{k_1}, L_2^{k_2}, \ldots, L_2^{k_r} \rangle,$$

for some $k_1, \ldots, k_r \in K$. Now, $L_2 = L^{L_1}$ and so

$$L_1 = \langle (L^{L_1})^{k_1}, \ldots, (L^{L_1})^{k_r} \rangle = \langle (L^{k_1})^{L_1}, \ldots, (L^{k_r})^{L_1} \rangle$$
$$= \langle L^{k_1}, \ldots, L^{k_r} \rangle^{L_1}.$$

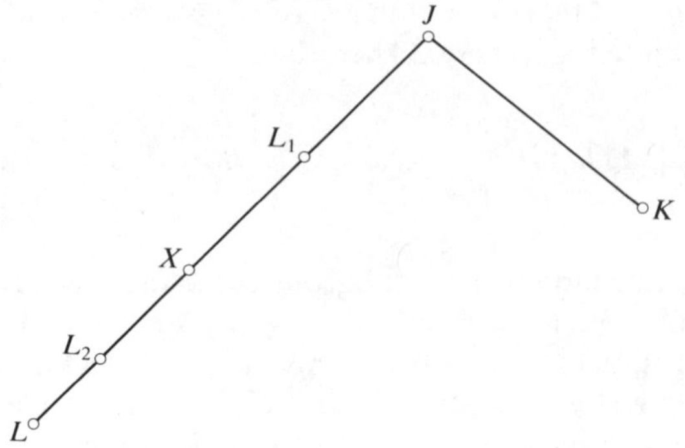

However, by Theorem 1.6.12(ii) and induction on r,

$$\langle L^{k_1}, \ldots, L^{k_r} \rangle \; sn \; L_1$$

and hence $L_1 = \langle L^{k_1}, \ldots, L^{k_r} \rangle$. Thus our induction (and a second induction on the number of conjugates of H under K involved in the generation of L_1) show that $L_1 \in \mathfrak{Max}\text{-}\lhd^2$ and therefore so does J. For, $J = L_1 K$ and the class $\mathfrak{Max}\text{-}\lhd^2$ is closed with respect to forming quotients and extensions.

It follows that if H_i denotes the ith normal closure of H in J, then H_1 is generated by finitely many conjugates of H_2. Therefore, as above (with H replacing L), H_1 is generated by finitely many conjugates of H. Hence $H_1 \in \mathfrak{Max}\text{-}\lhd^2$ by what we have already proved. Thus H_2 is generated by finitely many conjugates of H_3 and, continuing, H_i is generated by finitely many conjugates of H_{i+1}, for all i. Therefore

$$H_1 \in \mathrm{N}_0 \mathfrak{X} = \mathfrak{X}$$

and similarly $K^H \in \mathfrak{X}$. Then $J \in \mathfrak{X}$ as required. □

Now we turn our attention to the minimal condition for subnormal subgroups. Roseblade and Robinson obtained the analogue to Theorem 1.7.1 independently in 1964 and 1965, respectively ([123] and [113]).

Theorem 1.7.4. *If $\mathfrak{X} = \mathrm{N}_0 \mathfrak{X} \leqq \mathfrak{Min}\text{-}sn$, then $\mathfrak{X}$ is a subnormal coalition class.*

We can quickly establish the subnormality of the join here using Theorem 1.6.11 and a corollary of

Lemma 1.7.5. *Let H be a subgroup of G and suppose that there is an integer d such that, for every finitely generated subgroup F of H, there is a subgroup $X \lhd^d G$ with $F \leqq X \leqq H$. Then $H \lhd^d G$.*

Proof. For any subgroup S of G, let S_i denote the ith normal closure of S in G.

We show, by induction on i, that for any i and any element g of H_i, there is a finitely generated subgroup F of H such that $g \in F_i$. This is clear if $i \leqq 1$. Suppose that it is true for some $i-1$, $i \geqq 2$ and let $g \in H_i$. Then

$$g = h_1^{y_1} h_2^{y_2} \cdots h_s^{y_s} \tag{1}$$

where $h_1, \ldots, h_s \in H$ and $y_1, \ldots, y_s \in H_{i-1}$. By induction there is a finitely generated subgroup F of H, without loss of generality containing $h_1, \ldots, h_s$, such that

$$y_1, y_2, \ldots, y_s \in F_{i-1}.$$

(Note that $S \leqq T$ implies that $S_i \leqq T_i$ for all i.) Thus $g \in F_i$ by (1).

Now take $i = d$. By hypothesis there is a subgroup $X \lhd^d G$ such that $F \leqq X \leqq H$. Thus

$$g \in F_d \leqq X_d = X \leqq H,$$

that is $H_d \leqq H$. Therefore $H = H_d \lhd^d G$. □

An *ascending chain* of subgroups of a group G is a set $(H_\alpha)_{\alpha < \lambda}$ of subgroups, where λ is an ordinal, $H_\alpha \leqq H_{\alpha+1}$ whenever $\alpha + 1 < \lambda$, and $H_\beta = \bigcup_{\alpha < \beta} H_\alpha$ for all limit ordinals $\beta < \lambda$. (This concept was involved in the definition of $\mathfrak{Max}$-*sn* in §1.3, but only with $\lambda = \omega$, the first limit ordinal, and then the idea was understood to be intuitively obvious.) If each H_α is normal in $H_{\alpha+1}$, then the chain is called an *ascending series.*

Corollary 1.7.6. *Let d be an integer and suppose that a subgroup H of a group G is the union of an ascending chain of subgroups X_α with all $X_\alpha \lhd^d G$. Then $H \lhd^d G$.*

Proof of Theorem 1.7.4, *part* 1. Let H and K be subnormal subgroups of G with $H, K \in \mathfrak{Min}$-*sn* and $J = \langle H, K \rangle$. We show that J *sn* G.

Let $\bar{H} = H^{(\omega)}$ (the ωth term of the derived series of H) and $\bar{K} = K^{(\omega)}$. Using the fact that $\mathfrak{Min}$ (the class of groups satisfying the minimal condition for subgroups) is P-closed, that is closed under forming extensions (see p. 13), it follows that $H/\bar{H}$ and $K/\bar{K}$ are soluble groups satisfying Min. Such groups are finite extensions of direct products of finitely many groups isomorphic to the quasicyclic group C_{p^∞}, for various primes p (see [119], p. 68). Therefore there are ascending chains of subgroups

$$\bar{H} \leqq L_1 \leqq L_2 \leqq \cdots \leqq H = \bigcup_{i \geqq 1} L_i,$$

$$\bar{K} \leqq M_1 \leqq M_2 \leqq \cdots \leqq K = \bigcup_{i \geqq 1} M_i,$$

and an integer r with all the factors $L_i/\bar{H}$, $M_i/\bar{K}$ generated by at most r

elements. Since $\bar{H}$ is perfect, $\bar{H} \leqq L_i'$ and so L_i/L_i' and similarly M_i/M_i' are generated by at most r elements, for each i.

Now from Theorem 1.6.11 we see that there is a subgroup $X_i \lhd^d G$ with

$$\langle L_i, M_i \rangle \leqq X_i \leqq J,$$

and d depends only on r and the defects of H and K in G. Every finitely generated subgroup of J lies in some $\langle L_i, M_i \rangle$ and therefore, by Lemma 1.7.5, we have $J \lhd^d G$ as required.

It remains to show that if

$$H, K \in \mathfrak{X} = \mathrm{N}_0 \mathfrak{X} \leqq \mathfrak{Min}\text{-}sn,$$

then $J \in \mathfrak{X}$. For this we need some preliminary results and so we interrupt the proof of Theorem 1.7.4.

Lemma 1.7.7. *Let $H \lhd^m J$, $K \lhd^n J$ and let $J = \langle H, K \rangle = HK$. Then, given integers $a, b \geqq 0$, there is an integer λ ($\geqq 0$), depending only on m, n, a and b, such that*

$$J^{(\lambda)} \leqq H^{(a)} K^{(b)}.$$

Proof. If H and K are normal in J, then $H^{(a)}$ and $K^{(b)}$ are also normal in J. Moreover, $J/H^{(a)}K^{(b)}$, as the product of two soluble normal subgroups, is soluble of derived length $\leqq a+b$. Thus we proceed by induction on $m+n$ and assume that one of m and n, say m, is at least 2. Let $H_1 = H^J$. Then $H_1 = H(H_1 \cap K)$. By induction there is $\lambda_1 = \lambda_1(m, n, a, b)$ with

$$H_1^{(\lambda_1)} \leqq H^{(a)}(H_1 \cap K)^{(b)}.$$

Finally, since $J = H_1 K$, induction also gives $\lambda = \lambda(n, \lambda_1, b)$ such that

$$J^{(\lambda)} \leqq H_1^{(\lambda_1)} K^{(b)} \leqq H^{(a)} K^{(b)}. \quad \square$$

Remark. We shall extend Theorem 1.6.4 in Theorem 3.1.1, showing that the hypothesis $\langle H, K \rangle = HK$ is unnecessary in Lemma 1.7.7.

The next result was originally proved by Černikov in [25]. Here we follow Roseblade [123].

Lemma 1.7.8. *Let H, K be subnormal subgroups of a group and suppose that $H \cong C_{p^\infty}$, $K \cong C_{q^\infty}$ for primes p and q (not necessarily distinct). Then $\langle H, K \rangle$ is abelian.*

Proof. Let $J = \langle H, K \rangle$. Then H^J is a p-group and K^J is a q-group (by induction up the normal closure series). Thus if $p \neq q$, then

$$[H, K] \leqq H^J \cap K^J = 1.$$

Therefore we may suppose that $p = q$. By induction on defect, we may also suppose that H^J and K^J are abelian.

Let $h \in H$ and $k \in K$. Then $h^{p^l} = 1$, for some l. Also there is an element k_1 in K with $k_1^{p^l} = k$. Now we use the fact that $[H, K] \leqq \zeta_1(J)$ and the commutator identities

$$[ab, c] = [a, c]^b [b, c],\ [a, bc] = [a, c][a, b]^c$$

to obtain

$$1 = [h^{p^l}, k_1] = [h, k_1]^{p^l} = [h, k_1^{p^l}] = [h, k].$$

Therefore J is abelian. □

As we shall see shortly, the above two lemmas, together with what we have already proved of Theorem 1.7.4, suffice to show that

any subjunctive subclass of $\mathfrak{Min}$-*sn is subnormally coalescent.*

However, to remove the hypothesis of $\mathrm{s_n}$-closure on this subclass (that is to complete the proof of Theorem 1.7.4) requires a further important result due to Roseblade. First an easy result.

Lemma 1.7.9. *Suppose that $\mathscr{S}$ is a set of minimal normal subgroups of the group G and that S is the subgroup generated by the members of $\mathscr{S}$. Then S is the direct product of some members of $\mathscr{S}$.*

Proof. Call a subset $\mathscr{T}$ of $\mathscr{S}$ *independent* if $\langle \mathscr{T} \rangle$ is the direct product of the members of $\mathscr{T}$. Then order the independent subsets of $\mathscr{S}$ by inclusion. By Zorn's lemma there is a maximal independent subset $\mathscr{T}$. If $S > \langle \mathscr{T} \rangle = T$, say, then there is a group H in $\mathscr{S}$ such that $H \nleqq T$. Since $T \lhd G$ and H is a minimal normal subgroup of G, we must have $T \cap H = 1$. Therefore $TH = T \times H$, contradicting the maximality of $\mathscr{T}$. Thus $S = T$. □

Roseblade's result is

Theorem 1.7.10. *If G satisfies* Min-*sn and H is subnormal in G, then H has only finitely many conjugates in G.*

Proof. Suppose the theorem is false. Then there is a group G, satisfying Min-*sn*, with a subnormal subgroup G_0 minimal subject to containing a subnormal subgroup with infinitely many conjugates. We may assume that $G_0 = G$. Similarly, we may assume that H is a subnormal subgroup of G minimal subject to $|G\colon N_G(H)|$ being infinite.

Let G^* be the unique minimal subnormal subgroup of finite index in G. Then $G^* \lhd G$ and hence $G_1 = G^*H$ *sn* G. If $G_1 < G$, our choice of G shows that $|G_1\colon N_{G_1}(H)|$ is finite. Since $|G\colon G_1|$ is finite, it follows that $|G\colon N_G(H)|$ is finite, giving a contradiction. Thus $G = G^*H$.

Let $H_1 = H^G$. Then $H_1 < G$ and so $|H_1 : N_{H_1}(H)|$ is finite. Therefore, since every subgroup of finite index contains a normal subgroup of finite index, H_1^* normalizes H so that

$$[H_1^*, H] \leqq H. \tag{2}$$

Clearly H_1^* is characteristic in H_1 and hence $H_1^* \lhd G$. Thus the centralizer $C_G(H_1/H_1^*)$ has finite index in G and $[G^*, H_1] \leqq H_1^*$. With (2) this gives

$$[G^*, H, H] \leqq [G^*, H_1, H] \leqq H.$$

Then $H \lhd H[G^*, H] \lhd G^*H = G$ and therefore $H \lhd H_1$.

By choice of H, every proper normal subgroup L of H has only finitely many conjugates in G and so G^* normalizes L. If Q is the product of all such L, then G^* normalizes Q and therefore $Q \lhd G = G^*H$. Hence we may assume that $Q = 1$ and H is simple. Thus by Lemma 1.7.9, H_1 is the direct product of certain conjugates of H. Since G satisfies Min-*sn*, H_1 must be the direct product of finitely many conjugates of H. However, such a direct product has only finitely many normal subgroups (either by an easy exercise or by looking ahead to §2.3). This contradicts our supposition that $|G : N_G(H)|$ is infinite. □

Remark. If H *sn* $G \in \mathfrak{Min}$-*sn* and $|G : G^*| = d$, then $N_G(H) \geqq G^*$ and so

H has at most d conjugates in G.

Proof of Theorem 1.7.4, *part* 2. We have

$$H, K \in \mathfrak{X} = \mathrm{N}_0\mathfrak{X} \leqq \mathfrak{Min}\text{-}sn,$$

H and K are subnormal in $J = \langle H, K \rangle$ and we must show that $J \in \mathfrak{X}$. In fact it is sufficient to prove that

$$J \in \mathfrak{Min}\text{-}sn. \tag{3}$$

For then, by Theorem 1.7.10 and induction on defect, we may assume that H^J and $K^J \in \mathfrak{X}$, whence $J \in \mathrm{N}_0\mathfrak{X} = \mathfrak{X}$.

To prove (3) we recall that $H^{(\omega)}$ is perfect and hence $H^{(\omega)}K$ and $H^{(\omega)}K^{(\omega)}$ are subgroups by Theorem 1.6.6. Using Lemma 1.7.7 it follows that some term of the derived series of $H^{(\omega)}K$ is contained in $H^{(\omega)}K^{(\omega)}$. Since the reverse inclusion is clear, we have

$$N = H^{(\omega)}K^{(\omega)} \lhd H^{(\omega)}K.$$

Similarly N is normalized by H and hence $N \lhd J$. The class $\mathfrak{Min}$-*sn* is subjunctive and hence, by Proposition 1.6.3, N satisfies Min-*sn*. The class $\mathfrak{Min}$-*sn* is also closed under forming extensions. Thus we may assume that $N = 1$, that is H and K are soluble. (Note that we only need $H, K \in \mathfrak{Min}$-*sn* to prove (3).) Therefore H and K satisfy the minimal condition for *all* subgroups and so they have normal subgroups H^* and K^*, respectively, of finite index, each a direct product of finitely many C_{p^∞}-subgroups (for various primes p).

By Lemma 1.7.8, $M = \langle H^*, K^* \rangle^J$ is abelian and J/M is finite, by Theorem 1.3.3. It follows that M is the product of only finitely many conjugates of $\langle H^*, K^* \rangle = H^*K^*$ and so M satisfies Min. Then J satisfies Min and (3) is established. This completes the proof of Theorem 1.7.4. □

Remarks

1. The final two paragraphs above (proving (3)) and Proposition 1.6.3 combine to show that

   any subjunctive subclass $\mathfrak{X}$ of $\mathfrak{Min}$-sn is subnormally coalescent.

   For, M will belong to $\mathfrak{X}$, and then so will HM, KM and hence J. The result is weaker than Theorem 1.7.4, but avoids using Theorem 1.7.10.
2. Any group satisfying the minimal condition for subnormal subgroups of defect $\leqq 2$ actually satisfies Min-*sn*. (See [119], p. 184.) Thus there is no need to consider the analogue of Theorem 1.7.3.

Our final application of Theorem 1.6.11 to subnormal coalescence concerns groups of finite rank. A group G has *finite rank r* if every finitely generated subgroup of G can be generated by at most r elements and r is the least integer with this property. Clearly every finitely generated abelian group has finite rank. On the other hand the additive group $\mathbb{Q}$ of rationals has rank 1, but is not finitely generated. The class of groups of finite rank is easily seen to be S- and Q-closed. (For any class $\mathfrak{X}$, Q$\mathfrak{X}$ is the class of all quotients of $\mathfrak{X}$-groups.) However, P-closure is not quite so immediate.

Lemma 1.7.11 [8]. *Let $N \lhd G$, where N and G/N have finite ranks r, s, respectively. Then G has finite rank $\leqq r+s$.*

Proof. Let $H = \langle h_1, h_2, \ldots, h_k \rangle$ be a finitely generated subgroup of G. Then $H/H \cap N \cong HN/N$ has finite rank $\leqq s$ and so

$$h_i = w_i(a_1, \ldots, a_s)x_i$$

for certain words w_i and elements $a_1, \ldots, a_s$ of H and x_i of $H \cap N$. Moreover $\langle x_1, \ldots, x_k \rangle = \langle b_1, \ldots, b_r \rangle$ for some $b_1, \ldots, b_r \in H \cap N$. Therefore $H = \langle a_1, \ldots, a_s, b_1, \ldots, b_r \rangle$ as required. □

Thus the class of groups of finite rank is P-closed and therefore N_0-closed.

For any class $\mathfrak{X}$ of groups, L$\mathfrak{X}$ denotes the class of all groups G such that every finite subset of G lies in an $\mathfrak{X}$-subgroup of G. So L is the 'local' closure operation. Then Drukker *et al.* [32] proved

Theorem 1.7.12. *For $i \geqq 1$, let $\mathfrak{X}_i$ be* L- *and* $\mathrm{S_n}$*-closed classes of groups and let*

$$\mathfrak{X} = \bigcup_{i \geqq 1} \mathfrak{X}_i .$$

Suppose that the product of a normal $\mathfrak{X}_i$-subgroup and a normal $\mathfrak{X}_j$- subgroup always belongs to $\mathfrak{X}_k$, where k depends only on i and j. Then the $\mathfrak{X}$-groups of finite rank form a subnormal coalition class.

Proof. Let H, K be subnormal $\mathfrak{X}$-subgroups of G with defects m, n and having finite ranks r, s, respectively. Then H is an $\mathfrak{X}_i$-group and K is an $\mathfrak{X}_j$-group for some i and j. Thus if $L \leqq H$ and $M \leqq K$ with L and M finitely generated, then L and M can be generated by at most r and s elements, respectively, and so the same is true of L/L' and M/M'. Therefore, by Theorem 1.6.11, there is a subgroup X of G such that

$$\langle L, M \rangle \leqq X \leqq \langle H, K \rangle = J,$$

say,

$$X \lhd^d G$$

and

$$X \in (\mathrm{N}_2\,\mathrm{S_n})^t((H) \cup (K)),$$

where d and t depend only on m, n, r and s. Since every finitely generated subgroup of J lies in one like $\langle L, M \rangle$, it follows from Lemma 1.7.5 that $J \lhd^d G$. Also $(H) \cup (K) \leqq \mathfrak{X}_i \cup \mathfrak{X}_j$ and therefore by hypothesis there is an integer k, depending only on m, n, r and s, such that $X \in \mathfrak{X}_k$. Thus

$$J \in \mathrm{L}\,\mathfrak{X}_k = \mathfrak{X}_k \leqq \mathfrak{X}.$$

Finally, since the product of two normal subgroups of finite ranks r_1, r_2 has rank $\leqq r_1 + r_2$ (from Lemma 1.7.11), we see that X has finite rank bounded by a function of m, n, r and s and therefore the same is true of J. □

Some special cases deserve mention.

Corollary 1.7.13. *The following are all subnormal coalition classes:*

(i) *groups of finite rank;*
(ii) *nilpotent groups of finite rank;*
(iii) *soluble groups of finite rank;*
(iv) *locally nilpotent groups of finite rank;*
(v) *locally soluble groups of finite rank.*

Proof. (i) Take each $\mathfrak{X}_i$ in the theorem to be the class of all groups.

(ii) For each i take $\mathfrak{X}_i$ to be the nilpotent groups of class $\leqq i$. The hypotheses of the theorem are then satisfied, by Theorem 1.6.1.

(iii) For each i take $\mathfrak{X}_i$ to be the soluble groups of derived length $\leqq i$.

(iv) In this case we take each $\mathfrak{X}_i$ to be the class of locally nilpotent groups and use the Hirsch–Plotkin theorem (2.1.1(iii)) which says that this class is N_0-closed.

(v) Here we take $\mathfrak{X}_i$ to be the class of locally soluble groups of rank $\leqq i$. Then the product of a normal $\mathfrak{X}_i$-subgroup and a normal $\mathfrak{X}_j$-subgroup has rank $\leqq i+j$ and is locally soluble [120] (Corollary 1 of Theorem 10.3.8) and so Theorem 1.7.12 applies. (The method of (ii) fails here because the class of locally soluble groups is *not* N_0-closed. See Theorem 2.1.2(iii).) □

Further examples of subnormal coalition classes have been found by Van Werkhooven [146].

2
THE JOIN OF MANY SUBNORMAL SUBGROUPS

§2.1 Internal structure of joins of finitely many subnormal subgroups

In the previous chapter we were primarily concerned with conditions sufficient to guarantee that a join J of two subnormal subgroups H and K is subnormal, though in §1.6 and §1.7 (on subnormal coalescence) we discovered some internal properties of H and K which are inherited by J. The intention now is to continue this investigation of the internal structure of J. The concept of N_0-closure is clearly of fundamental importance. Indeed if J is to inherit properties from H and K, then it must do so when H and K are normal. Thus we begin with a résumé of the N_0-closed classes of groups. We shall omit proofs where they are already easily accessible.

Many classes are N_0-closed for elementary reasons. Thus the finite, finitely generated and countable groups, as well as the perfect groups, are all N_0-closed classes. Similarly any class $\mathfrak{X}$ which is closed with respect to forming extensions and quotients (that is P- and Q-closed) is obviously N_0-closed. Examples are the following:

(i) soluble groups,
(ii) SN^*-groups, that is groups G having an ascending series from 1 to G with abelian factors,
(iii) $\mathfrak{Max}$, $\mathfrak{Min}$,
(iv) $\mathfrak{Max}$-*sn*, $\mathfrak{Min}$-*sn*,
(v) $\mathfrak{Max}$-*n*, $\mathfrak{Min}$-*n* (the classes of groups satisfying the maximal/minimal conditions for *normal* subgroups),
(vi) π-groups (for any set π of primes),
(vii) locally finite groups,
(viii) groups of finite rank (Lemma 1.7.11),
(ix) locally soluble groups of finite rank [120] (Corollary 1 of Theorem 10.3.8).

The classes whose N_0-closure is not so obvious tend to be those related to nilpotency. Some of these are given in

Theorem 2.1.1. *The following classes are N_0-closed.*
(i) *Nilpotent groups.*
(ii) *Hypercentral groups (that is groups G having an ascending central series from 1 to G).*
(iii) *Locally nilpotent groups.*

(iv) L$\mathfrak{X}$*-groups, where* $\mathfrak{X}$ *is any* N_0*- and* S*-closed class of finitely generated groups. In particular the class of locally noetherian groups is* N_0*-closed.* (See also Theorem 2.2.5.)
(v) *Periodic divisible abelian groups.* (A group is *divisible* if every element has an nth root for each $n \geqq 1$. See also Theorem 2.2.8.)
(vi) *Divisible nilpotent groups.*
(vii) *Divisible hypercentral groups.*

Proof. (i) is Fitting's Theorem 1.6.1.
(ii) is due to Philip Hall [48] (Lemma 1). The proof is also given in [119] (page 51).
(iii) is the famous Hirsch–Plotkin theorem ([60], [109], [119]) and is a special case of (iv), choosing the class of finitely generated nilpotent groups for $\mathfrak{X}$.
(iv) was proved by Hall in his Cambridge lectures and a proof can be found in [119] (Theorem 2.3.1). Baer also has this result in [6] (§4, Satz 1) with the additional hypothesis of Q-closure of $\mathfrak{X}$.
(v) A periodic divisible abelian group is a direct product of C_{p^∞}-groups, for various p [72], and so the product of two such normal subgroups is certainly abelian, by Lemma 1.7.8. Thus it is also divisible and clearly periodic.
(vi) is a corollary of (i) and (vii).
(vii) is a consequence of (ii) and Theorem 9.23 of [120], which proves that a hypercentral group G is divisible provided G/G' is divisible. (A quotient of a divisible group is clearly divisible.) □

Remarks

1. We cannot omit 'periodic' in (v). For, the group G of unitriangular 3×3 matrices over the rational field $\mathbb{Q}$ is the product of two normal abelian subgroups, each isomorphic to $\mathbb{Q} \oplus \mathbb{Q}$. But G is not abelian.
2. Philip Hall proved in his Cambridge lectures that the class of perfect groups in which every subnormal subgroup is actually normal is N_0-closed.

It is probably of interest to record a few results in the opposite direction, namely classes which are not N_0-closed. Some of these are perhaps a little surprising.

Theorem 2.1.2. *The following classes are* **not** N_0*-closed.*
(i) *Finite supersoluble groups.*
(ii) *The Fitting groups, that is products of (arbitrarily many) normal nilpotent subgroups.*
(iii) *Locally soluble groups.*
(iv) *SI*-groups, that is hyperabelian groups (having an ascending series of normal subgroups, from* 1 *to the whole group, with abelian factors).*

(v) *$\overline{SI}$-groups, that is those groups for which the factors of every generalized composition series are abelian.* (See §2.3.)

Proof. (i) There is a non-supersoluble semidirect product G of an elementary abelian group of order 25 by the quaternion group of order 8, and G is the product of two normal supersoluble subgroups. (This example is due to Huppert [61]. See also [134] (p. 219, Exercise 9.2.29).)
(ii) This appears in [117] (pages 114–16). However, the example is the group G_1 of Theorem 1.5.3. We recall that

$$G_1 = G]\langle t_1 \rangle,$$

with t_1 an involution, V, H and K are elementary abelian 2-groups and $G = V]\langle H, K\rangle$. Routine calculations show that, for each x_n in the basis of H,

$$[G, x_n] = [V, x_n]\,[K, x_n]^V$$

and

$$[V, x_n, x_n] = 1 = [[K, x_n]^V, x_n]\,.$$

Therefore $[G, x_n, x_n] = 1$ and so $\langle x_n^G\rangle$ is abelian. Similarly $\langle y_n^G\rangle$ is abelian for each y_n in the basis of K. Thus G is a Fitting group.

If $T = \langle t_1\rangle$, then $T \lhd^4 G_1$ and hence, by Proposition 1.6.3, VT is nilpotent. Moreover $VT \lhd^2 G_1$ and so VT^{G_1} is a Fitting group. Thus G_1 is the product of the normal Fitting subgroups G and VT^{G_1}. However, G_1 is not itself a Fitting group. For otherwise $T_1 = T^{G_1}$ would be nilpotent and hence $VHT_1 = G_1$ would also be nilpotent, by Proposition 1.6.3. This is not the case, since G_1 has a proper self-normalizing subgroup, namely $\langle H, K\rangle$.
(iii) This remarkable result was due in the first instance to Philip Hall in his 1962/3 Cambridge lectures. Baumslag *et al.* gave a second proof in 1965 [12]. It is a corollary of the following result, also due to Hall, which appears in [120] (Theorem 8.19.1(i)):

There is a finitely generated group which is the product of two normal locally soluble hyperabelian subgroups, but which is not itself hyperabelian. (1)

(iv) is also shown by (1) above.
(v) is again due to Hall and appears in [120] as a consequence of Theorem 8.19.1(ii):

There is a finitely generated group G which is the product of two normal locally soluble subgroups, but which is not an SI-group (that is, there is no generalized series of normal subgroups from 1 *to G with abelian factors).*

On the other hand a *chief factor* (that is, a minimal normal subgroup of a quotient) of a locally soluble group is always abelian [97]; see also [119] (Corollary 1 of Theorem 5.27). Thus every locally soluble group is an $\overline{SI}$-group. □

Other examples of non-$\mathbf{N}_0$-closed classes follow from an important result by Hall and Hartley [50]:

Theorem 2.1.3. *For each group H there is a group G which is the product of two normal free subgroups such that H can be embedded in G.*

The proof of this also appears in [120] (Theorem 8.19.2). It follows that the only $\mathbf{s}$- and $\mathbf{N}_0$-closed class containing the free groups is the class of all groups. For any class $\mathfrak{X}$, we say that a group G is *residually-$\mathfrak{X}$* (and write $G \in \mathbf{R}\mathfrak{X}$) if there are normal subgroups $N_\lambda (\lambda \in \Lambda)$ of G such that $G/N_\lambda \in \mathfrak{X}$, all $\lambda \in \Lambda$, and $\bigcap_{\lambda \in \Lambda} N_\lambda = 1$. Thus G is residually nilpotent if and only if $\gamma_\omega(G)$ (the ωth term of the lower central series of G) is 1. By a famous result of Iwasawa ([65]; see also [120] (Theorem 9.11)) free groups are residually nilpotent. We have therefore

Theorem 2.1.2 (continued). *The following classes are* **not** *$\mathbf{N}_0$-closed.*

(vi) *SN-groups, that is those groups having a generalized series with abelian factors.* (*See* §2.4.)

(vii) *SI-groups, that is those groups having a generalized series of normal subgroups with abelian factors.* (See §2.4.)

(viii) *Residually soluble groups.*

(ix) *Residually nilpotent groups.*

All of these results follow from Theorem 2.1.3 because free groups belong to each of the four classes which are subgroup-closed.

More non-$\mathbf{N}_0$-closed classes appear in [12] and [120] (pages 92 and 93). Recall that if F is a free group with a countably infinite basis and if W is a subset of F, then the class of all groups G, for which the kernel of each homomorphism $F \to G$ contains W, is called a *variety*. Clearly varieties are always closed with respect to $\mathbf{L}$, $\mathbf{Q}$, $\mathbf{R}$ and $\mathbf{s}$. Also every class $\mathfrak{X}$ such that $\mathfrak{X} = \mathbf{Q}\mathfrak{X} = \mathbf{R}\mathfrak{X}$ is a variety (Birkhoff, Kogalovskiĭ, Šain; see [119]). A further interesting (and unpublished) result due to Philip Hall is

Theorem 2.1.4. *The only $\mathbf{N}_0$-closed varieties are* (i) *the groups of order* 1 *and* (ii) *the class of all groups.*

Proof. If $\mathfrak{B}$ $(\neq (1))$ is a variety, then $\mathfrak{B}$ contains a cyclic group of prime order p (say). Assuming that $\mathfrak{B}$ is $\mathbf{N}_0$-closed, it follows, by induction on order, that $\mathfrak{B}$ contains all finite p-groups which can be generated by elements of order p. If a group G has a normal subgroup N, then G can be embedded in the complete wreath product $N \,\bar{\wr}\, (G/N)$ [130]. Thus any finite p-group can be embedded in a repeated wreath product

$$(\cdots (C_p \wr C_p) \wr \cdots) \wr C_p$$

and hence $\mathfrak{B}$ contains all finite p-groups. Since free groups are residually finite p-groups ([65]; also [120] (Theorem 9.11)), $\mathfrak{B}$ contains all free groups and therefore all groups. □

We conclude our discussion of the operator N_0 by considering its application to the class $\mathfrak{A}$ of abelian groups. By Theorem 1.6.1, $\mathrm{N}_0\mathfrak{A} \leqq \mathfrak{N}$. In fact Philip Hall has shown that

$$\mathfrak{G} \cap \mathfrak{N} < \mathrm{N}_0\mathfrak{A} < \mathfrak{N}.$$

Clearly the first *strict* inclusion will follow from $\mathfrak{G} \cap \mathfrak{N} \leqq \mathrm{N}_0\mathfrak{A}$. Thus suppose for a contradiction that G is a finitely generated nilpotent group and $G \notin \mathrm{N}_0\mathfrak{A}$. Since G satisfies the maximal condition for subgroups, we may assume that every proper subgroup of G lies in $\mathrm{N}_0\mathfrak{A}$. Now G is not cyclic and so $G = HK$, where $H, K \lhd G$ and $H \neq G \neq K$. Then $H, K \in \mathrm{N}_0\mathfrak{A}$ and hence $G \in \mathrm{N}_0\mathfrak{A}$.

Secondly consider the group G presented by

$$\langle x_n \mid n \in \mathbb{Z}, x_n^2 = [x_l, x_m, x_n] = 1, \text{ all } l, m, n \in \mathbb{Z} \rangle .$$

Then G is nilpotent of class 2. On the other hand, it is easy to show that if A is an abelian subgroup of G, then $|AG'/G'| \leqq 2$. Since G/G' is infinite, it follows that G is not in the class $\mathfrak{X}$ of nilpotent groups generated by finitely many abelian subgroups. However,

$$\mathfrak{A} \leqq \mathfrak{X} = \mathrm{N}_0\mathfrak{X}$$

and so

$$\mathrm{N}_0\mathfrak{A} \leqq \mathfrak{X} < \mathfrak{N} .$$

N_1-*closure.* We say that a class $\mathfrak{X}$ is N_1-*closed* if the join of a *finite* number of subnormal $\mathfrak{X}$-subgroups is always an $\mathfrak{X}$-group. Thus the class of finite groups is N_1-closed (Theorem 1.3.3) and the N_1-closure of the class of finite nilpotent groups then follows without difficulty using Theorem 1.6.1. In fact for classes of finite groups, N_1- and N_0-closure coincide.

Clearly every subnormal coalition class is N_1-closed (see Theorems 1.6.2, 1.6.12, 1.7.1, 1.7.2, 1.7.3, 1.7.4, 1.7.12 and Corollary 1.7.13). On the other hand, there are N_1-closed classes which are not subnormally coalescent, besides the class of all groups.

Theorem 2.1.5 [139]. *A join of finitely many subnormal soluble subgroups is always soluble.*

This result will be proved in its proper context in Chapter 3 as an immediate consequence of Theorem 3.1.1. It is interesting to observe, however, that if a join of two subnormal $\mathfrak{X}$-subgroups, while being an $\mathfrak{X}$-group, is not necessarily subnormal, then it is not possible to establish the N_1-closure of $\mathfrak{X}$ by induction on the number of subnormal subgroups generating the join.

Other N_1-closed classes are determined by Van Werkhooven [146]. For example, if $\mathfrak{X}$ is N_1-closed, then so is the class $\mathfrak{X}\mathfrak{F}$, where $\mathfrak{F}$ denotes the class of finite groups. Wielandt has many results concerning N_1-closure of classes of finite groups. Thus, if π is a set of primes, then in [155] (Satz 3.1) it is shown that the class of finite groups possessing a Hall π-subgroup with a Sylow tower (corresponding to a fixed ordering of the primes in π) is N_1-closed. By choosing all orderings of π simultaneously, it follows that the class of finite groups with a nilpotent Hall π-subgroup is N_1-closed. Also, by choosing π to be the set of all primes, we see that the class of finite groups with a Sylow tower (corresponding to a fixed ordering of the primes) is N_1-closed.

Of course any class which is not N_0-closed cannot be N_1-closed. However, N_0-closure does not imply N_1-closure. For, a join of two subnormal nilpotent subgroups need not be nilpotent. This is shown by all the groups constructed in Theorems 1.5.1, 1.5.2 and 1.5.5. Thus, in each case VH is nilpotent (by Proposition 1.6.3) and subnormal in G. Moreover $G = \langle VH, K\rangle$. Since $\langle H, K\rangle$ is a proper self-normalizing subgroup, G cannot be nilpotent. We note that K in Theorem 1.5.5 can have order 2. Another striking situation is that of Theorem 1.5.2 and it is worthwhile giving more details.

Theorem 2.1.6. *There is a non-nilpotent group G which is generated by two elementary abelian 2-subgroups, each subnormal of defect two in G.*

Proof. Recall that in Theorem 1.5.2, $V = A \times B, J = \langle H, K\rangle, G = V]J, A, B,$ H and K are elementary abelian 2-groups, $[B, H] = A$, $[A, K] = B$, $[A, H] = [B, K] = 1$.
Thus $(AH)^G = VH[H, K]$ and $(AH)^{VH[H,K]} = (AH)^{H[H,K]} = AH$ since $[H, K]$ normalizes A and centralizes H. So $AH \lhd^2 G$ and similarly $BK \lhd^2 G$. Therefore G is generated by the elementary abelian 2-subgroups AH and BK, each subnormal of defect two in G. On the other hand, J is self-normalizing in G and so G is not nilpotent. □

Remarks

1. The above group G is not even hypercentral. This is because proper subgroups of hypercentral groups cannot be self-normalizing.
2. It is not hard to see that if G is the join of a finitely generated subnormal nilpotent subgroup K and a subnormal nilpotent subgroup H of defect $\leqq 2$, then G is nilpotent. For, by Lemma 1.4.3,

$$G = H^*K,$$

where H^* is generated by finitely many conjugates of H. Thus, by Theorem 1.6.1, H^* is nilpotent. Then G is nilpotent, by Proposition 1.6.3. (This argument appeared in [129] (§3.2).)

Following through the N_0-closed classes of Theorem 2.1.1, we postpone a discussion of the locally nilpotent, locally noetherian and periodic divisible abelian groups until §2.2, where results stronger than $\mathbf{N}_1$-closure will be established without difficulty. The class of divisible abelian groups, however, is not so well-behaved. The following result shows that neither the divisible nilpotent nor the divisible hypercentral groups form an $\mathbf{N}_1$-closed class.

Theorem 2.1.7. *There is a group generated by two subnormal torsion-free divisible abelian subgroups which is not hypercentral.*

Proof. We modify the example of Theorem 1.5.5. Thus let A be an infinite-dimensional $\mathbb{Q}$-space and let R be the exterior algebra on A. So, as a $\mathbb{Q}$-space, R is a torsion-free divisible abelian group under addition.

For a in A, write

$$\lambda_0(a) = \begin{pmatrix} 1 & a \\ 0 & 1 \end{pmatrix} \quad \text{and} \quad H = \{\lambda_0(a) \mid a \in A\}.$$

Then H (under multiplication) is isomorphic to A (under addition). For $q \in \mathbb{Q}$, write

$$\xi_q = \begin{pmatrix} 1 & 0 \\ q & 1 \end{pmatrix}$$

and $K = \{\xi_q \mid q \in \mathbb{Q}\}$. So K (under multiplication) is isomorphic to $\mathbb{Q}$ (under addition). With $J = \langle H, K \rangle$, we can show, as in Theorem 1.5.5, that $H^K, K^H \in \mathfrak{N}_2$. Now J acts on $V = R \oplus R$ according to

$$(x, y)^{\lambda_0(a)} = (x, xa + y),$$
$$(x, y)^{\xi_q} = (x + qy, y).$$

Let $G = V]J$. Then exactly as in Theorem 1.5.5, we see that $H, K \lhd^5 G$ and J is self-normalizing in G. So G is not hypercentral. However, with $V_2 = \{(0, y) \mid y \in R\}$, we have

$$G = \langle V_2 H, K \rangle.$$

Moreover $V_2 H = V_2 \times H$ is a torsion-free divisible abelian subgroup, subnormal in G, by Theorem 1.2.1. □

We note that one of the two subnormal subgroups generating G in the above theorem, namely K, has rank 1. By contrast, a join of two subnormal abelian subgroups of finite rank is always nilpotent, by Corollary 1.7.13(ii).

An interesting example due to Rex Dark (see [129]) is described in

Theorem 2.1.8. *There is an $\mathbf{N}_0$-closed class $\mathfrak{X}$ of finitely generated groups which is not subnormally coalescent. Thus it is not $\mathbf{N}_1$-closed.*

Proof. Let $Q = \{r/2^s \mid r, s \in \mathbb{Z}\}$, the additive group of dyadic rationals, and let

$$D = Q_1 \oplus Q_2$$

be the direct sum of two copies of Q. Then D has commuting automorphisms α and κ defined by

$$d^\alpha = 2d, \qquad \text{all } d \in D,$$

and

$$(x, y)^\kappa = (x, x + y), \tag{2}$$

all $(x, y) \in D$. Let A, K be the subgroups of the automorphism group of D generated by α, κ, respectively, and let

$$G = D]AK.$$

We take $H = DA$ and $\mathfrak{X} = \mathrm{N}_0((H) \cup (K))$. Since $H = \langle (1,0), (0,1), \alpha \rangle$ and K is cyclic, *$\mathfrak{X}$-groups are finitely generated.*

Equation (2) shows that $[D,K] \leqq Q_2$ and $[Q_2,K] = 1$. Therefore $K \lhd K^D = K^G$ and so

$$K \lhd^2 G.$$

Clearly $H \lhd G$ and thus G is generated by two subnormal $\mathfrak{X}$-subgroups. We show that $G \notin \mathfrak{X}$.

Let S be any subgroup of G such that
(i) $S \cap Q_1 \neq 1$ and $Q_2/S \cap Q_2$ is finite, and
(ii) $S \nleqq H$ and $S \nleqq DK$.
Suppose that $S = MN$, where $M,N \lhd S$ and $M,N \in \mathfrak{X}$. Then

$$\text{at least one of } M,N \text{ satisfies (i) and (ii).} \tag{3}$$

For, suppose that $M \nleqq DK$. In this case there is an element $d\alpha^m\kappa^n \in M$ with $d \in D$ and $m \neq 0$, and so $[S \cap Q_2, M]$ $(\leqq M \cap Q_2)$ has finite index in Q_2. Thus

$$Q_2/M \cap Q_2 \qquad \text{is finite.} \tag{4}$$

Also, by considering $[S \cap Q_1, M]$, we see that M contains an element $(x, y) \in D$ with $x \neq 0$. By (4) there is an integer $l \geqq 1$ such that $(0, ly) \in M$. Therefore $(lx, 0) \in M$ and so

$$M \cap Q_1 \neq 1. \tag{5}$$

Therefore to prove (3) we may assume that *either* $M \leqq H$ *or* $M \leqq DK$. Similar arguments apply to N and so without loss of generality $M \leqq H$ and $N \leqq DK$. Thus $M \nleqq DK$ (by (ii)) and (5) holds. Since $M \nleqq D$, it follows that $DM \cap A \neq 1$. Now by considering the conjugates of elements in $M \cap Q_1$ by elements of M, we deduce that $Q_1/M \cap Q_1$ is finite. Similarly $DN \cap K \neq 1$ and, since $[M \cap Q_1, DN \cap K] \leqq N \cap Q_2$, we have $Q_2/N \cap Q_2$ finite.

Thus $N \cap Q_2$ is not finitely generated and so the nilpotent group N cannot be finitely generated, contradicting $N \in \mathfrak{X}$. Therefore (3) follows.

Now let $\mathfrak{X}_0 = (H) \cup (K)$ and, for $n \geqq 0$, define $\mathfrak{X}_{n+1} = \mathrm{N}_2\mathfrak{X}_n$, that is $\mathfrak{X}_{n+1}$-groups are the product of two normal $\mathfrak{X}_n$-subgroups. Then $\mathfrak{X} = \bigcup_{n \geqq 0} \mathfrak{X}_n$. Hence if $G \in \mathfrak{X}$, some subgroup S satisfying (i) and (ii) is isomorphic to H or

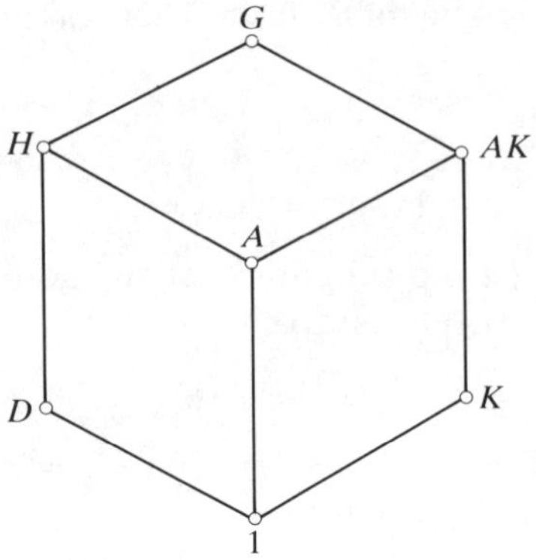

K. Clearly $S \nleqq K$ and therefore there is an isomorphism $\theta: H \to S$. However, $H' = G' = D$ and so

$$D^{\theta} = S' \leqq D.$$

Let $\alpha^{\theta} = d\alpha^{m}\kappa^{n}$, where $d \in D$. Then $m \neq 0 \neq n$, by (ii). Corresponding to (1,0) and (0,1) there must be two $\mathbb{Z}$-independent elements in S' satisfying

$$(x,y)^{d\alpha^{m}\kappa^{n}} = (2^{m}x, n2^{m}x + 2^{m}y) = (2x, 2y).$$

However, this equation has solutions ($\neq (0,0)$) only when $m = 1$ and then they have the form $(0,y)$. Thus $G \notin \mathfrak{X}$. □

§2.2 Internal structure of joins of arbitrarily many subnormal subgroups

In §2.4 and §2.6 we shall consider when the join of arbitrarily many subnormal subgroups is subnormal. However, the problem of determining the internal structure of such a join occurs most naturally at this point of our development of the theory of subnormal subgroups.

Proposition 2.2.1. *Let $\mathfrak{X}$ be a class of groups. Then every join of subnormal $\mathfrak{X}$-subgroups is an $\mathfrak{X}$-group if and only if every product of normal $\mathfrak{X}$-subgroups is an $\mathfrak{X}$-group.*

Proof. Suppose that every product of normal $\mathfrak{X}$-subgroups is an $\mathfrak{X}$-group and let H be a subnormal $\mathfrak{X}$-subgroup of defect m in a group G. Denoting the ith normal closure of H in G by H_i, we have

$$H_i = H_{i+1}^{H_{i-1}},$$

for $1 \leqq i \leqq m-1$, a product of conjugates of H_{i+1}, all normal in H_i. Therefore, by induction on i decreasing, we see that $H_1 \in \mathfrak{X}$. Thus a join of subnormal $\mathfrak{X}$-subgroups is a product of normal $\mathfrak{X}$-subgroups, namely their normal closures. Hence the join is an $\mathfrak{X}$-group.

The converse is immediate. □

For any class $\mathfrak{X}$ we have defined $\mathrm{N}\mathfrak{X}$ to be all those groups which are generated by subnormal $\mathfrak{X}$-subgroups. It follows from the above proposition that $\mathfrak{X} = \mathrm{N}\mathfrak{X}$ *if and only if every product of normal* $\mathfrak{X}$*-subgroups is an* $\mathfrak{X}$*-group.* Thus the closure operation N can be seen as an obvious generalization of both N_0 and N_1.

The $\mathfrak{X}$*-radical* $\rho_{\mathfrak{X}}(G)$ of a group G is defined to be the product of all the normal $\mathfrak{X}$-subgroups of G. Thus it follows that

if $\mathfrak{X} = \mathrm{N}\mathfrak{X}$, *then the* $\mathfrak{X}$*-radical of a group* G *contains all the subnormal* $\mathfrak{X}$*-subgroups of* G.

Also if $\mathfrak{X} = \mathrm{N}\mathfrak{X} = \mathrm{Q}\mathfrak{X}$ and $G/\rho_{\mathfrak{X}}(G)$ has trivial $\mathfrak{X}$-radical for every group G, then $\mathfrak{X}$ is said to be a *radical* class. Various characterizations of radical classes are given in [119] (pages 20–2).

Our main concern in this section is to determine the more important N-closed classes. Already from Proposition 2.2.1 it follows easily that *the class* SN^* (that is the class of groups G having an ascending series, from 1 to G, with abelian factors) *is* N*-closed.*

Proposition 2.2.2. *If* $\mathfrak{X} = \mathrm{L}\mathfrak{X} = \mathrm{N}_0\mathfrak{X}$, *then* $\mathfrak{X} = \mathrm{N}\mathfrak{X}$.

Proof. Let G be a product of normal $\mathfrak{X}$-subgroups N_λ, $\lambda \in \Lambda$. By Proposition 2.2.1, it is sufficient to prove that $G \in \mathfrak{X}$. Let X be a finite subset of G. Then X lies in a product H of finitely many of the N_λ. By hypothesis $H \in \mathfrak{X}$. Therefore $G \in \mathrm{L}\mathfrak{X} = \mathfrak{X}$. □

Then we obtain

Corollary 2.2.3. *The following are* N*-closed classes:*
(i) *locally finite groups;*
(ii) π*-groups, for any set* π *of primes.*

It follows from (ii) that subnormal π- and π'-subgroups commute elementwise. (π' is the set of primes excluded from π.) In particular the periodic elements of a nilpotent (or even locally nilpotent) group G form a subgroup G_0 and the p-elements (for p a prime) also form a subgroup G_p (the *p-component* of G); and G_0 is the direct product of the G_p.

Since the class of locally nilpotent groups is N_0-closed (Theorem 2.1.1(iii)), we also have

Theorem 2.2.4. *A join of subnormal locally nilpotent subgroups is locally nilpotent.*

Thus a join of subnormal nilpotent or hypercentral groups is locally nilpotent. Theorem 2.2.4 also follows from

Theorem 2.2.5. *Let* $\mathfrak{X} = \mathrm{N}_0\mathfrak{X} = \mathrm{S}\mathfrak{X}$ *be a class of finitely generated groups. Then the class* $\mathrm{L}\mathfrak{X}$ *is* N*-closed.*

Proof. By Theorem 2.1.1(iv), $\mathrm{L}\mathfrak{X}$ is N_0-closed. Thus the result follows from Proposition 2.2.2. □

As a special case we see that

the class of locally noetherian groups is N*-closed.*

For any class $\mathfrak{X}$, let $\mathrm{L}_n\mathfrak{X}$ denote the class of groups in which each finite set of elements lies in a subnormal $\mathfrak{X}$-subgroup. Then Corollary 1.6.15 says that

for any subjunctive class $\mathfrak{X}$, $\mathrm{L}_n\mathfrak{X}$ *is* N*-closed.*

The class of soluble groups is not N-closed; for example a direct product of soluble groups of unbounded derived lengths is not soluble. However, as we have already pointed out in §1.6, a join of subnormal soluble subgroups is locally soluble. This is also a corollary of Theorem 2.1.5.

An analogue of Theorem 2.2.5 is due to Rips and Robinson and is described in detail in [119] (pages 62–4). Accordingly we shall merely state the result here. For any class $\mathfrak{X}$, let $\mathrm{M}\mathfrak{X}$ denote the class of groups G in which each finite set of elements lies in a normal $\mathfrak{X}$-subgroup. Thus $\mathrm{M}\mathfrak{X} \leqq \mathrm{L}_n\mathfrak{X}$. However, M is *not* a closure operation. For example

$$(C_4 \times C_4 \times \cdots)]C_2,$$

with infinitely many copies of C_4 on each of which C_2 acts non-trivially, lies in $\mathrm{M}^2\mathfrak{F}$, but not in $\mathrm{M}\mathfrak{F}$. Thus let $\bar{\mathrm{M}}$ be the closure operation generated by M; so

$$\bar{\mathrm{M}}\mathfrak{X} = \bigcup_{\alpha} \mathrm{M}^{\alpha}\mathfrak{X},$$

where the union is taken over *all* ordinals α. Then

Theorem 2.2.6. *Let* $\mathfrak{X} = \mathrm{N}_0\mathfrak{X} = \mathrm{S}\mathfrak{X}$ *be a class of finitely generated groups. Then the class* $\bar{\mathrm{M}}\mathfrak{X}$ *is* N*-closed.*

It is easy to see that $\bar{\mathrm{M}}\mathfrak{X} \leqq \mathrm{L}_n\mathfrak{X} \leqq \mathrm{N}\mathfrak{X}$, for any class $\mathfrak{X}$. On the other hand, if $\mathfrak{X}$ satisfies the hypotheses of Theorem 2.2.6, then $\mathrm{N}\mathfrak{X} \leqq \mathrm{N}\bar{\mathrm{M}}\mathfrak{X} = \bar{\mathrm{M}}\mathfrak{X}$. Therefore we have

Corollary 2.2.7. *If* $\mathfrak{X} = \mathrm{N}_0\mathfrak{X} = \mathrm{S}\mathfrak{X}$ *is any class of finitely generated groups, then* $\bar{\mathrm{M}}\mathfrak{X} = \mathrm{N}\mathfrak{X} = \mathrm{L}_n\mathfrak{X}$.

We mention that $\bar{\mathrm{M}} \neq \mathrm{L}_n$, that is there are classes $\mathfrak{X}$ for which $\bar{\mathrm{M}}\mathfrak{X} < \mathrm{L}_n\mathfrak{X}$, and we leave this as an exercise for the interested reader. Similarly $\mathrm{L}_n \neq \mathrm{N}$, though this is very easy to see.

The only N_0-closed class of Theorem 2.1.1 which remains to be discussed in relation to N-closure is that of the periodic divisible abelian groups. In fact a periodic divisible hypercentral group is abelian [120] (Corollary 2 of Theorem 9.23). Moreover, by Lemma 1.7.8, subnormal C_{p^∞}- and C_{q^∞}-subgroups commute elementwise (even when $p = q$). Consequently a join of any number of periodic divisible abelian subnormal subgroups is abelian. Thus

Theorem 2.2.8. *The class of periodic divisible abelian groups is* N*-closed.*

An extension of this result to joins of ascendant subgroups (see p. 70 for definition) can be found in [119] (Lemma 4.46).

Semisimple groups

A direct product of non-abelian simple groups is said to be *semisimple.* Already we have seen that perfect subnormal subgroups permute with arbitrary subnormal subgroups (Theorem 1.6.6). Moreover we shall see in §4.5 that under suitable conditions perfect subnormal subgroups normalize and even centralize other subnormals. In particular disjoint perfect subnormal subgroups will be shown to commute elementwise (Corollary 4.5.4). That disjoint non-abelian simple subnormal subgroups commute elementwise follows without difficulty from Theorem 1.6.6. Then, using Zorn's lemma, the fact that the class of semisimple groups is N-closed is an immediate corollary. However, we choose to establish this result by more elementary considerations.

First we prove

Proposition 2.2.9. *Let G be the direct product of non-abelian simple groups $G_\lambda(\lambda \in \Lambda)$. Then every subnormal subgroup H of G is the direct product of a selection of the G_λ and hence $H \lhd G$. In particular a semisimple group can be expressed in only one way as a direct product of simple groups.*

Proof. Suppose that there is an element h in H whose projection in G_λ is $g_\lambda \neq 1$. Choose $x_\lambda \in G_\lambda$ such that $[g_\lambda, x_\lambda] \neq 1$. Then $[g_\lambda, x_\lambda] = [h, x_\lambda]$ and hence $[H, G_\lambda] \neq 1$. Therefore $[H, G_\lambda] = G_\lambda$ and so

$$H^{HG_\lambda} = HG_\lambda.$$

But $H \operatorname{sn} HG_\lambda$ so that $H = HG_\lambda$. Thus $H \geqq G_\lambda$. Therefore H is the direct product of those G_λ into which H projects non-trivially. □

Of course the final statement of the proposition is a special case of the Krull–Remak–Schmidt theorem on direct decompositions (see [72] and [171]). As a corollary we have

Theorem 2.2.10. *The class of semisimple groups is* N*-closed. Thus a join of subnormal non-abelian simple subgroups is their direct product; and a join of subnormal semisimple subgroups is just the direct product of the simple direct factors of these subgroups.*

Proof. Let G be the product of normal semisimple subgroups N_λ, $\lambda \in \Lambda$. By Proposition 2.2.1, it is sufficient to show that G is semisimple.

From Proposition 2.2.9, any direct product of non-abelian simple groups has a unique expression as such. Then the union of any ascending chain of direct products of non-abelian simple groups is similarly such a direct product. Thus by Zorn's lemma there is a maximal normal semisimple subgroup M of G. Suppose that $M \neq G$. Then there is some $N_\lambda \nleq M$. By Proposition 2.2.9,

$$N_\lambda = (N_\lambda \cap M) \times D,$$

where both factors are semisimple and $D \neq 1$. Moreover any conjugate (in G) of a simple direct factor of D lies in $N_\lambda \cap M$ or D; and the former possibility cannot occur because $N_\lambda \cap M \lhd G$. Therefore $D \lhd G$. Thus $M < \langle M,D \rangle = M \times D \lhd G$, contradicting the maximality of M. Hence $M = G$ and G is semisimple. □

Remarks

1. The first application of Proposition 2.2.9 above can be avoided by arguing as in Lemma 1.7.9. Thus consider those subsets $\mathscr{S}$ of the set of all simple direct factors of all the N_λ for which $\langle \mathscr{S} \rangle$ is the direct product of the members of $\mathscr{S}$. Ordering the sets $\mathscr{S}$ by inclusion and applying Zorn's lemma to those for which $\langle \mathscr{S} \rangle \lhd G$, we see that there is a maximal one, say $\mathscr{M}$. Take $M = \langle \mathscr{M} \rangle$ and proceed as before.
2. As a special case of the theorem, we see that if H is a non-abelian simple subnormal subgroup of G, then H^G is the direct product of the conjugates of H in G and H^G is a minimal normal subgroup of G, by Proposition 2.2.9. (This latter fact was proved by Wielandt in [154] (5).)
3. Let H,K be subnormal subgroups of G with H non-abelian and simple and $H \cap K = 1$. Then $[H,K] = 1$. For, we may assume that $G = \langle H,K \rangle$. Also H^G is a minimal normal subgroup of G by 2 above. Therefore $H^G \cap K^G = H^G$ or 1. In the former case $K^G = G$ and so $K = G$ and $H = 1$, a contradiction. Hence $H^G \cap K^G = 1$ and then $[H,K] = 1$ ([154] (2)).
4. Let G be the direct product of simple groups $G_\lambda \, (\lambda \in \Lambda)$ where, unlike Proposition 2.2.9, there is now no requirement that the G_λ be non-abelian. Suppose that H is a subnormal subgroup of G. Then an easy application of Zorn's lemma shows that H is complemented in G by the direct product of a selection of the G_λ and hence $H \lhd G$.
5. If a group has a minimal normal subgroup, then that subgroup must be *characteristically simple* (that is have no characteristic subgroups other

than itself and 1—a subgroup is *characteristic* if it is invariant under all automorphisms of the group). Now suppose that a characteristically simple group G has a minimal normal subgroup H. Then G is generated by H^α as α runs through the automorphisms of G. Therefore, by Lemma 1.7.9, G is the direct product of certain of the H^α. So H must be simple and G is either semisimple or elementary abelian.

§2.3 Subnormal composition factors

In this section we continue the investigation of N-closed classes, but now with particular reference to composition factors. A series

$$1 = G_0 \lhd G_1 \lhd \cdots \lhd G_l = G$$

of a group G, with l finite, is called a (finite) *composition* series of G if each factor G_{i+1}/G_i is simple. The famous Jordan–Hölder theorem (see [72] and [171]) says the following:

If a group G has a composition series of length l, then any series of G (of finite length and without repetitions) has length $\leqq l$, any two composition series have the same length l and there is a bijection between the two sets of composition factors (counting multiplicities) such that corresponding factors are isomorphic. (l is called the *composition length* of G.)

In this situation it follows that any series (without repeated terms) can be *refined* (by inserting extra terms if need be) to a composition series. Also the factors (with multiplicities) in any composition series of G are uniquely determined as a set of abstract groups called *the composition factors of G.*

The Jordan–Hölder theorem depends on the equally famous Zassenhaus lemma, a proof of which we include for the sake of completeness and for its applicability to other problems which we shall consider.

Lemma 2.3.1. *Let $A_1 \lhd A_2$ and $B_1 \lhd B_2$ be subgroups of a group. Then*

$$(A_2 \cap B_2)A_1/(A_2 \cap B_1)A_1 \cong (A_2 \cap B_2)B_1/(A_1 \cap B_2)B_1.$$

Proof. Clearly $A_1 \cap B_2 \lhd A_2 \cap B_2$ and there is an isomorphism

$$(A_2 \cap B_2)/(A_1 \cap B_2) \simeq (A_2 \cap B_2)A_1/A_1. \tag{1}$$

Also $A_2 \cap B_1 \lhd A_2 \cap B_2$ and hence

$$(A_1 \cap B_2)(A_2 \cap B_1) \lhd A_2 \cap B_2.$$

Restricting the isomorphism (1), we have

$$(A_1 \cap B_2)(A_2 \cap B_1)/(A_1 \cap B_2) \simeq (A_2 \cap B_1)A_1/A_1. \tag{2}$$

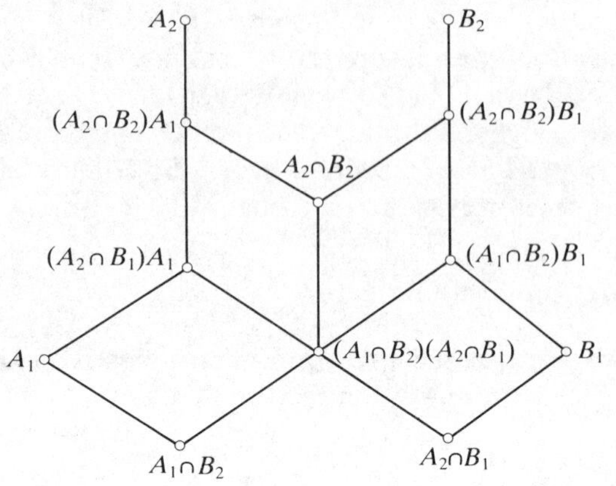

Then forming quotients from (1) and (2) we obtain

$$(A_2 \cap B_2)/(A_1 \cap B_2)(A_2 \cap B_1) \simeq (A_2 \cap B_2)A_1/(A_2 \cap B_1)A_1.$$

The result now follows by symmetry. □

The Jordan–Hölder theorem is an immediate corollary, since any two series have isomorphic refinements according to the Zassenhaus lemma. Thus the groups with a composition series are precisely the groups satisfying Max-*sn* and Min-*sn*. It is probably also worth recording the following result.

Theorem 2.3.2. *A group G has a composition series if and only if G has only finitely many subnormal subgroups.*

Proof. Suppose that G has a composition series. We show that G has only finitely many subnormal subgroups, by induction on the composition length l of G. Thus we assume that $l \geqq 2$. It is sufficient to show that G has only finitely many maximal normal subgroups.

Let K be a maximal normal subgroup of G. By induction there are only finitely many subgroups N of K such that $N \lhd G$. Therefore, again by induction, there are only finitely many maximal normal subgroups of G intersecting K non-trivially. Thus suppose that there is a maximal normal subgroup L of G such that $K \cap L = 1$. So $G = K \times L$ and K, L are simple. We may assume that K and L are not both abelian. Since any maximal normal subgroup of G different from K and L is isomorphic to both K and L, it follows that K and L are both non-abelian. Proposition 2.2.9 then shows that K and L are the only maximal normal subgroups of G.

The converse statement of the theorem is immediate. □

Remarks

1. Let $H \, sn \, G$ and suppose that

$$H = H_0 \lhd H_1 \lhd \cdots \lhd H_l = G \tag{3}$$

is a series of finite length l with each factor H_{i+1}/H_i simple. Then we call (3) a *composition series between H and G*. The Jordan–Hölder theorem extends to such series. Also the argument of Theorem 2.3.2 adapts easily to show that there is a composition series between H (sn G) and G if and only if there are only finitely many subnormal subgroups of G containing H.

2. A series

$$1 = G_0 \lhd G_1 \lhd \cdots \lhd G_l = G$$

of a group G, with l finite and all $G_i \lhd G$, is called a *chief* series of G if each factor G_{i+1}/G_i is a minimal normal subgroup of G/G_i. The groups G_{i+1}/G_i are the *chief factors* of G. The Jordan–Hölder theorem extends to chief series and a group G has a chief series if and only if G satisfies Max-n and Min-n (the maximal and minimal conditions for normal subgroups). Again the argument of Theorem 2.3.2 adapts easily to show that G has a chief series if and only if G has only finitely many normal subgroups.

Since a group G with a composition series satisfies Max-sn, Theorem 1.3.2 shows that the join J of any collection (necessarily finite) of subnormal subgroups H_λ ($\lambda \in \Lambda$) of G is subnormal in G. Thus J has a composition series. Thinking of composition factors as abstract groups, it is then an easy exercise (using the Jordan–Hölder theorem and induction on composition length) to show that the set (*not* counting multiplicities) of composition factors of J is the union of the sets of composition factors of the subgroups H_λ. In fact there is a generalization of this result to arbitrary groups. Accordingly if X is subnormal in a group G, $Y \lhd X$ and X/Y is simple, then we call X/Y a *subnormal composition factor of* G. Also we denote the set (without multiplicities) of abstract simple groups which occur as subnormal composition factors of G by $\mathscr{C}(G)$. It seems to be unknown whether $\mathscr{C}(G)$ can be empty if $G \neq 1$. However,

Theorem 2.3.3. *If G is generated by subnormal subgroups H_λ ($\lambda \in \Lambda$), then*

$$\mathscr{C}(G) = \bigcup_\lambda \mathscr{C}(H_\lambda).$$

Thus the class of groups with subnormal composition factors from a fixed set of abstract simple groups is N*-closed.*

This result combines with Corollary 1.6.7 to give

Corollary 2.3.4. *The class of groups in which all the subnormal subgroups are perfect is subnormally coalescent.*

Of course the subnormal subgroups of G are all perfect if and only if $\mathscr{C}(G)$ contains no abelian groups.

In order to prove Theorem 2.3.3 we need

Lemma 2.3.5. *Any two ascending series of subnormal subgroups of a group G, starting from a subnormal subgroup H, have isomorphic refinements.*

Proof. Let
$$H = A_0 \lhd A_1 \lhd \cdots A_\alpha \lhd A_{\alpha+1} \lhd \cdots,$$
$$H = B_0 \lhd B_1 \lhd \cdots B_\beta \lhd B_{\beta+1} \lhd \cdots$$
be ascending series with all $A_\alpha, B_\beta \; sn \; G$ and $G = \bigcup_\alpha A_\alpha = \bigcup_\beta B_\beta$. By Lemma 2.3.1,
$$(A_{\alpha+1} \cap B_{\beta+1})A_\alpha/(A_{\alpha+1} \cap B_\beta)A_\alpha \cong (A_{\alpha+1} \cap B_{\beta+1})B_\beta/(A_\alpha \cap B_{\beta+1})B_\beta.$$
Now $A_{\alpha+1} \cap B_\beta \; sn \; A_{\alpha+1}$. Therefore $(A_{\alpha+1} \cap B_\beta)A_\alpha \; sn \; A_{\alpha+1}$ (by Proposition 1.1.4) and hence $(A_{\alpha+1} \cap B_\beta)A_\alpha$ is subnormal in G. Since, for fixed α, the subgroups $(A_{\alpha+1} \cap B_\beta)A_\alpha$ form an ascending series of subnormal subgroups from A_α to $A_{\alpha+1}$ and, for fixed β, the subgroups $(A_\alpha \cap B_{\beta+1})B_\beta$ similarly form an ascending series of subnormal subgroups from B_β to $B_{\beta+1}$, the result follows. □

It is in fact true that ascending series of subgroups which are not necessarily subnormal also have isomorphic refinements (see [72]).

Proof of Theorem 2.3.3. Clearly $\mathscr{C}(G) \supseteq \bigcup_\lambda \mathscr{C}(H_\lambda)$. To establish the reverse inclusion, we may assume, by Proposition 2.2.1, that $H_\lambda \lhd G$, all λ. Let X/Y be a subnormal composition factor of G. Well-order Λ and let $N_\lambda = \prod_{\mu<\lambda} H_\mu$. Introduce a further index $\nu > \lambda$, all λ, and put $N_\nu = G$. Then the subgroups $N_\lambda (\lambda \in \{\Lambda, \nu\} = \Lambda^*$, say) form an ascending series from 1 to G of normal subgroups with each factor isomorphic to a quotient of some H_λ.

Let $X \lhd X_1 \lhd \cdots \lhd X_r = G$ be a series. By Lemma 2.3.5, the series $(N_\lambda)_{\lambda\in\Lambda^*}$ and
$$1 \lhd Y \lhd X \lhd X_1 \lhd \cdots \lhd X_r = G$$
have isomorphic refinements. Since X/Y admits no non-trivial refinements, X/Y must be isomorphic to a subnormal composition factor of H_λ, for some λ, and hence the theorem follows. □

If D is an intersection of subnormal subgroups H_λ $(\lambda \in \Lambda)$, then clearly
$$\mathscr{C}(D) \neq \bigcap_\lambda \mathscr{C}(H_\lambda)$$
in general. Though obviously $\mathscr{C}(D) \subseteq \bigcap_\lambda \mathscr{C}(H_\lambda)$ if $|\Lambda|$ is finite. However, there

are groups G with subnormal subgroups $H_\lambda (\lambda \in \Lambda)$ and

$$D = \bigcap_\lambda H_\lambda$$

such that $\mathscr{C}(D) \nsubseteq \mathscr{C}(G)$. For example let p be prime and let R be the localization of $\mathbb{Z}$ at p. Write $G = GL(n, R)$, where $n \geqq 3$, and define

$$D = \{1 + ae_{12} | a \in pR\}.$$

Here e_{12} denotes the matrix with (1, 2)-coefficient equal to 1 and with zeros elsewhere. Wilson shows in [168] (pages 174–5) that if D_m is the mth normal closure of the subgroup D in G, then

$$\bigcap_{m \geqq 1} D_m = D.$$

Moreover, from Theorem 2*(b) of the same paper, every subnormal composition factor of G has order p. Since D is abelian and torsion-free, D has subnormal composition factors of every prime order.

Let L be a subgroup of a group G, and denote by $\mathscr{C}(G: L)$ the set of abstract subnormal composition factors X/Y of G with $L \leqq Y$. We shall obtain a much stronger result than Theorem 2.3.3 (see Theorem 2.3.7). This is motivated by Sätze 6 and 9 of Wielandt's paper [152]:

If G possesses a composition series and H, K sn G, then

$$\mathscr{C}(H{:}H \cap K) = \mathscr{C}(\langle H, K \rangle{:}K). \tag{4}$$

Indeed the hypothesis that G has a composition series will be shown to be redundant. The relation $\subseteq$ in (4) is fairly straightforward and was generalized in the following way by Wielandt [158] (8.5).

Theorem 2.3.6. *Let $H_1, H_2, \dots, H_n$ be finitely many subnormal subgroups of G. Then, for each i,*

$$\mathscr{C}(H_i{:}D) \subseteq \bigcup_{j \neq i} \mathscr{C}(G{:}H_j)$$

where $D = H_1 \cap \cdots \cap H_n$.

Proof. We suppose that $n \geqq 2$ and $i = 1$. Let X/Y be a subnormal composition factor of H_1 such that

$$D \leqq Y \lhd X \leqq H_1.$$

Let $D_1 = H_2 \cap H_3 \cap \cdots \cap H_n$. Thus $H_1 \cap D_1 = D$ and there are series

$$1 \lhd Y \lhd X \lhd U_1 \lhd \cdots \lhd U_r = G,$$
$$1 \lhd D_1 = V_0 \lhd V_1 \lhd \cdots \lhd V_s = G.$$

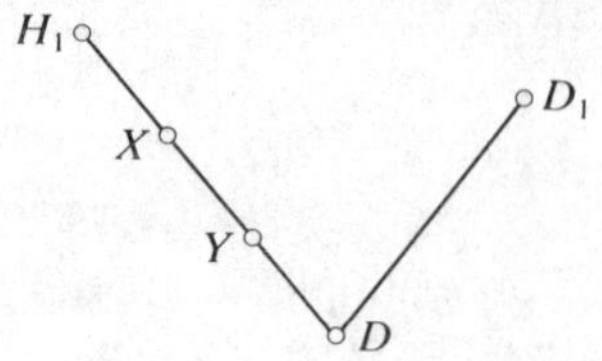

By Lemma 2.3.5, these series have isomorphic refinements, hence either

$$X/Y \cong (X \cap D_1)/(Y \cap D_1)$$

or

$$X/Y \cong (X \cap V_{k+1})V_k/(Y \cap V_{k+1})V_k,$$

for some k, $0 \leqq k \leqq s-1$. The first possibility cannot hold and so $X/Y \in \mathscr{C}(G:D_1)$. Then once more by Lemma 2.3.5 we obtain

$$X/Y \in \mathscr{C}(G:H_2) \cup \mathscr{C}(H_2:D_1). \tag{5}$$

If $n = 2$, the result now follows. On the other hand, if $n \geqq 3$, then we can argue by induction on n to obtain

$$\mathscr{C}(H_2:D_1) \subseteq \bigcup_{j \geqq 3} \mathscr{C}(G:H_j).$$

Thus, again from (5), the result follows. □

Remarks

1. Without difficulty we deduce that $\mathscr{C}(G:D) = \bigcup_{i=1}^{n} \mathscr{C}(G:H_i)$.
2. The case $n = 2$ shows that if H, K *sn* G, then

$$\mathscr{C}(H:H \cap K) \subseteq \mathscr{C}(G:K).$$

   By taking $G = \langle H, K \rangle$ we obtain the relation $\subseteq$ in (4).
3. Wielandt also pointed out that Theorem 2.3.6 does not extend to infinitely many subnormals. For, let $G = \langle g \rangle$ be an infinite cyclic group and let $H_n = \langle g^{2^n} \rangle$, for $n \geqq 1$. Then $D = \bigcap_{n \geqq 1} H_n = 1$ and the theorem obviously fails.
4. Theorem 2.3.6 shows that if π is a set of primes and G is a finite group, then the subnormal subgroups of G with index a π-number form a lattice. Similarly if $\mathscr{S}$ is a set of finite simple groups, then the subnormal subgroups H of G for which $\mathscr{C}(G:H) \subseteq \mathscr{S}$ also form a lattice.

In order to obtain the relation $\supseteq$ in (4), Wielandt needed a more complicated argument. We describe his interesting proof here. Thus suppose that G has a composition series and $H, K \; sn \, G = \langle H, K \rangle$. We prove that

$$\mathscr{C}(H:H \cap K) \supseteq \mathscr{C}(G:K) \tag{6}$$

by induction on the length l of a composition series between H and G. We may assume that $l \geqq 1$. Choose a maximal normal subgroup G^* of G containing H and put $G^* \cap K = K^*$.

Case (*i*). Suppose that $K^* = H \cap K$. Then $H \cap K \lhd K$. If $H \lhd HK$, then (6) follows without difficulty. Therefore we assume that $H^K \neq H$. Thus there is an element k in K such that $H^* = \langle H, H^k \rangle > H$. We note that $H^* \; sn \, G$, by Theorem 1.3.2. Let 1 denote a composition series between K and G and let $\{1\}$ denote the set (without multiplicities) of abstract composition factors of the series 1. Similarly we use 2, 3 and 4 as indicated in the diagram. Then

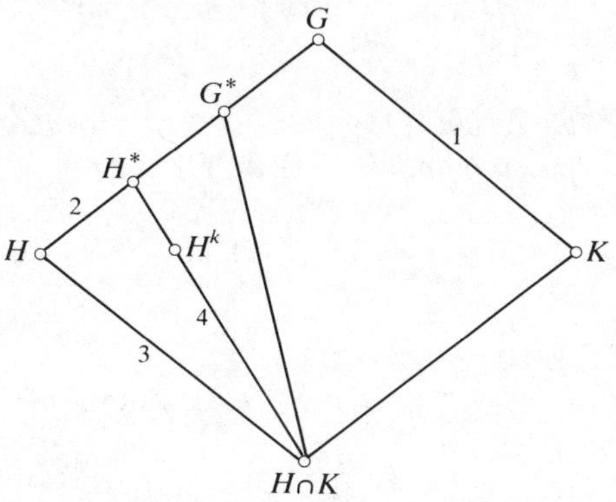

$\{1\} \subseteq \{2\} \cup \{3\}$, by induction on l applied to $G = \langle H^*, K \rangle$. Also by induction, applied to $H^* = \langle H, H^k \rangle$, $\{2\} \subseteq \{4\}$. Since conjugation by k takes a composition series between $H \cap K$ and H to one between $H \cap K$ and H^k, we have $\{4\} = \{3\}$. Thus (6) holds in this case.

Case (*ii*). (Diagram on p. 62.) Suppose that $K^* > H \cap K$. Then $\bar{H} = \langle H, K^* \rangle > H$. So $\bar{H} \; sn \, G$ and $\{1\} \subseteq \{2\}$ by induction applied to $\langle \bar{H}, K \rangle$. Also

$$\{2\} \subseteq \{2\} \cup \{3\} = \{4\}$$

and $\{4\} \subseteq \{5\}$ by induction applied to $\langle H, K^* \rangle$. Therefore again (6) holds.

It is clear that the above argument will not adapt to arbitrary groups G, first because composition series do not exist in general (and composition lengths cannot be replaced by subnormal defects in this instance) and secondly because a join of subnormal subgroups is not always subnormal. Thus it is perhaps a

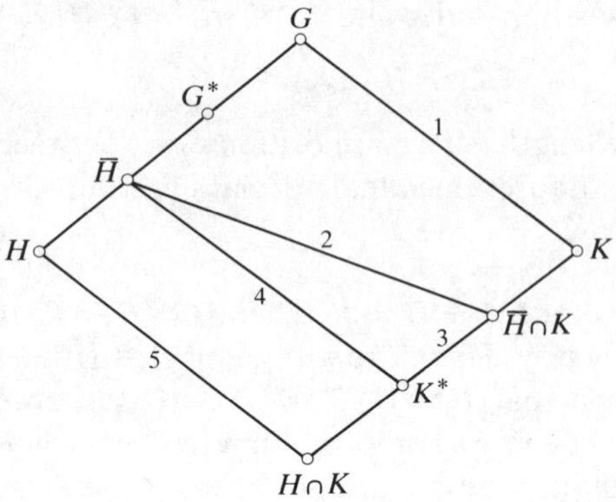

little surprising that the hypothesis that G has a composition series can be shown to be unnecessary in (6). We do this in

Theorem 2.3.7 (Stonehewer [144]). *Let G be generated by subnormal subgroups H_λ ($\lambda \in \Lambda$, possibly infinite). Then, for each λ,*

$$\mathscr{C}(G:H_\lambda) = \bigcup_{\mu \neq \lambda} \mathscr{C}(H_\mu : H_\lambda \cap H_\mu).$$

Remarks

1. Using Theorem 2.3.6 we see that, if $|\Lambda|$ is finite, then

$$\bigcup_\lambda \mathscr{C}(G:H_\lambda) = \bigcup_\lambda \mathscr{C}(H_\lambda : D),$$

where $D = \bigcap_\lambda H_\lambda$. Also if H, $K \operatorname{sn} G$, then

$$\mathscr{C}(\langle H, K \rangle : K) = \mathscr{C}(H : H \cap K),$$

and so (4) holds for arbitrary groups.

2. With the notation of Theorem 2.3.7, we have

$$\mathscr{C}(G) = \bigcup_\lambda \mathscr{C}(H_\lambda),$$

that is Theorem 2.3.3.

In order to prove Theorem 2.3.7, we need one of the main results of Chapter 3, namely Theorem 3.1.1. This says that if the group G is generated by finitely many subnormal subgroups $H_1, \ldots, H_n$ and $d_1, \ldots, d_n$ are integers $\geqq 0$, then there is an integer $d \geqq 0$ such that

$$G^{(d)} \leqq H_1^{(d_1)} H_2^{(d_2)} \cdots H_n^{(d_n)}. \tag{7}$$

(As a corollary of (7) we shall have Theorem 2.1.5 which says that a join of finitely many subnormal soluble subgroups is soluble.)

Proof of Theorem 2.3.7. The group G is generated by subnormal subgroups H_λ ($\lambda \in \Lambda$). Let X/Y be a subnormal composition factor of G with

$$H_\lambda \leqq Y \lhd X \leqq G,$$

for some λ. We have to prove that $X/Y \in \mathscr{C}(H_\mu : H_\lambda \cap H_\mu)$, for some $\mu \neq \lambda$. We reduce first to the case when

$$\Lambda \text{ is a finite set.} \tag{8}$$

To see this choose an element x from $X \setminus Y$. Then there are elements $\lambda_1, \ldots, \lambda_n$ in Λ such that $x \in \langle H_{\lambda_1}, \ldots, H_{\lambda_n} \rangle = J$, say, and we may assume that $H_{\lambda_1} = H_\lambda$. To simplify notation put $H_i = H_{\lambda_i}$, $i = 1, 2, \ldots, n$. Now by Theorem 1.6.14, there is a subnormal subgroup L of G such that $x \in L \leqq J$. Therefore $X \cap L \, sn \, G$ and so $(X \cap L) Y \, sn \, G$, by Proposition 1.2.1. Since $x \in X \cap L$ and $x \notin Y$, we must have $(X \cap L) Y = X$. Hence $(X \cap J) Y = X$ and therefore

$$X/Y \cong (X \cap J)/(Y \cap J),$$

a subnormal composition factor of J above H_1 ($= H_\lambda$). Replacing G by J, it follows that we may assume (8).

Thus we have $G = \langle H_1, H_2, \ldots, H_n \rangle$ with each $H_i \, sn \, G$ and X/Y is a subnormal composition factor of G with $H_1 \leqq Y \lhd X \leqq G$. It now suffices to prove the theorem in the case when

$$G \text{ is soluble.} \tag{9}$$

For, let $P_2 = P_{H_2}(H_1)$, the permutizer of H_1 in H_2. Then $H_1 \cap H_2 \leqq P_2$. Put $Q_2 = H_1 P_2$. Since $P_2 \, sn \, G$, by Theorem 1.6.9, we have $Q_2 \, sn \, G$, by Theorem 1.2.5. Thus there are series

$$H_1 \lhd \cdots \lhd Q_2 \lhd \cdots \lhd G$$

and

$$H_1 \lhd \cdots \lhd Y \lhd X \lhd \cdots \lhd G,$$

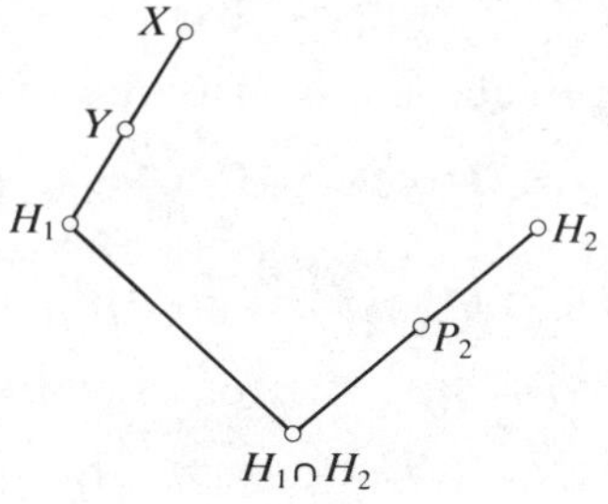

and so, by Lemma 2.3.5, we may assume that either $H_1 \leqq Y \lhd X \leqq Q_2$ or

$$Q_2 \leqq Y \lhd X \leqq G. \tag{10}$$

In the first case

$$X/Y \in \mathscr{C}(Q_2:H_1) = \mathscr{C}(P_2:H_1 \cap P_2) \subseteq \mathscr{C}(H_2:H_1 \cap H_2),$$

as required. Therefore we assume that (10) holds. By Theorem 1.6.8, there is an integer d_2 such that $H_2^{(d_2)} \leqq P_2$. There is no loss of generality in taking $H_1 = Q_2$ and so we may assume that $H_2^{(d_2)} \leqq H_1$. Repeating this process with $H_3, H_4, \ldots, H_n$ in turn replacing H_2, we obtain

$$H_i^{(d_i)} \leqq H_1,$$

for $2 \leqq i \leqq n$ and for certain integers d_i. However, according to (7) there is an integer d such that

$$G^{(d)} \leqq H_1 H_2^{(d_2)} \cdots H_n^{(d_n)}$$

and so $G^{(d)} \leqq H_1$. Let $N = G^{(d)}$. Since $H_1 \cap (H_i N) = (H_1 \cap H_i)N$ and

$$\mathscr{C}(H_i N:(H_1 \cap H_i)N) \subseteq \mathscr{C}(H_i:H_1 \cap H_i),$$

for all i, it follows that we may replace G by G/N. Then we have established (9) and so we may suppose that G is soluble.

It is now clear that $X/Y \simeq C_p$, a cyclic group of prime order p. The next step is to show that

we may assume that G is finitely generated. (11)

For, choose $x \in X \setminus Y$. Then $x \in \langle K_1, K_2, \ldots, K_n \rangle = M$, say, where K_i is a finitely generated subgroup of H_i, for each i. So M is finitely generated. Let $H_i^* = H_i \cap M$. Thus $H_i^* \geqq K_i$, $M = \langle H_1^*, \ldots, H_n^* \rangle$ and H_i^* *sn* M. Put $X_1 = X \cap M$, $Y_1 = Y \cap M$. Since $x \in X_1 \setminus Y_1$, we have

$$X_1/Y_1 \simeq X_1 Y/Y = X/Y$$

and hence $C_p \in \mathscr{C}(M:H_1^*)$. If the theorem holds for finitely generated groups, then

$$C_p \in \mathscr{C}(H_j^*:H_1^* \cap H_j^*) \tag{12}$$

for some $j \geqq 2$. Let

$$H_1 \cap H_j = A_0 \lhd A_1 \lhd \cdots \lhd A_m = H_j$$

be a series with all A_{i+1}/A_i abelian. Intersecting with M, we obtain

$$H_1^* \cap H_j^* = A_0 \cap M \lhd A_1 \cap M \lhd \cdots \lhd A_m \cap M = H_j^*$$

and

$$(A_{i+1} \cap M)/(A_i \cap M) \simeq (A_{i+1} \cap M)A_i/A_i \leqq A_{i+1}/A_i.$$

By (12), $C_p \in \mathscr{C}((A_{i+1} \cap M)/(A_i \cap M))$, for some i, and so

$$C_p \in \mathscr{C}(A_{i+1}/A_i) \subseteq \mathscr{C}(H_j:H_i \cap H_j),$$

showing that the theorem will hold for all groups. Therefore (11) follows.

Now G is a finitely generated soluble group. Let

$$H_1 = B_0 \lhd B_1 \lhd \cdots \lhd B_l = G$$

be a series with each factor B_{i+1}/B_i abelian. Choose i as large as possible such that B_{i+1}/B_i has C_p as a subnormal composition factor. Then for all $j \geqq i+1$, B_{j+1}/B_j must be periodic. Since G is finitely generated, induction over j decreasing shows that $|G:B_{i+1}|$ is finite and B_{i+1} is finitely generated. Therefore we may assume that $X = B_{i+1}$ and $Y = H_1$. Then $|G:H_1|$ is finite. We have already seen (while proving (9)) that it is in order to replace G by G factored by any normal subgroup contained in H_1. Thus we may assume that H_1 is core-free in G (that is the intersection of the conjugates of H_1 in G is 1) and so G is finite.

Finally, for $1 \leqq i \leqq n$, let $G_i = \langle H_1, H_2, \ldots, H_i \rangle$. Since all the subgroups G_i are subnormal in G, Lemma 2.3.5 gives

$$X/Y \in \mathscr{C}(G_{i+1}: G_i),$$

for some $i \geqq 1$. Therefore, by (6) applied to the finite group G_{i+1} generated by the subnormal subgroups H_{i+1} and G_i,

$$X/Y \in \mathscr{C}(H_{i+1}: G_i \cap H_{i+1}) \subseteq \mathscr{C}(H_{i+1}: H_1 \cap H_{i+1})$$

as required. Thus

$$\mathscr{C}(G:H_\lambda) \subseteq \bigcup_{\mu \neq \lambda} \mathscr{C}(H_\mu: H_\lambda \cap H_\mu).$$

The reverse inclusion follows from Theorem 2.3.6. This completes the proof of Theorem 2.3.7. □

Let G be a finite group and let $\mathscr{S}$ be a set of abstract simple groups. By Theorem 2.3.6, there is a unique subnormal subgroup G^* of G minimal such that $\mathscr{C}(G:G^*) \subseteq \mathscr{S}$. Clearly $G^* \lhd G$. Now suppose that G (still finite) is generated by subnormal subgroups $H_1, H_2, \ldots, H_n$. Then

$$H_i/(H_i \cap G^*) \simeq H_i G^*/G^*$$

and hence $\mathscr{C}(H_i: H_i \cap G^*) \subseteq \mathscr{S}$. Therefore $H_i^* \leqq H_i \cap G^*$ and so

$$\langle H_1^*, H_2^*, \ldots, H_n^* \rangle \leqq G^*.$$

Conversely, let $\langle H_1^*, \ldots, H_n^* \rangle \leqq Y \lhd X \, sn \, G$ with X/Y simple. Then $G = \langle H_1, \ldots, H_n, Y \rangle$ and hence, by Theorem 2.3.7,

$$X/Y \in \bigcup_{i=1}^{n} \mathscr{C}(H_i: H_i \cap Y) \tag{13}$$

$$\subseteq \bigcup_{i=1}^{n} \mathscr{C}(H_i: H_i^*) \subseteq \mathscr{S}.$$

Therefore $G^* \leqq \langle H_1^*, H_2^*, \ldots, H_n^* \rangle$. Consequently

$$G^* = \langle H_1^*, H_2^*, \ldots, H_n^* \rangle. \tag{14}$$

This result is due to Wielandt [153] (Satz 2.4). Clearly by far the greater part of the argument lies in establishing (13), that is Theorem 2.3.7. However, this theorem does not require G or the number of generating subnormal subgroups to be finite, and so we can obtain what is essentially a generalization of (14) to arbitrary groups:

Corollary 2.3.8 (Stonehewer [144]). *Let G be a group generated by subnormal subgroups H_λ ($\lambda \in \Lambda$). If X/Y is a subnormal composition factor of G, then*

$$X/Y \in \bigcup_\lambda \mathscr{C}(H_\lambda : H_\lambda \cap Y).$$

Proof. Since $G = \langle Y, H_\lambda | \lambda \in \Lambda \rangle$, we may take H_λ in Theorem 2.3.7 to be Y here. □

In conclusion we remark that there is no result like (14) with radicals replacing residuals. For, with $\mathscr{S} = \{C_2\}$ and G the dihedral group of order 12, then $G = HK$ where H and K are dihedral of order 6 (and therefore normal in G). However, denoting the join of those subnormal subgroups of G with composition factors in $\mathscr{S}$ by G_*, we have

$$H_* = K_* = 1, \qquad \text{whereas} \qquad G_* \cong C_2.$$

§2.4 Seriality of joins and the class $\mathfrak{S}^\infty$

Theorem 1.6.14 shows that if J is the join of a set of subnormal subgroups of a group G, then

every finitely generated subgroup of J lies in a subnormal subgroup of G contained in J. (1)

Hickin and Phillips used this result to prove that J is a *serial* subgroup of G [58]. First we must define this concept.

Definition. A subgroup H of a group G is said to be *serial* in G if there is a set

$$\{\Lambda_\sigma, V_\sigma | \sigma \in \Omega\} \tag{2}$$

of subgroups of G, where Ω is a totally ordered indexing set, such that

(i) $H \leqq V_\sigma \lhd \Lambda_\sigma$, all $\sigma \in \Omega$;
(ii) $\Lambda_\sigma \leqq V_\tau$ if $\sigma < \tau$;
(iii) $G \backslash H = \bigcup_{\sigma \in \Omega} (\Lambda_\sigma \backslash V_\sigma)$.

The set (2) is said to be a *generalized series between H and G*.

Thus the terms Λ_σ, V_σ form a nested set of subgroups between H and G. The quotients Λ_σ/V_σ are called the *factors* of the series. Moreover each element of G outside H lies in just one of the layers $\Lambda_\sigma \backslash V_\sigma$. From (ii) we have

$$\Lambda_\sigma \leqq \bigcap_{\sigma<\tau} V_\tau = V, \qquad \text{say.}$$

If $\Lambda_\sigma < V$, then there is an element $x \in V \backslash \Lambda_\sigma$ and so $x \in \Lambda_\rho \backslash V_\rho$ for some $\rho \in \Omega$. Thus $\sigma < \rho$ and hence $x \in V \leqq V_\rho$. This contradiction shows that

$$\Lambda_\sigma = \bigcap_{\sigma<\tau} V_\tau. \tag{3}$$

Similarly

$$V_\sigma = \bigcup_{\tau<\sigma} \Lambda_\tau. \tag{4}$$

(If Ω has a greatest or least element σ, then (3) and (4) are intepreted as $\Lambda_\sigma = G$ and $V_\sigma = H$, respectively.)

When Ω is a finite set, then H *sn* G and so subnormality is a special case of seriality.

A generalized series between 1 and G is called a generalized series *of* G. Such a series which contains all intersections and unions of its terms is said to be *complete*. If $\mathscr{S}_1$ and $\mathscr{S}_2$ are generalized series of G, then $\mathscr{S}_2$ is called a *refinement* of $\mathscr{S}_1$ if each term of $\mathscr{S}_1$ is also a term of $\mathscr{S}_2$. If $\mathscr{S}_1$ has no refinements other than itself, then it is called a *generalized composition series of G*. One uses Zorn's lemma to show that every group has such series. Similarly if $\mathscr{S}$ is a generalized series of *normal* subgroups of G and if $\mathscr{S}$ has no refinements (consisting of normal subgroups of G) other than itself, then $\mathscr{S}$ is called a *generalized chief series of G*. Again by Zorn's lemma, every group has such series. Clearly generalized composition and chief series are complete. We use Kuroš's terminology and say that G is an $\overline{SN}$-group if every factor Λ_σ/V_σ of every generalized composition series of G is abelian. Similarly, G is an $\overline{SI}$-group if every factor Λ_σ/V_σ of every generalized chief series of G is abelian. (See Theorem 2.1.2.) A generalized series with all factors belonging to a class $\mathfrak{X}$ is called an *$\mathfrak{X}$-series*.

Our main concern with generalized series at this point is

Theorem 2.4.1 [58]. *Let H be a subgroup of G. Then the following are equivalent.*

(i) *H is serial in G.*

(ii) *For each finite subset F of H, there is a serial subgroup L of G with* $F \leqq L \leqq H$.

(iii) *If K is a finitely generated subgroup of G and* $H < \langle H, K\rangle$, *then* $H^K < \langle H, K\rangle$.

If H is a join of subnormal subgroups of G, then, by (1), Theorem 2.4.1 (ii) holds and so we have

Theorem 2.4.2 [58]. *A join of arbitrarily many subnormal subgroups is serial.*

In order to prove Theorem 2.4.1, we need a lemma. First, suppose that G is a group and F is a subset of $G\backslash 1$. A finite sequence

$$A_1 \lhd B_1 \leqq A_2 \lhd B_2 \leqq \cdots \leqq A_n \lhd B_n \tag{5}$$

of subgroups of G is called a *separating chain for* F if $F \subseteq \bigcup_1^n (B_i \backslash A_i)$.

Lemma 2.4.3. *Let H be a subgroup of G and suppose that, for each finite subset F of $G\backslash H$, there is a separating chain* (5) *with $H \leqq A_1$. Then H is serial in G.*

Before proving this result we show how it is used to establish Theorem 2.4.1.

Proof of Theorem 2.4.1. To see that (i) implies (ii) we simply take $L = H$.

We show that (ii) implies (iii). Suppose, for a contradiction, that K is finitely generated and $H < \langle H, K \rangle$, but $K \leqq H^K$. Then there is a finite subset F of H such that $K \leqq F^K$, and hence $K \leqq L^K$, where L is serial in G and $F \leqq L \leqq H$. Thus $\langle L, K \rangle = L^K$. However,

$$K \nleqq H \Rightarrow K \nleqq L \Rightarrow L < \langle L, K \rangle,$$

and L is serial in $\langle L, K \rangle$ (by an elementary property of serial subgroups analogous to Proposition 1.1.2(i)). Therefore, since K is finitely generated, there is a subgroup $N \ntrianglelefteq \langle L, K \rangle$ with $L \leqq N$. Then $L^K \leqq N^K = N < \langle L, K \rangle$, giving the required contradiction.

Finally (i) follows from (iii). For, it suffices to prove that the hypotheses of Lemma 2.4.3 hold. Thus suppose, for a contradiction, that this is not the case and let $\mathscr{C}$ be the set of all finite, non-empty subsets of $G\backslash H$ for which the hypotheses of the lemma fail. Choose $F \in \mathscr{C}$ with $|F|$ as small as possible and let $K = \langle F \rangle$. So

$$K \nleqq H \Rightarrow H^K \ntrianglelefteq \langle H, K \rangle, \tag{6}$$

by hypothesis. Let $F_1 = H^K \cap F$.

Case 1. If $F_1 = \varnothing$, then

$$F \subseteq \langle H, K \rangle \backslash H^K,$$

contradicting $F \in \mathscr{C}$.

Case 2. Suppose that $F_1 \neq \varnothing$. By (6), $K \nleqq H^K$ and so $F \nsubseteq H^K$. Therefore $|F_1| < |F|$ and by our choice of F there must be a finite chain

$$H \leqq A_1 \lhd B_1 \leqq \cdots \leqq A_n \lhd B_n$$

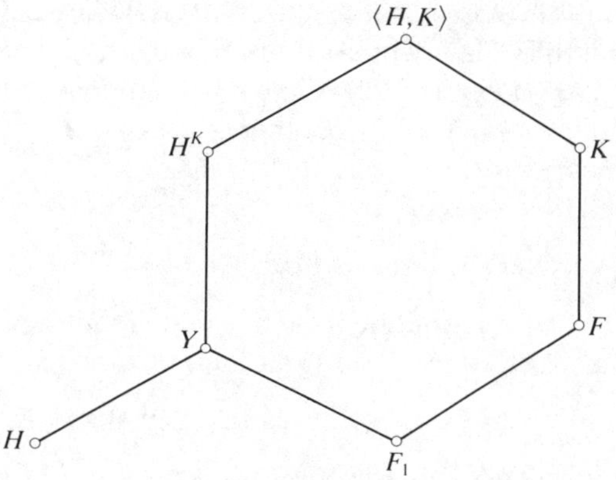

with $F_1 \subseteq \bigcup_1^n (B_i \backslash A_i)$. Let $Y = \langle H, F_1 \rangle$. Then

$$H \leqq A_1 \cap Y \lhd B_1 \cap Y \leqq \cdots \leqq A_n \cap Y \lhd B_n \cap Y \leqq H^K \lhd \langle H, K \rangle \quad (7)$$

and $F \backslash F_1 \subseteq \langle H, K \rangle \backslash H^K$. Hence (7) is a separating chain for F, contradicting $F \in \mathscr{C}$. Therefore the hypotheses of Lemma 2.4.3 hold and so H is serial in G. □

The proof of Lemma 2.4.3 uses Mal'cev's work on local systems. If X is a set, a *local system* $\mathscr{L}$ on X is a set of subsets of X such that every finite subset of X is contained in some element of $\mathscr{L}$. Now let M be a subset of a group G with $1 \in M$ and let $\prec$ be a relation on M. We shall be interested in the following properties:

(i) $x \prec y \prec z \Rightarrow x \prec z$;
(ii) $x \prec y$ or $y \prec x$;
(iii) $x \prec 1 \Rightarrow x = 1$;
(iv) $x \prec y$, $z \prec y$ and $xz^{-1} \in M \Rightarrow xz^{-1} \prec y$;
(v) $y \nprec x$ and $x^y \in M \Rightarrow y \nprec x^y$.

Proof of Lemma 2.4.3. Let F be a finite subset of $G \backslash H$. So there is a separating chain (5) for F with $H \leqq A_1$. Write $S = H \cup F$ and put $A_0 = 1$, $B_0 = H$. Then define a relation $\prec$ on S by:

$$x \prec y \Leftrightarrow x = 1 \quad \text{or} \quad r \leqq s \quad \text{where } x \in B_r \backslash A_r \text{ and } y \in B_s \backslash A_s.$$

One easily verifies (taking $M = S$) that S and $\prec$ satisfy properties (i)–(v) above. Now define a function f_S: $S \times S \to \{0, 1\}$ by

$$f_S(x, y) = \begin{cases} 0 & \text{if } x \prec y \\ 1 & \text{otherwise.} \end{cases}$$

The sets S (as F runs through *all* finite subsets of $G\backslash H$) form a local system $\mathscr{L}$ of G. Then according to P. Hall (who formulated a method due to McLain [91]; see for example [120] (Lemma 8.22)) there is a function $f: G \times G \to \{0, 1\}$ such that, given any finite subset C of $G \times G$, there exists $S \in \mathscr{L}$ for which $C \subseteq S \times S$ and f and f_S coincide on C.

We define a relation $\prec$ on G by:

$$x \prec y \Leftrightarrow f(x, y) = 0.$$

Again (taking $M = G$) it is routine to verify that G and $\prec$ satisfy properties (i)–(v). Now define a relation $\sim$ on G as follows:

$$x \sim y \Leftrightarrow x \prec y \quad \text{and} \quad y \prec x.$$

Then $\sim$ is an equivalence relation. According to (iii), $\{1\}$ is an equivalence class. Let Ω be the set of all equivalence classes *except* $\{1\}$. If $\sigma, \tau \in \Omega$, we write

$$\sigma < \tau \quad \text{if} \quad \sigma \neq \tau \text{ and there are } x \in \sigma,\ y \in \tau \text{ with } x \prec y.$$

One checks easily that $<$ is a well-defined total ordering of Ω.

Finally we see that Ω has a least element, namely $\sigma_0 = H\backslash 1$. For $\sigma \neq \sigma_0$, we define

$$\Lambda_\sigma = \{x \mid x \prec y, \text{ some } y \in \sigma\}$$

and

$$V_\sigma = \bigcup_{\tau < \sigma} \Lambda_\tau.$$

Then it is straightforward to show that $\{\Lambda_\sigma, V_\sigma \mid \sigma \in \Omega, \sigma \neq \sigma_0\}$ is a generalized series from H to G. □

When H is serial in a group G and the indexing set Ω is well-ordered, then H is called an *ascendant* subgroup of G. In this case Ω may be taken to be a set of ordinals

$$\{\alpha \mid \alpha < \gamma\}. \tag{8}$$

Then (3) shows that $\Lambda_\alpha = V_{\alpha+1}$ if $\alpha + 1 < \gamma$, and $\Lambda_\alpha = G$ if $\alpha + 1 = \gamma$. Thus the subgroups Λ_α are superfluous, except possibly for $\Lambda_{\gamma-1}$ which, if it exists, equals G. In any case we put $V_\gamma = G$ and write

$$H = V_0 \lhd V_1 \lhd \cdots V_\alpha \lhd V_{\alpha+1} \lhd \cdots V_\gamma = G.$$

Observe from (4) that if α is a limit ordinal, then

$$V_\alpha = \bigcup_{\beta < \alpha} V_\beta. \tag{9}$$

Similarly if Ω is inversely well-ordered, then a serial subgroup H of G is said to be *descendant*. This time Ω with reverse ordering may be taken to be a set of ordinals (8). Then the subgroups V_α are superfluous (defining $\Lambda_\gamma = H$) and we write

$$H = \Lambda_\gamma \cdots \lhd \Lambda_{\alpha+1} \lhd \Lambda_\alpha \cdots \lhd \Lambda_1 \lhd \Lambda_0 = G.$$

When α is a limit ordinal, we have, by analogy with (9),

$$\Lambda_\alpha = \bigcap_{\beta<\alpha} \Lambda_\beta.$$

Clearly a subnormal subgroup is both ascendant and descendant and hence (together with seriality) we have three generalizations of the concept of subnormality. In the light of Theorem 2.4.2 it is natural to ask if a join of an arbitrary number of subnormal subgroups is always ascendant or descendant. In fact a join of two subnormal subgroups need be neither ascendant nor descendant. For, in the group G of Theorem 1.5.2, J is self-normalizing and so cannot be ascendant. Also $J^G = G$ and so J cannot be descendant either.

Serial, ascendant and descendant subgroups have not proved to be very tractable and so we turn our attention now to the consideration of conditions sufficient to guarantee that a join J of arbitrarily many subnormal subgroups of a group G is actually subnormal. A necessary and sufficient condition has already been given in Theorem 1.3.10:

J sn G if and only if the set of subnormal subgroups of G lying in J contains at least one maximal member. (10)

Denote by $\mathfrak{S}^\infty$ the class of groups in which all joins of arbitrarily many subnormal subgroups are subnormal. An equivalent formulation of (10) is given in the following result [111] (Lemma 8.1).

Theorem 2.4.4. *$G \in \mathfrak{S}^\infty$ if and only if the union of each ascending chain of subnormal subgroups of G is subnormal in G.*

Proof. Suppose that the union of each ascending chain of subnormal subgroups of G is subnormal and let J be a join of subnormal subgroups of G. Then it is clear by Zorn's lemma that the set of subnormal subgroups of G lying in J contains a maximal member and so $G \in \mathfrak{S}^\infty$ by (10). The converse is immediate. □

Remark. It follows that $\mathfrak{Max}$-*sn* $\leq \mathfrak{S}^\infty$, but this was already clear from (10), that is Theorem 1.3.10.

The remainder of the theory of $\mathfrak{S}^\infty$-groups which we shall present (here and in §2.6) is almost entirely due to Robinson ([111] and [113]), though some proofs have been modified. We note the elementary facts that $\mathrm{Q}\mathfrak{S}^\infty = \mathrm{s_n}\mathfrak{S}^\infty = \mathfrak{S}^\infty$. We saw in Corollary 1.2.4 that in $\mathfrak{N}\mathfrak{A}$-groups the join of two (and hence finitely many) subnormal subgroups is always subnormal, that is $\mathfrak{N}\mathfrak{A} \leq \mathfrak{S}$. It is not surprising that $\mathfrak{S}^\infty$ is a much smaller class than $\mathfrak{S}$. In fact it is easy to see that there are metabelian groups which are not in $\mathfrak{S}^\infty$. We give three examples:

1. Let G_m be the dihedral group of order 2^m, for each $m \geqq 2$. Then G_m has a subgroup H_m of order 2 which is subnormal in G_m of defect $m-1$. Let
$$G = G_2 \times G_3 \times \cdots$$
and let $H = H_2 \times H_3 \times \cdots$. Thus each H_m is subnormal in G, but H is not subnormal in G. (Otherwise $H \lhd^n G$, for some n, and hence $H \cap G_m = H_m \lhd^n G_m$, which is clearly not the case if $m-1 > n$.)
2. Let p be an odd prime and let A_m be a cyclic group of order p^m, for each $m \geqq 2$. In the automorphism group of A_m there is a unique cyclic subgroup H_m of order p^{m-1} (see [134] (5.7.12)). Form the natural semidirect product $G_m = A_m]H_m$. Then it is easy to check that H_m is subnormal of defect m in G_m. Put
$$G = G_2 \times G_3 \times \cdots$$
and $H = H_2 \times H_3 \times \cdots$. As in example 1, each H_m is subnormal in G, but H is not subnormal in G.
3. Let p be any prime, $A \cong C_p$, $H \cong C_{p^\infty}$ and $G = A \wr H$, the standard wreath product formed by taking H in its regular representation. Since every cyclic subgroup of a soluble p-group of finite exponent is subnormal (see Theorem 2.5.7), it follows that H is generated by subnormal subgroups of G. On the other hand, if H is subnormal in G, then H^G is abelian, by Lemma 1.7.8. However, H is its own centralizer in G and so this would imply that $H \lhd G$, which is false. Thus H is not subnormal in G.

A positive result concerning the class $\mathfrak{S}^\infty$ is analogous to Corollary 1.4.5:

Theorem 2.4.5. $\mathfrak{S}^\infty(\mathfrak{Max}\text{-}sn) = \mathfrak{S}^\infty$.

Proof. Let $N \lhd G$ with $N \in \mathfrak{S}^\infty$ and $G/N \in \mathfrak{Max}\text{-}sn$. By Theorem 2.4.4 it is sufficient to prove that the union of an ascending chain of subnormal subgroups of G is subnormal. Thus let
$$H_1 \leqq H_2 \leqq \cdots H_\alpha \leqq \cdots$$
$(\alpha < \gamma)$ be such an ascending chain and let $H = \bigcup_{\alpha<\gamma} H_\alpha$. Since $G/N \in \mathfrak{Max}\text{-}sn$ and $HN = \bigcup_{\alpha<\gamma} H_\alpha N$ with all $H_\alpha N$ *sn* G, we must have $HN = H_\alpha N$, for some α, and so $H = H_\alpha(H \cap N)$. Also
$$H \cap N = \bigcup_{\beta<\gamma} (H_\beta \cap N).$$
Each $H_\beta \cap N$ is subnormal in N and therefore $H \cap N$ *sn* N, since $N \in \mathfrak{S}^\infty$. It now follows that H is the join of two subnormal subgroups of G, namely

H_α and $H \cap N$. Moreover H_α normalizes $H \cap N$ and hence $H\, sn\, G$, by Theorem 1.2.1. □

Corollary 2.4.6. *Finitely generated $\mathfrak{N}^2$-groups lie in $\mathfrak{S}^\infty$.*

For comparison with Theorem 2.4.5, we recall Theorem 1.4.2:

$$\mathfrak{S}\mathfrak{G}^{\mathrm{S_n}} = \mathfrak{S}.$$

Clearly $\mathfrak{Max}$-*sn* $\leqq \mathfrak{S}^\infty$ $(\leqq \mathfrak{S})$, but it appears to be unknown whether $\mathfrak{G}^{\mathrm{S_n}} \leqq \mathfrak{S}^\infty$. In fact it is easy to see that this will be the case if and only if $\mathfrak{G}^{\mathrm{S_n}} \leqq \mathfrak{Max}$-*sn*. For the necessity, let $G \in \mathfrak{G}^{\mathrm{S_n}}$. Suppose that

$$H_1 \leqq H_2 \leqq \cdots$$

is an ascending chain of subnormal subgroups of G. If $G \in \mathfrak{S}^\infty$, then $H = \bigcup_\alpha H_\alpha$ is subnormal in G and therefore H is finitely generated. Thus $H = H_n$, some n, and G will satisfy Max-*sn*.

Now consider a group G satisfying Min-*sn* and let H be subnormal in G. If G^* is the unique minimal subgroup of finite index in G, then

$$N = N_G(H) \geqq G^*,$$

by Theorem 1.7.10. Thus $H \lhd HG^* \lhd^d G$, where $d = |G:G^*|$. It follows from Corollary 1.7.6 that the union of each ascending chain of subnormal subgroups of G is subnormal and hence $G \in \mathfrak{S}^\infty$, by Theorem 2.4.4. Therefore

$$\mathfrak{Min}\text{-}sn \leqq \mathfrak{S}^\infty.$$

By analogy with Theorem 2.4.5 it is natural to ask if $\mathfrak{S}^\infty(\mathfrak{Min}\text{-}sn) = \mathfrak{S}^\infty$. In fact this is not the case, because, by example 3 above,

$$C_p \wr C_{p^\infty} \notin \mathfrak{S}^\infty.$$

There is a result in this direction, however. If G satisfies Min-*sn* and G has no proper subgroups of finite index, then every subnormal subgroup of G is actually normal, by Theorem 1.7.10. We denote the class of groups in which all subnormal subgroups are normal by $\mathfrak{T}$. (These are the groups in which normality is transitive.) Thus

$$\mathfrak{Min}\text{-}sn \leqq \mathfrak{T}\mathfrak{F}. \tag{11}$$

Also we shall prove

Theorem 2.4.7. $\mathfrak{S}^\infty \mathfrak{T} \leqq \mathfrak{S}^\infty$.

Then, using (11),
$$\begin{aligned} \mathfrak{S}^\infty(\mathfrak{Min}\text{-}sn) &\leqq \mathfrak{S}^\infty(\mathfrak{T}\mathfrak{F}) \leqq (\mathfrak{S}^\infty\mathfrak{T})\mathfrak{F} \\ &\leqq \mathfrak{S}\mathfrak{F}, \quad \text{by Theorem 2.4.7} \\ &= \mathfrak{S}, \quad \text{by Theorem 1.4.2.} \end{aligned}$$

Therefore

Corollary 2.4.8. $\mathfrak{S}^\infty(\mathfrak{Min}\text{-}sn) \leqq \mathfrak{S}$.

Another special case of Theorem 2.4.7 is

Corollary 2.4.9. $\mathfrak{S}^\infty\mathfrak{A} \leqq \mathfrak{S}$.

Proof of Theorem 2.4.7. Let $N \lhd G$, $N \in \mathfrak{S}^\infty$ and $G/N \in \mathfrak{T}$. Suppose that H, K are subnormal subgroups of G with $\langle H, K \rangle = J$. We show that $J \, sn \, G$ by induction on the defect r of H in J. We may assume that $r \geqq 1$.

Since $HN \, sn \, G$ and $G/N \in \mathfrak{T}$, it follows that $HN \lhd G$. Therefore $H^K \leqq HN$ and so

$$H^K = H(H^K \cap N). \tag{12}$$

If $x \in H^K \cap N$, then there is a finitely generated subgroup F of K such that $x \in H^F \cap N$. By induction on r and a second induction on $|F|$, $H^F \, sn \, G$. Hence $H^F \cap N \, sn \, N$. Now

$H^K \cap N = \langle H^F \cap N | \text{ all finitely generated subgroups } F \text{ of } K \rangle \; sn \, N$ since $N \in \mathfrak{S}^\infty$. Therefore $H^K \cap N \, sn \, G$. Then, by (12), $H^K sn \, G$ and so $J \, sn \, G$. □

For further results in this direction see §3.5.

The main result on the class $\mathfrak{S}^\infty$ (Theorem 2.6.1) involves the theory of Baer groups, which we therefore introduce in the next section.

§2.5 Baer groups

A group which is generated by subnormal abelian subgroups is called a *Baer* group. Since

$$\mathrm{N}\mathfrak{A} = \mathrm{N}\mathfrak{N} = \mathrm{N}(\mathfrak{G} \cap \mathfrak{N}),$$

Baer groups can also be thought of as joins of subnormal nilpotent subgroups or as joins of subnormal finitely generated nilpotent subgroups. By Theorem 1.6.2, a join of finitely many subnormal finitely generated nilpotent subgroups is always subnormal and nilpotent. Therefore every finitely generated subgroup of a Baer group is subnormal and nilpotent. This is sufficiently important to be stated as

Theorem 2.5.1

(i) *A group G is a Baer group if and only if every finitely generated subgroup of G is subnormal.*

(ii) *Every finitely generated subgroup of a Baer group is nilpotent.*

In particular the finitely generated Baer groups are just the finitely generated nilpotent groups and the finite Baer groups are just the finite nilpotent groups. Also there is the following connection with the class $\mathfrak{S}^{\infty}$.

Corollary 2.5.2. *All the subgroups of a group G are subnormal if and only if G is a Baer group in the class* $\mathfrak{S}^{\infty}$.

(An example of a metabelian Baer group of rank 2 which is *not* in $\mathfrak{S}^{\infty}$ will be given in Theorem 3.5.5.)

From the definition of Baer groups it is by no means clear that the class is closed with respect to forming subgroups, but Theorem 2.5.1(i) shows that this is in fact the case. Since Q- and N-closure are immediate, we obtain

Theorem 2.5.3. *The class of Baer groups is closed with respect to taking subgroups, quotients and subnormal joins.*

Clearly $\mathfrak{N} \leqq \mathrm{N}\mathfrak{A}$ and it is easy to see from example 1 on page 72 that the inclusion is strict. On the other hand, every Baer group is locally nilpotent, by Theorem 2.5.1(ii). The group $C_{2^\infty}]C_2$ (with C_2 acting by inversion) is locally nilpotent, but the C_2-subgroup is not subnormal. Therefore

$$\mathfrak{N} < \mathrm{N}\mathfrak{A} < \mathrm{L}\mathfrak{N}.$$

We now consider a way in which Baer groups can arise. Let N be a group and H a subgroup of Aut N (the automorphism group of N). Suppose that

$$1 = N_m \lhd N_{m-1} \lhd \cdots \lhd N_0 = N \tag{1}$$

is a series of subgroups of N, each invariant under H, such that all the factors N_{i-1}/N_i are fixed elementwise by H. Then H is said to *stabilize* the series (1) and H is called a *stability* group. Clearly

$$[N, \overbrace{H, H, \ldots, H}^{m}] = 1$$

and so H is subnormal in the natural semidirect product $G = N]H$. Conversely if H (a subgroup of Aut N) is subnormal in $N]H$, then there exists a series

$$H = H_m \lhd H_{m-1} \lhd \cdots \lhd H_0 = G.$$

Putting $N_i = N \cap H_i$, we obtain a series (1) and

$$[N_{i-1}, H] \leqq N \cap [H_{i-1}, H] \leqq N \cap H_i = N_i.$$

Thus H normalizes the subgroups N_i and centralizes the factors N_{i-1}/N_i, that is H stabilizes the series (1).

The main result on stability groups is due to P. Hall [46].

Theorem 2.5.4. *Suppose that H stabilizes a series*

$$1 = N_m \lhd N_{m-1} \lhd \cdots \lhd N_0 = N.$$

Then H is nilpotent of class $\leqq \binom{m}{2}$.

The connection with Baer groups is the immediate

Corollary 2.5.5. *If H is a group of automorphisms of N and if H is subnormal in* $N]H$, *then* H^N *is a Baer group.*

We need a useful commutator identity in order to prove Theorem 2.5.4. For $n \geqq 3$ and elements $x_1, \ldots, x_n$ of a group, we define inductively $[x_1, \ldots, x_n] = [[x_1, \ldots, x_{n-1}], x_n]$. Then

Lemma 2.5.6. *If a, b, c are elements of a group, then*

$$[a, b^{-1}, c]^b [b, c^{-1}, a]^c [c, a^{-1}, b]^a = 1.$$

Proof. Observe that $[a, b^{-1}, c]^b = (a^{-1}b^{-1}ac^{-1}a^{-1})(b^{-1}c^{-1}ba^{-1}b^{-1})^{-1}$, etc. □

Proof of Theorem 2.5.4. Let $G = N]H$. Since the cases $m \leqq 1$ are trivial, we suppose that $m \geqq 2$. Let $C = C_H(N_1)$. Thus $C \lhd H$ and H/C is isomorphic to the group of automorphisms of N_1 induced by H. By induction on m, we may assume that H/C is nilpotent of class at most $\binom{m-1}{2}$. It is sufficient to show that $C \leqq \zeta_{m-1}(H)$, the $(m-1)$th term of the upper central series of H.

For a fixed $x \in N$, the map $c \mapsto [c, x]$ is a homomorphism from C to $[C, N] \leqq N_1$. Therefore the set C_i of all $c \in C$ such that $[c, N] \leqq N_i$ is a subgroup of C, normalized by H. We claim that the series

$$1 = C_m \lhd C_{m-1} \lhd \cdots \lhd C_1 = C$$

is actually centralized by H. For, $[H, N, C_i] \leqq [N_1, C] = 1$. It follows that if $h \in H$, $x \in N$ and $c \in C_i$, then $[h, x^{-1}, c] = 1$ and Lemma 2.5.6 gives

$$[x, c^{-1}, h]^c [c, h^{-1}, x]^h = 1.$$

Now $[x, c^{-1}] \in N_i$ and so $[x, c^{-1}, h]^c \in N_{i+1}$. Hence $[c, h^{-1}, x] \in N_{i+1}$. This is true for all $x \in N$ and therefore $[c, h^{-1}] \in C_{i+1}$, that is $[C_i, H] \leqq C_{i+1}$. Thus our claim is established and $C \leqq \zeta_{m-1}(H)$, as required. □

Remark. Hall shows in [46] (Lemma 3) that if the subgroups N_i are all normal in N, then H is nilpotent of class $\leqq m-1$ (assuming $m \geqq 1$).

The next result shows that a large class of Baer groups is to be found among the soluble p-groups.

Theorem 2.5.7. *For any prime p, soluble p-groups of finite exponent are Baer groups. More precisely, let $G(\neq 1)$ be a soluble p-group of finite exponent and let l be the minimal length of a series of G with elementary abelian factors. Then*

$$[G, \overbrace{\langle g \rangle, \langle g \rangle, \ldots, \langle g \rangle}^{d}] = 1$$

for all $g \in G$, where $d = 1 + p + p^2 + \cdots + p^{l-1}$. Hence $\langle g \rangle \lhd^d G$.

Proof. Let $1 = G_0 \lhd G_1 \lhd \cdots \lhd G_l = G$ be a series of G with elementary abelian factors. Replacing each G_i by its core in G (that is the intersection of its conjugates), we may assume that $G_i \lhd G$, for all i. Let $g \in G$. Since the result is clear if $l = 1$, we suppose that $l \geqq 2$. Then, by induction on l,

$$[G, \overbrace{\langle g \rangle, \langle g \rangle, \ldots, \langle g \rangle}^{d_1}] \leqq G_1, \tag{2}$$

where $d_1 = 1 + p + p^2 + \cdots + p^{l-2}$. Now $g_1 = g^{p^{l-1}} \in G_1$. Let $\gamma = (g-1)^{p^{l-1}}$ (in the group ring $\mathbb{Z}G$). Thus if $a \in G_1$, then

$$\begin{aligned} 1 = [a, g_1] = a^{g_1 - 1} = a^{\gamma} \quad &(\text{since } G_1 \text{ is elementary abelian}) \\ = [a, \overbrace{g, g, \ldots, g}^{p^{l-1}}]. & \end{aligned}$$

Hence

$$[G_1, {}_{p^{l-1}}\langle g \rangle] = 1. \tag{3}$$

(*Note*: If $H \leqq G$ and $H^g = H$, then $[H, \langle g \rangle] = [H, g]$.) Since $d = d_1 + p^{l-1}$, the theorem follows from (2) and (3). □

This and other results about soluble p-groups of finite exponent appear in §7.1 of [120].

We introduce the notation

$$\mathfrak{M} = \mathbf{P}(\mathfrak{Max}\text{-}sn \cup \mathfrak{Min}\text{-}sn).$$

Thus a group G belongs to the class $\mathfrak{M}$ if and only if there is a series of finite length

$$1 = G_0 \lhd G_1 \lhd \cdots \lhd G_l = G \tag{4}$$

where each factor G_{i+1}/G_i satisfies Max-*sn* or Min-*sn*. In Theorem 2.6.1 we shall prove that $\mathfrak{M}\mathfrak{S}^{\infty} = \mathfrak{S}^{\infty}$ and we shall need the following result.

Theorem 2.5.8. *Baer $\mathfrak{M}$-groups are nilpotent.*

In order to prove this Theorem, we must consider a rather special situation.

Lemma 2.5.9. *Let $G = AX$ where A is a normal divisible abelian subgroup and X is a subnormal nilpotent subgroup of G. Suppose that X is periodic. Then* $[A, X] = 1$.

Proof. By Proposition 1.6.3, G is nilpotent and so $[A, {}_sX] = 1$ for some s. Choose s minimal and suppose for a contradiction that $s \geqq 2$. Let $A_0 = A$ and $A_i = [A_{i-1}, X]$ for $i \geqq 1$. Then

$$A_i = \prod_{x \in X} [A_{i-1}, x]$$

and each factor in the product is a homomorphic image of A_{i-1}. It follows, by induction on i, that each A_i is divisible. Denote the homomorphism $a \mapsto [a, x]$ from A_{s-2} to A_{s-1} by x^*. Then $x \mapsto x^*$ defines a homomorphism from X to the additive group $\mathrm{Hom}(A_{s-2}, A_{s-1})$ of homomorphisms from A_{s-2} to A_{s-1}. Since divisible groups of finite exponent have order 1, this latter group is torsion-free and so $X/C_X(A_{s-2})$ must be trivial, giving the desired contradiction. □

Remarks

1. The above lemma can fail if X is not periodic. For example let A_1, A_2 be C_{2^∞}-groups and $a_1 \mapsto a_2$ be an isomorphism from A_1 to A_2. Take $A = A_1 \times A_2$, $X = \langle x \rangle \cong C_\infty$ (an infinite cyclic group) and let x act on A according to

$$a_1^x = a_1 a_2, \qquad \text{all } a_1 \in A_1,$$
$$a_2^x = a_2, \qquad \text{all } a_2 \in A_2.$$

   Now form the natural semidirect product $G = A]X$. Then $X \lhd^2 G$, but $[A, X] \neq 1$.
2. If X/X' is periodic but X is not necessarily nilpotent, then we can at least say that A normalizes H (Robinson [119], Lemma 3.13).

Proof of Theorem 2.5.8. Let $G \in \mathfrak{M}$. Call the smallest integer l, for which there is a series (4) with each G_{i+1}/G_i satisfying Max-*sn* or Min-*sn*, the *$\mathfrak{M}$-length* of G. Now suppose, in addition, that G is a Baer group. We proceed by induction on l. The case $l = 0$ is trivial and so we assume that $l \geqq 1$ and the usual induction hypothesis holds. There is a subgroup $N \lhd G$ of $\mathfrak{M}$-length $l - 1$ and

$$G/N \in \mathfrak{Max}\text{-}sn \cup \mathfrak{Min}\text{-}sn.$$

The class of Baer groups is subgroup closed (Theorem 2.5.3) and so, by induction, N is nilpotent. Also G/N is a Baer group. If G/N satisfies Max-*sn*, then G/N is finitely generated, by theorem 2.5.1(i). Therefore G is the product of N and a finitely generated (and hence subnormal and nilpotent) subgroup, showing that G is nilpotent (Proposition 1.6.3). Thus we may assume that

$$G/N \text{ satisfies Min-}sn.$$

Let $\bar{G} = G/N$. The minimal subgroup $\bar{G}^*$ of finite index in $\bar{G}$ normalizes each subnormal subgroup of $\bar{G}$, by Theorem 1.7.10. In particular every finitely generated subgroup of $\bar{G}^*$ is normal and so every subgroup of $\bar{G}^*$ is normal. Then $\bar{G}^*$ is a Dedekind group and nilpotent of class $\leqq 2$ (see Theorem 6.1.1). Hence $\bar{G}$, as the product of $\bar{G}^*$ and a finitely generated subnormal nilpotent subgroup, is nilpotent (Proposition 1.6.3). Thus $\bar{G}$ satisfies Min and has a normal periodic divisible abelian subgroup of finite index. Then it is easy to see that we may assume that

$$G/N \cong C_{p^\infty} \tag{5}$$

for some prime p.

Now G/N' is a Baer $\mathfrak{M}$-group. If it is nilpotent, then so is G, by a well-known result of P. Hall [46] (Theorem 7). Therefore, replacing G by G/N', it is in order to assume that N is abelian. A divisible subgroup of an abelian group is a direct factor (see [72]) and hence the abelian $\mathfrak{M}$-group N has a finitely generated subgroup L with

$$N/L \cong C_{p_1^\infty} \times C_{p_2^\infty} \times \cdots \times C_{p_s^\infty}.$$

According to (5) there is an ascending chain

$$X_1 \leqq X_2 \leqq \cdots$$

of finitely generated subgroups such that $G = NX$, where

$$X = \bigcup_{i \geqq 1} X_i. \tag{6}$$

Also each $X_i/X_i \cap N \cong X_i N/N$ is a finite group. Therefore L has only finitely many conjugates under X_i and consequently L^{X_i} is finitely generated. By Theorem 2.5.1, X_i is subnormal in G and X_i is nilpotent. The group $X_i N/L^{X_i}$ is now the product of the normal divisible abelian subgroup N/L^{X_i} and the subnormal nilpotent subgroup $X_i L^{X_i}/L^{X_i}$. Moreover $X_i N/L^{X_i}$ is periodic, since G/N and N/L are periodic. It follows from Lemma 2.5.9 that

$$[N, X_i] \leqq L^{X_i}.$$

Therefore $[N, X_i]$ is finitely generated. From $[N, X'] = 1$, we deduce

$$[N, X_i]^X = [N, X_i^X] = [N, X_i[X_i, X]] = [N, X_i] = A,$$

say.

If the torsion subgroup of A has order m, then the subgroups $A_j = A^{mj}$, $j \geqq 1$, are normal in G and $\bigcap_{j \geqq 1} A_j = 1$. (Here A^n is the subgroup consisting of the nth powers of the elements of A.) So the group $X/X \cap N$ ($\cong C_{p^\infty}$) acts on each finite quotient A/A_j and hence must act trivially. Thus $[A, X] \leqq A_j$, all j, and therefore $[N, X_i, X] = [A, X] = 1$. Then from (6) we obtain

$$[N, X, X] = 1$$

and so $X \lhd^2 G$. Also $X' \leqq N$ and consequently $[X', X, X] = 1$, that is X is nilpotent. Therefore G is nilpotent, by Proposition 1.6.3. □

We saw in §1.6 that perfect subnormal subgroups of arbitrary groups behave remarkably well. In Baer groups they excel themselves. Thus

Theorem 2.5.10. *If H is a perfect subnormal subgroup of a Baer group G, then $H \lhd G$.*

Proof. We may assume that $G = \langle H, g \rangle$. So $G = H\langle g \rangle$, since $\langle g \rangle \,\mathrm{sn}\, G$ (Theorem 1.6.6). Let

$$H = H_n \lhd H_{n-1} \lhd \cdots \lhd H_0 = G$$

be the normal closure series of H in G. If $n \geqq 2$, then $H_{n-1} = H^{H_{n-2}}$ is perfect. However, $H_{n-1} = H(H_{n-1} \cap \langle g \rangle)$ and so H_{n-1}/H is cyclic and perfect, that is $H_n = H_{n-1}$. It follows that $H = H_1 \lhd G$. □

Remarks

1. There are non-trivial perfect Baer groups, for example the McLain groups [120] (Theorem 6.21). One of these groups appears in Theorem 3.1.6.
2. Philip Hall showed that Theorem 2.5.10 extends to perfect ascendant subgroups of Baer groups in a fairly routine way.

While Baer groups are rich in subnormal abelian subgroups, there are Baer groups ($\neq 1$) with no non-trivial *normal* abelian subgroups. Their construction is due to R. S. Dark. First we introduce the following concept.

Definition. A group G is said to be *prime* if $[A, B] \neq 1$ whenever A and B are non-trivial normal subgroups of G.

Clearly a prime group cannot have a non-trivial normal abelian subgroup. Dark's result is

Theorem 2.5.11 [45]. *For each prime p there is a Baer p-group ($\neq 1$) which is prime.*

Proof. We divide the argument into three parts.

1. *The construction.* We shall form a semidirect product G of a group H by a group W of automorphisms of H.

Let p be a prime and, for each positive integer m, let U_m be the group of upper unitriangular $(m+1) \times (m+1)$ matrices over the field of p elements. For

$1 \leqq k \leqq m$, define ε_k and E_k as follows: ε_k is the element of U_m with 0 everywhere above the diagonal except for 1 in the $(k, k+1)$-position; and E_k is the subgroup of U_m consisting of those matrices with (i, j)-coefficient 0 unless either $i = j$ or $i \leqq k \leqq j$. Elementary matrix calculations show that, for $1 \leqq k \leqq m$,

$$\varepsilon_k^p = 1, \tag{7}$$

$$\langle \varepsilon_k^{U_m} \rangle = E_k \tag{8}$$

and

$$E_k \text{ is abelian.} \tag{9}$$

Let F be a free group with basis $\{x_i \mid i \geqq 1\}$ and let Φ_m be the set of all maps ϕ from the set of positive integers into the set $\{0, 1, 2, \ldots, m\}$. For $\phi \in \Phi_m$ there is a homomorphism $F \to U_m$ defined by

$$x_i \mapsto \begin{cases} 1 & \text{if } \phi(i) = 0 \\ \varepsilon_{\phi(i)} & \text{otherwise.} \end{cases}$$

Denote the kernel of this homomorphism by K_ϕ and let $R = \bigcap_{\phi \in \Phi} K_\phi$, where $\Phi = \bigcup_{m \geqq 1} \Phi_m$. Thus $R \lhd F$ and we define $H = F/R$.

If $y_i = Rx_i$, then $H = \langle y_i \mid i \geqq 1 \rangle$,

$$y_i^p = 1 \tag{10}$$

and

$$\langle y_i^H \rangle \text{ is abelian,} \tag{11}$$

for all $i \geqq 1$, by (7), (8) and (9). Also, for $\phi \in \Phi_m$, $K_\phi \geqq R$ and so there is a homomorphism $H \to U_m$ defined by

$$y_i \mapsto \begin{cases} 1 & \text{if } \phi(i) = 0 \\ \varepsilon_{\phi(i)} & \text{otherwise.} \end{cases}$$

We shall also use the symbol ϕ to denote this homomorphism.

If σ is any permutation of the set of positive integers, then the automorphism of F defined by $x_i \mapsto x_{i\sigma}$ $(i \geqq 1)$ permutes the subgroups K_ϕ, $\phi \in \Phi_m$, for fixed m, and so leaves R invariant. It follows that σ determines an automorphism of H defined by $y_i \mapsto y_{i\sigma}$; we shall also use the symbol σ to denote this automorphism. Now, for each $k \geqq 1$, let σ_k be the automorphism of H corresponding to the permutation

$$(1, 1+p^{k-1}, 1+2p^{k-1}, \ldots, 1+(p-1)p^{k-1})\,(2, 2+p^{k-1}, 2+2p^{k-1}, \ldots,$$
$$2+(p-1)p^{k-1}) \cdots (p^{k-1}, 2p^{k-1}, 3p^{k-1}, \ldots, p^k).$$

Define $W = \langle \sigma_k \mid k \geqq 1 \rangle$ and form the natural semidirect product

$$G = H]\,W.$$

Put $W_k = \langle \sigma_1, \sigma_2, \ldots, \sigma_k \rangle$ and regard W_k as a group of permutations of $\{1, 2, \ldots, p^k\}$. Then

$$W_k \cong \overbrace{C_p \wr C_p \wr \cdots \wr C_p}^{k} \tag{12}$$

where C_p is the cyclic group of order p in its regular representation. (The permutational wreath product is defined in [43] (pp. 81–3) and in [120] (pp. 18–19).) It follows that W is isomorphic to the wreath power $\mathrm{Wr}\, C_p^{\Lambda}$, where Λ is the set of positive integers in their natural order, as defined by P. Hall in [49] (see also [120], pages 18–19). Thus G is a locally finite p-group.
2. *Showing that G is prime.* We claim that

$$\text{if } 1 \neq A \lhd G, \text{ then } H \cap A \neq 1. \tag{13}$$

In order to see this, observe first that

$$\text{if } i \neq j, \text{ then } \langle y_i^H \rangle \neq \langle y_j^H \rangle. \tag{14}$$

For, define $\phi \in \Phi_2$ by

$$\phi(k) = \begin{cases} 1 & \text{if } k = i \\ 2 & \text{if } k = j \\ 0 & \text{otherwise.} \end{cases}$$

Then $E_1 = \langle y_i^H \rangle^{\phi}$ and $E_2 = \langle y_j^H \rangle^{\phi}$, by (8), and $E_1 \neq E_2$ by definition.
Suppose, for a contradiction to (13), that $1 \neq A \lhd G$, but $H \cap A = 1$. Thus

$$[H, A] = 1. \tag{15}$$

However, if $y\sigma \in A$, where $y \in H$ and $\sigma \in W$, then $\sigma \neq 1$ since $H \cap A = 1$. Therefore there is a positive integer i such that $i\sigma \neq i$ and hence

$$\langle y_i^H \rangle^{y\sigma} = \langle y_i^H \rangle^{\sigma} = \langle y_{i\sigma}^H \rangle.$$

Thus, by (14), $\langle y_i^H \rangle^{y\sigma} \neq \langle y_i^H \rangle$, contradicting (15), and so (13) must hold.
Let A, B be non-trivial normal subgroups of G. By (13) there are non-trivial elements $a \in H \cap A$ and $b \in H \cap B$. Choose an integer k such that

$$a, b \in \langle y_i \mid 1 \leqq i \leqq p^{k-1} \rangle.$$

Then $b^{\sigma_k} \in \langle y_i \mid p^{k-1} < i \leqq 2p^{k-1} \rangle \cap B$. To prove that G is prime, it suffices to show that $[a, b^{\sigma_k}] \neq 1$. This in turn will follow from:

if $1 \leqq r < s$, $1 \neq c \in \langle y_1, y_2, \ldots, y_r \rangle$ *and* $1 \neq d \in \langle y_{r+1}, y_{r+2}, \ldots, y_s \rangle$,

$$\textit{then } [c, d] \neq 1. \tag{16}$$

To establish (16), note first that if $x \in F$ and $y = Rx$, then $y \neq 1$ if and only if $x \notin R$, that is if and only if $x \notin K_\phi$ for some $\phi \in \Phi$. Hence $y \neq 1$ if and only if $y^\phi \neq 1$ for some $\phi \in \Phi$. In particular there are elements $\phi \in \Phi$ such that $c^\phi \neq 1$. Choose such a $\phi \in \Phi_m$ with m as small as possible. Then $c^\phi \in U_m$. We claim that

$$c^\phi \in E_1. \tag{17}$$

For, suppose not. Then $m \geqq 2$. If $z \in U_m$, we write z^α for the element of U_{m-1} obtained from z by deleting the first row and column. Clearly $\alpha: U_m \to U_{m-1}$ is a homomorphism. Define $\phi_1 \in \Phi_{m-1}$ by

$$\phi_1(i) = \begin{cases} 0 & \text{if } \phi(i) = 0 \text{ or } 1 \\ \phi(i) - 1 & \text{otherwise.} \end{cases}$$

The homomorphisms ϕ_1 and $\phi\alpha$ from H to U_{m-1} coincide on the generators y_i of H and therefore $\phi_1 = \phi\alpha$ and $c^{\phi_1} = c^{\phi\alpha}$. Since (17) is assumed to be false, $c^{\phi\alpha} \neq 1$ and hence $c^{\phi_1} \neq 1$, contradicting the minimality of m. Thus we have established (17). In a similar way we can show that

$$c^\phi \in E_m. \tag{18}$$

Denote the matrix with 1 in the (i,j)-position and 0s elsewhere by e_{ij}. Then (17) and (18) show that

$$c^\phi = 1 + \lambda e_{1,\,m+1}, \tag{19}$$

and $0 < \lambda < p$ since $c^\phi \neq 1$. From the hypothesis of (16), $c \in \langle y_1, y_2, \ldots, y_n \rangle$. Define $\phi_2 \in \Phi_m$ by

$$\phi_2(i) = \begin{cases} \phi(i) & \text{if } 1 \leqq i \leqq r \\ 0 & \text{otherwise.} \end{cases}$$

Thus $c^{\phi_2} = c^\phi$ and so without loss of generality we may assume that

$$\phi(i) = 0, \qquad \text{unless } 1 \leqq i \leqq r. \tag{20}$$

Similarly we can find an integer n and an element $\psi \in \Phi_n$ such that

$$d^\psi = 1 + \mu e_{1,\,n+1} \tag{21}$$

with $0 < \mu < p$ and

$$\psi(i) = 0, \qquad \text{unless } r < i \leqq s. \tag{22}$$

Now (20) and (22) allow us to define a map $\theta \in \Phi_{m+n}$ by

$$\theta(i) = \begin{cases} 0 & \text{if } \phi(i) = \psi(i) = 0 \\ \phi(i) & \text{if } 0 < \phi(i) \leqq m \\ m + \psi(i) & \text{if } 0 < \psi(i) \leqq n. \end{cases}$$

From (19) and (21) it follows that

$$c^\theta = 1 + \lambda e_{1,\,m+1} \qquad \text{and} \qquad d^\theta = 1 + \mu e_{m+1,\,m+n+1}.$$

Therefore $[c, d]^\theta = [c^\theta, d^\theta] = 1 + \lambda\mu e_{1,m+n+1}$. Since $\lambda\mu \not\equiv 0 \bmod p$, we have $[c, d]^\theta \neq 1$ and this proves (16).

3. *Showing that G is a Baer group.* We show that G is generated by subnormal abelian subgroups. From (11) it is clear that H is generated by subnormal abelian subgroups of G. Therefore it is sufficient to show that

$$\langle \sigma_k \rangle \text{ sn } H \langle \sigma_k \rangle \tag{23}$$

and

$$H\langle\sigma_k\rangle \; sn \; G, \tag{24}$$

for $k \geqq 1$. By (12), $W_k = \langle\sigma_1, \sigma_2, \ldots, \sigma_k\rangle$ is a finite p-group. Also $W_k{}^W$ is the direct product of the conjugates of W_k in W, since the permutational wreath product is associative. Thus $\langle\sigma_k\rangle \; sn \; W_k \lhd W_k{}^W \lhd W$ and so (24) holds.

Finally, if $i > p^k$, then σ_k fixes y_i. Putting $N = \langle y_i^H \mid 1 \leqq i \leqq p^k\rangle$, it follows that $[H, \sigma_k] \leqq N$ and so $N\langle\sigma_k\rangle \lhd H\langle\sigma_k\rangle$. From (10) and (11) we see that N is the product of p^k elementary abelian normal subgroups. Then N is a soluble p-group of finite exponent and the same is true of $N\langle\sigma_k\rangle$. Thus, by Theorem 2.5.7, $\langle\sigma_k\rangle \; sn \; N\langle\sigma_k\rangle$ and (23) follows. □

§2.6 The class $\mathfrak{S}^\infty$ (continued)

We recall the notation $\mathfrak{M} = \mathrm{P}(\mathfrak{Max}\text{-}sn \cup \mathfrak{Min}\text{-}sn)$. The main result on the class $\mathfrak{S}^\infty$ is then

Theorem 2.6.1. $\mathfrak{M}\mathfrak{S}^\infty = \mathfrak{S}^\infty$.

We shall need some preliminary results, the first of which will allow us to apply Theorem 2.5.4 (that is the nilpotency of stability groups).

Lemma 2.6.2. *Let $\mathfrak{X} = \mathrm{Q}\mathfrak{X} = \mathrm{s_n}\mathfrak{X}$ and $\mathfrak{Y} = \mathrm{s_n}\mathfrak{Y}$ be subclasses of $\mathfrak{S}^\infty$. Suppose that in every semidirect product $G = N]J$, where $N \in \mathfrak{X}$, $J \in \mathfrak{Y}$ and J is the union of an ascending chain of subnormal subgroups of G, J is subnormal in G. Then $\mathfrak{X}\mathfrak{Y} \leqq \mathfrak{S}^\infty$.*

Proof. Let $G \in \mathfrak{X}\mathfrak{Y}$. Then there is a subgroup $N \lhd G$ with $N \in \mathfrak{X}$ and $G/N \in \mathfrak{Y}$. Suppose that J is the union of an ascending chain of subnormal subgroups H_α of G,

$$J = \bigcup_{\alpha<\gamma} H_\alpha.$$

By Theorem 2.4.4, it is sufficient to prove that J is subnormal in G.

Clearly $N \cap H_\alpha$ is subnormal in N and $N \cap J = \bigcup_{\alpha<\gamma} (N \cap H_\alpha)$. Since $N \in \mathfrak{X} \leqq \mathfrak{S}^\infty$, it follows that $N \cap J \; sn \; N$. Let

$$N \cap J = N_m \lhd N_{m-1} \lhd \cdots \lhd N_0 = N$$

be the normal closure series of $N \cap J$ in N. Then J normalizes $N \cap J$ and N and hence also each N_i. Therefore we have a chain

$$J = N_m J \leqq \cdots \leqq N_{i+1} J \leqq N_i J \leqq \cdots \leqq N_0 J = NJ$$

of subgroups of G. As a join of subnormal subgroups of $G/N (\in \mathfrak{S}^\infty)$, NJ/N is subnormal in G/N, that is $NJ \; sn \; G$. Suppose that $N_i J \; sn \; G$ for some i. Then, by induction on i, it suffices to show that $N_{i+1} J \; sn \; N_i J$.

Use bars to denote subgroups modulo N_{i+1}. Since

$$N_i \cap N_{i+1}J = N_{i+1}(N_i \cap J) = N_{i+1},$$

we have $\overline{N_i J} = \overline{N}_i]\overline{J}$, $\overline{N}_i \in s_n Q\mathfrak{X} = \mathfrak{X}$,

$$\overline{J} \cong J/N_{i+1} \cap J = J/N \cap J \cong NJ/N \in s_n \mathfrak{Y} = \mathfrak{Y}$$

and $\overline{J} = \bigcup_{\alpha < \gamma} \overline{H}_\alpha$ with each $\overline{H}_\alpha$ *sn* $\overline{N_i J}$. So the result follows from the hypotheses. □

During the proof of Theorem 2.5.8, we saw that an abelian $\mathfrak{M}$-group A has a finitely generated subgroup B such that

$$A/B \cong C_{p_1^\infty} \times C_{p_2^\infty} \times \cdots \times C_{p_s^\infty}. \tag{1}$$

This leads to

Lemma 2.6.3. *Let $G = A]J$, where A is an abelian $\mathfrak{M}$-group and J sn G. Then there is an integer $d = d(A)$ such that $J \lhd^d G$.*

Proof. As we have just remarked, there is a finitely generated subgroup B of A such that (1) holds. Let q be a prime different from $p_1, \ldots, p_s$. Suppose that $b \in B$ and $b = a^q$, for some $a \in A$. There is a positive integer m, prime to q, such that $a^m \in B$ and so $a \in B$. Therefore

$$B^q \leqq B \cap A^q \leqq B^q$$

and hence $B^q = B \cap A^q$.

We choose infinitely many values $q_1, q_2, \ldots$ for q and let

$$X = \bigcap_{i \geqq 1} A^{q_i} \lhd G.$$

Then $B \cap X = \bigcap_{i \geqq 1} B^{q_i}$ is finite and so X is periodic, since $X/X \cap B \cong XB/B$ is periodic. By Lemma 1.7.11, A has finite rank, say r_1, and hence A/A^{q_i} has order dividing $q_i^{r_1}$. Thus $JA^{q_i} \lhd^{r_1} JA$. It is easy to see that $\bigcap_{i \geqq 1} JA^{q_i} = JX$ and so, by Proposition 1.1.2(ii), $JX \lhd^{r_1} JA = G$.

The periodic subgroup of A has the form $A_1 \times A_2$, where A_1 is divisible and A_2 is finite. Thus A splits over $A_1 \times A_2$ (see [72], Volume 1, pages 202–5), that is $A = A_1 \times A_2 \times A_3$, where A_3 is torsion-free. Therefore $X = A_1 \times A_2^*$, where $A_2^* \leqq A_2$. Now A_1 is J-invariant and $[A_1, J]$ is smaller than A_1 and (as in the proof of Lemma 2.5.9) divisible. It follows easily that $[A_{1,r_2} J] = 1$, where r_2 is the rank of A_1. Hence $J \lhd^{r_2} A_1 J$. Finally

$$A_1 J \lhd^{r_3} A_1 A_2^* J = XJ,$$

where $r_3 = |A_2|$. Thus $J \lhd^{r_1+r_2+r_3} G$. □

Now we can prove an important special case of Theorem 2.6.1.

Theorem 2.6.4. $\mathfrak{M} \leqq \mathfrak{S}^\infty$.

Proof. We have already seen in §2.4 that $\mathfrak{Min}$-*sn* $\leqq \mathfrak{S}^\infty$; and $\mathfrak{Max}$-*sn* $\leqq \mathfrak{S}^\infty$ follows from Theorem 1.3.10 (or even Theorem 1.3.2). Therefore we may argue by induction on $\mathfrak{M}$-length l (as defined in the proof of Theorem 2.5.8), assuming $l \geqq 2$ and the usual induction hypothesis.

Let $\mathfrak{X}$ be the class of all $\mathfrak{M}$-groups of $\mathfrak{M}$-length $\leqq l-1$ and let

$$\mathfrak{Y} = \mathfrak{Max}\text{-}sn \cup \mathfrak{Min}\text{-}sn.$$

So $\mathfrak{X}, \mathfrak{Y} \leqq \mathfrak{S}^\infty$ and we must show that $\mathfrak{X}\mathfrak{Y} \leqq \mathfrak{S}^\infty$. Thus according to Lemma 2.6.2, we consider a semidirect product $G = N]J$, where $N \in \mathfrak{X}$, $J \in \mathfrak{Y}$ and J is the union of an ascending chain of subnormal subgroups H_α of G, that is

$$J = \bigcup_{\alpha<\gamma} H_\alpha.$$

It suffices to show that J is subnormal in G.

Let $C = C_J(N)$. So $C \lhd G$. Also if $H_\alpha \lhd^r G$, $r \geqq 1$, then

$$[N, \overleftrightarrow{H_\alpha, H_\alpha, \ldots, H_\alpha}^{r}] \leqq N \cap H_\alpha = 1.$$

Thus defining $N_0 = N$ and $N_i = [N, {}_i H_\alpha]$, $i \geqq 1$, we see that $H_\alpha C/C$ is a subgroup of Aut N stabilizing the series

$$N = N_0 \geqq N_1 \geqq \cdots \geqq N_r = 1 .$$

By Theorem 2.5.4, $H_\alpha C/C$ must therefore be nilpotent. Let B/C be the *Baer radical* of G/C, that is the join of all the subnormal nilpotent subgroups of G/C. Then $H_\alpha \leqq B$, for all $\alpha < \gamma$, and hence $J \leqq B$. However, B/C is a Baer group and an $\mathfrak{M}$-group (since $B \lhd G$ and $\mathfrak{M}$ is $\mathrm{s_n}$-closed) and so B/C is nilpotent, by Theorem 2.5.8. Therefore J/C is subnormal in B/C, that is J *sn* $B \lhd G$, as required. □

All is now ready for our main result.

Proof of Theorem 2.6.1. We have to show that $\mathfrak{M}\mathfrak{S}^\infty = \mathfrak{S}^\infty$. By Theorem 2.6.4, $\mathfrak{M} \leqq \mathfrak{S}^\infty$. Therefore according to Lemma 2.6.2, it is sufficient to consider a semidirect product $G = N]J$, where $N \in \mathfrak{M}$ and J is the union of an ascending chain of subnormal subgroups H_α ($\alpha < \gamma$) of G, and show that J is subnormal in G. (We may assume that $J \in \mathfrak{S}^\infty$, but we do not require this fact.)

Let $C = C_J(N) \lhd G$. As in Theorem 2.6.4, $H_\alpha C/C$ is nilpotent. Since we may replace G by G/C, we may even assume that H_α is nilpotent. Then if B is the Baer radical of G, each $H_\alpha \leqq B$ and so $J \leqq B \lhd G$. Moreover, $B \cap N$, as a subgroup of the Baer group B, is also a Baer group (Theorem 2.5.3); and as a normal subgroup of N, $B \cap N \in \mathfrak{M}$. Therefore, by Theorem 2.5.8, $B \cap N$ is nilpotent. Replacing G by B, this means that we may assume that N is nilpotent.

Let $Z = \zeta_1(N)$. By induction on the class of N, it is sufficient to show that J is subnormal in ZJ; in other words we may assume that $N = A$ (say) is abelian. Finally, according to Lemma 2.6.3, there is an integer $d = d(A) \geqq 1$ such that $H_\alpha \lhd^d AH_\alpha$, that is

$$[A, {}_dH_\alpha] \leqq A \cap H_\alpha = 1,$$

all $\alpha < \gamma$. Therefore $[A, {}_dJ] = 1$ and so $J \lhd^d AJ$ as required. □

To conclude this section, we use the above results to prove a theorem outstanding from §1.4. Recall that $\mathfrak{S}$ is the class of all groups in which the join of two subnormal subgroups is always subnormal. Then

Theorem 1.4.6. $\mathfrak{M}\mathfrak{S} = \mathfrak{S}$.

Proof. Let H, K be subnormal subgroups of $G \in \mathfrak{M}\mathfrak{S}$ and $J = \langle H, K \rangle$. We proceed by induction on the defect m of H in G to show that J is subnormal in G. Thus suppose that $m \geqq 2$ and assume the usual induction hypothesis.

There is a normal subgroup N of G with $N \in \mathfrak{M}$ and $G/N \in \mathfrak{S}$. Since NJ is subnormal in G, we may assume that $G = NJ$. If $x \in N \cap J$, then, by Theorem 1.6.14, there is a subgroup F subnormal in G such that $x \in F \leqq J$. So

$$x \in N \cap F \ sn\ N.$$

Therefore $N \cap J$ is subnormal in N, since $N \in \mathfrak{M} \leqq \mathfrak{S}^\infty$, by Theorem 2.6.4.

Consider the normal closure series

$$N \cap J = N_r \lhd N_{r-1} \lhd \cdots \lhd N_0 = N$$

of $N \cap J$ in N. Now J normalizes N and therefore J normalizes each N_i. Thus it is sufficient, by induction on i, to show that $N_{i+1}J$ is subnormal in N_iJ, for $r-1 \geqq i \geqq 0$. Since $N_{i+1} \lhd N_iJ$, we may assume that $G = N]J$ and $C_J(N) = 1$ (first replacing G by N_iJ/N_{i+1} and N by N_i/N_{i+1} and then factoring by $C_J(N)$). Then H and K, being subnormal subgroups of G, each stabilize a series of N and are nilpotent, by Theorem 2.5.4. Therefore H and K lie in the Baer radical B of G and so $J \leqq B \lhd G$. Without loss of generality $G = B$ and thus N is nilpotent, by Theorem 2.5.8. Now by induction on the class of N we may assume that $N = A$ (say) is abelian.

Let J_0 be a finite subset of J. Clearly $H^G \in \mathfrak{M}\mathfrak{S}$ and $H \lhd^{m-1} H^G$. Therefore, by induction on m, $H^{J_0}\ sn\ H^G \lhd G$. By Lemma 2.6.3, there is an integer $d = d(A) \geqq 1$ such that $H^{J_0} \lhd^d AH^{J_0}$, that is

$$[A, {}_dH^{J_0}] \leqq A \cap H^{J_0} \leqq A \cap J = 1.$$

Thus $[A, {}_d H^J] = 1$ and so H^J *sn* $AH^J \lhd G$. Then $J = H^J K$ is subnormal in G, by Theorem 1.2.1. □

A significant improvement of Theorem 1.4.6 will be established in Chapter 3 (see Theorem 3.5.1).

§2.7 The commutator of two subnormal subgroups

Suppose that H is a nilpotent subnormal subgroup of the group G. Then $[G, {}_n H] = 1$ for some n. Conversely, if $H \leqq G$ and $[G, {}_n H] = 1$ for some n, it is clear that H is a nilpotent subnormal subgroup of G. Thus H^G in this case is generated by nilpotent subnormal subgroups and is a Baer group.

Suppose now that H_1 and H_2 are subgroups of a group G and that $[H_2, {}_n H_1] = 1$ for some n. In view of the above, the question arises as to what implication this condition has for H_2. Here, of course, H_1 need not be nilpotent and so $H_1^{H_2}$ need not be nilpotent. However, $H_1^{H_2} = H_1 [H_1, H_2]$ and we can prove that $[H_1, H_2]$ is a Baer group and therefore locally nilpotent. This we state as

Theorem 2.7.1. *Suppose that $J = \langle H_1, H_2 \rangle$ and that $[H_2, {}_n H_1] = 1$. Then $[H_2, H_1]$ is contained in the Baer radical of J.*

Proof. The hypothesis clearly implies that $[L, {}_n H_1] = 1$ where $L = H_2^{H_1}$, and so if $x \in H_1$, we have $[L, {}_n \langle x \rangle] = 1$. It follows that $\langle x \rangle$ *sn* $\langle L, x \rangle$ and hence $\langle x^L \rangle$ is generated by nilpotent subnormals and is a Baer group. Thus $[L, \langle x \rangle]$ is a Baer group. Moreover $[L, \langle x \rangle]$ is a normal subgroup of L and $[L, H_1]$ is generated by such. The result follows. □

Alternatively this result may be deduced from the theorem of Hall [46] that, under the hypotheses of Theorem 2.7.1, $[H_2, \gamma_{m+1}(H_1)] = 1$, where $m = n(n-1)/2$. (Compare Theorem 2.5.4.) In fact Theorem 2.7.1 will be superseded by Lemma 4.5.12.

Hall raised the question in 1958 [46] as to whether for two subgroups H_1 and H_2 of a group J, the hypotheses $[H_1, {}_n H_2] = 1 = [H_2, {}_m H_1]$ would force $[H_1, H_2]$ to be nilpotent. However, the examples on p. 22 of a pair of nilpotent subnormals whose join fails to be nilpotent show that $[H_1, H_2]$ need not be nilpotent, even in the case $[H_1, {}_3 H_2] = [H_2, {}_3 H_1] = 1$. Roseblade [128] showed that his result (Theorem 1.6.4) on the derived series of the join of a pair of subnormal subgroups has as a corollary the fact that, under the hypotheses of Hall's question, the commutator $[H_1, H_2]$ is always soluble.

3

THE DERIVED AND LOWER CENTRAL SERIES OF A JOIN OF SUBNORMAL SUBGROUPS

Our objective in this chapter is to carry forward our study of the properties of the derived series of a join of subnormal subgroups which has already played a decisive part in Chapter 1. We recall that the main purely group-theoretical result on subnormal coalescence (Theorem 1.6.11) depended on the fact that although the product of two subnormal subgroups need not be a subgroup, nevertheless the product set contains the λth term (for some finite λ) of the derived series of the join of those subgroups (Theorem 1.6.4). Furthermore, if the join of two subnormal subgroups happens to be their product, then the product of any term of the derived series of one subgroup with any term of the derived series of the other contains some term of the derived series of their join (Lemma 1.7.7). This fact was used in our study of subnormal coalition classes contained in $\mathfrak{Min}$-*sn* (Theorem 1.7.4).

Both of these results foreshadow the main result of the first part of this chapter, Theorem 3.1.1, that if we have a join of a finite number of subnormal subgroups of a group and take the product (set) of arbitrary terms of the derived series of each of these subgroups, then this set contains some term of the derived series of the join. This result, which is proved by purely group-theoretic means, reduces many questions on properties of joins of subnormal subgroups of an arbitrary group to the same questions for soluble groups—a fact that we have aready put to good use in Chapter 2 in investigating subnormal composition factors.

The question naturally arises as to whether anything analogous holds for the lower central series of a join of subnormal subgroups. It turns out that the answer to this question is in the affirmative provided we are prepared to impose some restriction on the ambient join J—for example J/J' being of finite rank (a corollary of Theorem 3.3.1). However, in order to prove this theorem, which leads to the most comprehensive results so far obtained in connection with the join problem, we first of all reduce to the case of soluble groups by Theorem 3.1.1 and then a further reduction allows us to reinterpret the problem as a question about properties of the integral group ring of a nilpotent group of finite rank. It is in this form that the solution is finally achieved using techniques developed by P. Hall.

§3.1 The derived series of a join of subnormal subgroups

The major theorem is due to Roseblade [128].

Theorem 3.1.1. *Suppose that* $J = \langle H_1, H_2, \ldots, H_n \rangle$ where $H_i \lhd^{h_i} G$ *for* $i = 1, 2, \ldots, n$. *Then there exists* $\lambda_1 = \lambda_1(h, r)$ *such that*

$$J^{(\lambda_1)} \leqq H_1^{(r_1)} H_2^{(r_2)} \cdots H_n^{(r_n)}$$

and

$$J^{(\lambda_1)} \lhd^{\lambda_1} G$$

whenever $h_1 + h_2 + \cdots + h_n \leqq h$ *and* $r_1 + r_2 + \cdots + r_n \leqq r$.

Wielandt (see Theorem 4.4.1) anticipated this result by proving that if J, in addition to satisfying the hypotheses of Theorem 3.1.1, is a group with finite composition series, then

$$J^{(\omega)} = H_1^{(\omega)} \cdots H_n^{(\omega)}$$

where, for a group $X, X^{(\omega)}$ denotes $\bigcap_{r=1}^{\infty} X^{(r)}$.

Of course as an immediate consequence of Theorem 3.1.1 we have Theorem 2.1.5 (stated, but not proved, on p. 46):

the join of finitely many soluble subnormal subgroups is soluble.

This result was first proved by Stonehewer in [139] in the following form: if $H_i \lhd^{h_i} G$ and $H_i^{(r)} = H_i^{(r+1)}$ for $i = 1, \ldots, n$, then there exists $\lambda_2 = \lambda_2(h_1, \ldots, h_n, r)$ such that if J is the join of the H_i, then $J^{(\lambda_2)} = H_1^{(r)} H_2^{(r)} \cdots H_n^{(r)}$.

We have already seen in §1.5 that joins of finitely many subnormal subgroups of a group G are not in general subnormal. Theorem 3.1.1 shows, however, that such joins are only a 'soluble step' away from being subnormal in that some bounded term of the derived series of the join is subnormal in G. Thus in the extreme case where J is perfect we have $J \text{ sn } G$ and $J = H_1 H_2 \cdots H_n$, although we shall see in Theorem 3.1.3 that this does not mean that $H_1, \ldots, H_n$ are perfect. It is the fact that, in general, some term of the derived series of the join J is actually subnormal in G that we apply to the join problem in §3.2.

It is convenient to split off the pair case of Theorem 3.1.1 as

Theorem 3.1.2. *Suppose that* $J = \langle H_1, H_2 \rangle$, *where* $H_i \lhd^{h_i} G, i = 1, 2$. *Then there exists* $\lambda_3 = \lambda_3(h, r)$ *such that*

$$J^{(\lambda_3)} \leqq H_1^{(r_1)} H_2^{(r_2)}$$

and

$$J^{(\lambda_3)} \lhd^{\lambda_3} G$$

whenever $h_1 + h_2 \leqq h$ *and* $r_1 + r_2 \leqq r$.

The three major ingredients in the proof of Theorem 3.1.2 have already been established in Chapter 1 (and two of them have been mentioned at the start of the introduction to this chapter). They are Theorem 1.6.4, Lemma 1.7.7 and Theorem 1.2.5, which we now list in that order as (1), (2) and (3) below, contenting ourselves with a slightly weaker form of the last result in the interests of symmetry.

Suppose that $J = \langle H_1, H_2 \rangle$ with $H_i \lhd^{h_i} J$, $i = 1, 2$. Then there exists $\lambda_4 = \lambda_4(h)$ such that whenever $h_1 + h_2 \leqq h$,

$$J^{(\lambda_4)} \leqq H_1 H_2. \tag{1}$$

If further $J = H_1 H_2$, then there exists $\lambda_5 = \lambda_5(h, r)$ such that, whenever $h_1 + h_2 \leqq h$ and $r_1 + r_2 \leqq r$, we have

$$J^{(\lambda_5)} \leqq H_1^{(r_1)} H_2^{(r_2)}. \tag{2}$$

Finally, suppose that $J = H_1 H_2$ with $H_i \lhd^{h_i} G$ for $i = 1, 2$. Then for $h_1 + h_2 \leqq h$

$$J \lhd^{h!} G. \tag{3}$$

The proof of Theorem 3.1.2 consists in using (1) to give a suitable permutable product of subnormal subgroups to which we can apply (2) and (3). We do this as follows. With $H_i \lhd^{h_i} G\,(i = 1, 2)$ and $J = \langle H_1, H_2 \rangle$, set $M = J^{(\lambda_4)}$. Then $M \leqq H_1 H_2$ by (1). We form the subgroup MH_2. Clearly

$$MH_2 = (H_1 H_2) \cap (MH_2) = (H_1 \cap MH_2) H_2.$$

Thus if $U = (H_1 \cap MH_2)$, we have that U permutes with H_2. Furthermore, since $M \lhd J$, we have $U \lhd^{h_2} M$, so that $U \lhd^{h_1 + h_2} G$. We now apply (3) to the pair U, H_2 to obtain

$$UH_2 = MH_2 \lhd^{(h_1 + 2h_2)!} G.$$

For $t \geqq \lambda_4$ we have $J^{(t)} \lhd MH_2$ and so

$$J^{(t)} \lhd^{(2h)!} G \qquad (t \geqq \lambda_4) \tag{4}$$

whenever $h_1 + h_2 \leqq h$. Applying (2) to the pair U, H_2 we obtain

$$(UH_2)^{(m)} \leqq U^{(r_1)} H_2^{(r_2)} \leqq H_1^{(r_1)} H_2^{(r_2)},$$

where $m = \lambda_5(2h, r)$ and $r_1 + r_2 \leqq r$. Since $J^{(\lambda_4)} \leqq UH_2$, we deduce that

$$J^{(\lambda_4 + m)} \leqq H_1^{(r_1)} H_2^{(r_2)}. \tag{5}$$

Taking (4) and (5) together we obtain the result — $\lambda_3(h, r)$ may be defined as $(2h)! + \lambda_4(h) + \lambda_5(2h, r)$. This completes the proof of Theorem 3.1.2. □

Before moving on to the proof of Theorem 3.1.1 for arbitrary n, we make some remarks about, and deduce some corollaries of, the pair case (Theorem 3.1.2). Wielandt's version of Theorem 3.1.1 in the limiting case for groups with a finite composition series naturally suggests the question of an analogue of

Theorem 3.1.1 for limit ordinals. However, as Stonehewer points out in [139], by Theorem 2.1.3 *any group X can be embedded in the product of a pair of normal free subgroups* and so taking for X any non-trivial perfect group we see that there is no possibility of such an analogue. If $J = \langle H_1, H_2 \rangle$ with $H_1 \, sn \, J$, $H_2 \, sn \, J$ and J is perfect, then we have at once from Theorem 3.1.2 that $J = H_1 H_2 = H_1^{(r_1)} H_2^{(r_2)}$ for any positive integers r_1, r_2. But this does not imply that H_1 or H_2 is perfect or even that they have any non-trivial perfect subgroups. Indeed Roseblade [128] has modified the argument of Theorem 2.1.3 to prove

Theorem 3.1.3. *Any torsion-free group can be embedded in a perfect group which is the product of a pair of normal free subgroups.*

Of course, free groups have no non-trivial perfect subgroups. It is an open question whether the analogue of Theorem 3.1.3 holds for an arbitrary group.

Proof of Theorem 3.1.3. Suppose that G is any torsion-free group. By a result of Higman *et al.* [59] we may embed G in an infinite group L which has only two classes of conjugate elements and hence is perfect. Let F be the free group with basis $x_\lambda \, (1 \neq \lambda \in L)$ and let X be the split extension of F by L, where the action of L on F is given by $x_\lambda^\mu = x_{\lambda^\mu} (\lambda, \mu \in L)$.

Let Y be the subgroup of X generated by all the elements $\lambda^{-1} x_\lambda$ with $1 \neq \lambda \in L$. As in the proof of the Hall–Hartley theorem (Theorem 2.1.3), Y is normal in X and is free. Also $X = FY$.

Now let $K = L^X$. Then X/K is generated by the symbols $x_\lambda \, (1 \neq \lambda \in L)$ and $\mu \, (\mu \in L)$ subject to the relations $\mu = 1$ and $x_\lambda^\mu = x_{\lambda^\mu} (\lambda, \mu \in L)$. Since all the non-identity elements of L are conjugate, it follows at once that X/K is infinite cyclic. Hence $X' \leqq K$. But since L is perfect, so also is K. We deduce that $X' = K$ is a perfect group containing L and therefore G as a subgroup. Now $X' = F'[F, Y]Y'[F, Y]$. Since F is normal, $F'[F, Y] \leqq F$ and so, as a subgroup of a free group, $F'[F, Y]$ is free. Similarly $Y'[F, Y]$ is free. Thus X' is the product of a pair of normal free subgroups as required. □

Subgroups with trivial intersection are said to be *disjoint*. We note that the normal free subgroups in the above proof are not disjoint. That this is no accident is demonstrated by a corollary of Theorem 3.1.2.

Theorem 3.1.4. *Suppose that H_1 and H_2 are disjoint subnormal subgroups of a group and that $J = \langle H_1, H_2 \rangle$. Then*

$$J^{(\lambda)} = H_1^{(\lambda)} H_2^{(\lambda)}$$

for all limit ordinals λ.

Proof. That $H_1^{(\lambda)} H_2^{(\lambda)} \leqq J^{(\lambda)}$ for any ordinal is clear. Suppose that λ is a limit ordinal and that $J^{(\mu)} = H_1^{(\mu)} H_2^{(\mu)}$ for all limit ordinals $\mu < \lambda$. If λ is a limit of limit ordinals, then $J^{(\lambda)} = \bigcap_{\mu<\lambda} J^{(\mu)}$ and this equals $H_1^{(\lambda)}H_2^{(\lambda)}$ since $H_1 \cap H_2 = 1$. If λ is not such a limit, then $\lambda = \mu + \omega$ for some limit ordinal μ. Therefore

$$J^{(\lambda)} = J^{(\mu+\omega)} \leqq H_1^{(\mu+n)}H_2^{(\mu+n)}$$

for all $n \geqq 0$. To see this we simply apply Theorem 3.1.2 to the pair $H_1^{(\mu)}, H_2^{(\mu)}$ whose join is $J^{(\mu)}$ by hypothesis if $\mu = 0$ and by induction if not. It now follows that

$$J^{(\lambda)} \leqq \bigcap_n H_1^{(\mu+n)}H_2^{(\mu+n)} = H_1^{(\lambda)}H_2^{(\lambda)},$$

again since $H_1 \cap H_2 = 1$. □

It is convenient at this point to mention a further corollary of Theorem 1.6.4 (that is (1) of this section).

Theorem 3.1.5. *Suppose that* $J = \langle H_1, H_2 \rangle$. *Then for all* $n \geqq 1$ *there exists* $\lambda_6 = \lambda_6(n)$ *such that*

$$[H_1, H_2]^{(\lambda_6)} \leqq [H_2, {}_nH_1][H_1, {}_nH_2].$$

This result has some bearing on the work of Hall [46] on sufficient conditions for a group to be nilpotent. It shows, for example, that $[H_1, H_2]$ is soluble whenever there exists an n such that $[H_2, {}_nH_1]$ and $[H_1, {}_nH_2] = 1$. Hall showed further that each of these conditions implies that $[H_1, H_2]$ is locally nilpotent and gave many conditions which were sufficient to ensure that it is nilpotent. However, as is pointed out on p. 88, there are examples of a pair of nilpotent subnormal subgroups H_1, H_2 such that $[H_1, H_2]$ is not nilpotent, even when $[H_1, {}_3H_2] = [H_2, {}_3H_1] = 1$. Roseblade [128] gives an example (3.1.6 below) to show that $[H_1, H_2]$ *may be perfect and non-trivial even when* $[H_1, {}_3H_2]$ *is trivial*, although, of course, in this case $[H_2, {}_nH_1] \neq 1$ for all n, by Theorem 3.1.5.

Proof of Theorem 3.1.5. Suppose that $F = H_1 * H_2$ is the free product of the groups H_1 and H_2. Suppose $n > 0$ and let K_1, K_2 be the nth normal closures of H_1, H_2, respectively, in F. It follows from (1) that if $\lambda_6(n) = \lambda_4(2n)$, then

$$F^{(\lambda_6)} \leqq K_1K_2.$$

Hence in the group F

$$[H_1, H_2]^{(\lambda_6)} \leqq [H_1, H_2] \cap K_1K_2.$$

Recalling the remark after Proposition 1.1.1,

$$K_1 = H_1[H_2, {}_nH_1] \quad \text{and} \quad K_2 = H_2[H_1, {}_nH_2].$$

Therefore

$$[H_1, H_2] \cap K_1K_2 = [H_1, H_2] \cap H_1[H_2, {}_nH_1]H_2[H_1, {}_nH_2].$$

Since $[H_1, H_2], [H_2, {}_nH_1], H_2$ and $[H_1, {}_nH_2]$ all lie in H_2^F and $H_1 \cap H_2^F = 1$, it follows that

$$\begin{aligned}[H_1,H_2] \cap K_1K_2 &= [H_1,H_2] \cap [H_{2,n}H_1]H_2[H_{1,n}H_2] \\ &= [H_1,H_2] \cap [H_{2,n}H_1][H_{1,n}H_2] \\ &= [H_{2,n}H_1][H_{1,n}H_2].\end{aligned}$$

Therefore

$$[H_1,H_2]^{(\lambda_6)} \leqq [H_{2,n}H_1][H_{1,n}H_2]. \tag{6}$$

Furthermore any group J generated by subgroups H_1 and H_2 is a homomorphic image of F and so, since commutators of subgroups are preserved under homomorphisms, it follows that (6) must hold when H_1 and H_2 are viewed as subgroups of J. □

We now give Roseblade's example in

Theorem 3.1.6. *There exists a group G and a cyclic subgroup H of G such that* $[G,{}_3H] = 1$ *and yet* $[G,H] = H^G$ *is non-trivial and perfect.*

Proof. Let K be the field of p-elements, p a prime, and let Λ be the set of rational numbers in their natural order. Let V be a vector space over K with basis elements v_λ $(\lambda \in \Lambda)$. For $\lambda < \mu$, we write $\zeta_{\lambda,\mu}$ for the automorphism of V defined by

$$v_\lambda^{\zeta_{\lambda,\mu}} = v_\lambda + v_\mu$$

and $\zeta_{\lambda,\mu}$ fixes all v_ν $(\nu \neq \lambda)$. Then the group

$$M = \langle \zeta_{\lambda,\mu} | \lambda < \mu \rangle$$

is one of McLain's characteristically simple groups [120]. The normal closure in M of any $\zeta_{\lambda,\mu}$ is abelian and yet $M = M'$. Hence M has no non-trivial cyclic homomorphic images.

Let T be the group of automorphisms of V induced by order-preserving maps $\theta:\lambda \mapsto \lambda'$ $(\lambda \in \Lambda)$ of Λ onto itself. Then

$$\zeta_{\lambda,\mu}^{\theta} = \zeta_{\lambda',\mu'}$$

and so T normalizes M. We form the group $G = MT$ and set $H = \langle \zeta_{0,1} \rangle$. Since H^M is abelian, we get $[G,{}_3H] = 1$. Since M has no non-trivial cyclic homomorphic image, the fact that $[H,G] = M$ is perfect will follow from showing that $M = H^G$. But this is clear, as McLain pointed out in [89], since if $\lambda < \mu$ are rational, there exists some $\theta:\lambda \mapsto \lambda'$ such that $0' = \lambda$ and $1' = \mu$. □

In order to prove Theorem 3.1.1 we need two lemmas. The first is a consequence of (1) and is a sort of 'swap-over' lemma which helps us cope with the fact that the product of subgroups involved in the theorem is not necessarily a subgroup.

Lemma 3.1.7. *Suppose that* $X_i \lhd^{x_i} G$ *for* $1 \leqq i \leqq n$. *Then there exists*

$\lambda_7 = \lambda_7(x)$ *such that*

$$\left(\prod_{i=1}^{n} X_i\right) X_1^{(\lambda_7)} \leqq \prod_{i=1}^{n} X_i$$

whenever $x_1 + x_2 + \cdots + x_n \leqq x$.

Proof. We define $\lambda_7(x) = x\lambda_4(x)$ where λ_4 comes from (1) on p. 91. We may assume that each x_i is greater than zero and that $n > 1$. We write $s_i = (x-i)\lambda_4(x)$ for $0 \leqq i \leqq n$. For $0 \leqq i \leqq n-2$, let $Y_i = \langle X_{n-i}, X_1^{(s_i+1)} \rangle$. For these i

$$\begin{aligned} X_{n-i} X_1^{(s_i)} = X_{n-i} X_1^{(s_i+1)(\lambda_4(x))} &\leqq X_{n-i} Y_i^{(\lambda_4(x))} \\ = Y_i^{(\lambda_4(x))} X_{n-i} &\leqq X_1^{(s_i+1)} X_{n-i}, \end{aligned}$$

by (1) applied to the pair $X_1^{(s_i+1)}, X_{n-i}$. The result follows by induction on i. □

Secondly we need a straightforward inductive generalization of Theorem 1.2.1.

Lemma 3.1.8. *Suppose that* $Y_i \lhd^{y_i} G$ *for* $1 \leqq i \leqq n$. *If each* Y_i *is normal in* $Y = \langle Y_1, \ldots, Y_n \rangle$, *then*

$$Y \lhd^{y_1 y_2 \cdots y_n} G.$$

We may now proceed with the proof of Theorem 3.1.1. Suppose then that $H_i \lhd^{h_i} G$ for $i = 1, 2, \ldots, n$, that $J = \langle H_1, H_2, \ldots, H_n \rangle$ and $r_1, r_2, \ldots, r_n$ are non-negative integers. Let

$$U = \prod_{i=1}^{n} H_i^{(r_i)}. \tag{7}$$

We must show that there exists $\lambda_1 = \lambda_1(h, r)$ such that $J^{(\lambda_1)} \lhd^{\lambda_1} G$ and $J^{(\lambda_1)} \leqq U$, whenever $h_1 + h_2 + \cdots + h_n \leqq h$ and $r_1 + r_2 + \cdots + r_n \leqq r$.

We define $\lambda_1(h, r)$ by induction on h. If $h = 0$, then all H_i are equal to G and it is clear that we may set $\lambda_1(0, r) = r$. Suppose therefore that $h > 0$ and that $\lambda_1(h-1, r)$ has been defined for all r.

If any $h_i = 0$, then obviously $G^{(r)} \leqq U$ and of course $G^{(r)} \lhd G$. Since requiring $\lambda_1(h, r) \geqq r$ is no restriction for us, we may assume that no h_i is zero. One consequence of this is the trivial

$$n \leqq h. \tag{8}$$

We set

$$K_i = \langle H_j \mid j \neq i \rangle, \qquad 1 \leqq i \leqq n.$$

Since $h_i \geqq 1$, there is a subgroup L_i with $H_i \lhd L_i \lhd^{h_i - 1} G$. Then $J \leqq \langle L_i, K_i \rangle$. Applying induction to L_i and H_j $(j \neq i)$, we see that for any $\mu_1 \geqq 0$

$$J^{(\mu_2)} \leqq \langle L_i, K_i \rangle^{(\mu_2)} \leqq L_i \prod_{j \neq i} H_j^{(\mu_1)} \leqq L_i K_i^{(\mu_1)}, \tag{9}$$

where $\mu_2 = \lambda_1(h-1, (h-1)\mu_1)$. We have used (8) here.

Now

$$J^{(\mu_2+1)} \leqq [J^{(\mu_2)}, J] = \prod_{i=1}^{n} [J^{(\mu_2)}, H_i], \tag{10}$$

by Lemma 1.6.5, and the factors on the right are normal in their join. For any $\mu_3 \geqq 0$, let $\mu_4 = 1 + h\mu_3 + \mu_2$. Then from (8) and (10)

$$\left.\begin{aligned} J^{(\mu_4)} &\leqq [J^{(\mu_2)}, J]^{(h\mu_3)} \leqq \prod_{i=1}^{n} [J^{(\mu_2)}, H_i]^{(\mu_3)} \\ &\leqq \prod_{i=1}^{n} [L_i K_i^{(\mu_1)}, H_i]^{(\mu_3)} \qquad \text{(by (9))} \\ &\leqq \prod_{i=1}^{n} \langle H_i, K_i^{(\mu_1)} \rangle^{(\mu_3)} \qquad (\text{since } H_i \lhd L_i) \\ &\leqq J. \end{aligned}\right\} \tag{11}$$

For any $\mu_5 \geqq 0$, let $\mu_1 = \lambda_1(h-1, (h-1)\mu_5)$. Induction applied to the H_j $(j \neq i)$ then gives

$$K_i^{(\mu_1)} \leqq \prod_{j \neq i} H_j^{(\mu_5)} \qquad \text{and} \qquad K_i^{(\mu_1)} \lhd^{\mu_1} G. \tag{12}$$

Now we apply Theorem 3.1.2 to the pair $H_i, K_i^{(\mu_1)}$. Thus with $\mu_3 = \lambda_3(h + \mu_1, r)$,

$$\langle H_i, K_i^{(\mu_1)} \rangle^{(\mu_3)} \leqq H_i^{(r_i)} K_i^{(\mu_1)} \qquad \text{and} \qquad \langle H_i, K_i^{(\mu_1)} \rangle^{(\mu_3)} \lhd^{\mu_3} G. \tag{13}$$

Therefore from (11)

$$\begin{aligned} J^{(\mu_4)} &\leqq \prod_{i=1}^{n} (H_i^{(r_i)} K_i^{(\mu_1)}) = \left(\prod_{i=1}^{n} H_i^{(r_i)} \right) \left(\prod_{i=1}^{n} K_i^{(\mu_1)} \right) \quad (\text{since } H_j \leqq K_i \text{ for } j \neq i) \\ &\leqq \left(\prod_{i=1}^{n} H_i^{(r_i)} \right) \left[\prod_{i=1}^{n} \left(\prod_{j \neq i} H_j^{(\mu_5)} \right) \right] \qquad \text{(by (12))}. \end{aligned}$$

Now choose $\mu_5 = r + \lambda_7(h)$ and apply Lemma 3.1.7 repeatedly to give

$$J^{(\mu_4)} \leqq \prod_{i=1}^{n} H_i^{(r_i)} = U.$$

From (11) we have

$$J^{(\mu_4)} \lhd \prod_{i=1}^{n} [J^{(\mu_2)}, H_i]^{(\mu_3)}. \tag{14}$$

Also $[J^{(\mu_2)}, H_i]^{(\mu_3)}$ is a characteristic subgroup of $[J^{(\mu_2)}, H_i] \lhd J^{(\mu_2)} \lhd J$ and

therefore

$$[J^{(\mu_2)}, H_i]^{(\mu_3)} \lhd^2 \langle H_i, K_i^{(\mu_1)} \rangle^{(\mu_3)} \qquad \text{(using (11))}$$
$$\lhd^{\mu_3} G \qquad \text{(by (13))}.$$

The factors on the right of (14) are normal in $J^{(\mu_2)}$ and therefore in their join. Thus Lemma 3.1.8 yields

$$\prod_{i=1}^{n} [J^{(\mu_2)}, H_i]^{(\mu_3)} \lhd^{(2+\mu_3)^h} G$$

(using (8)). Finally from (14)

$$J^{(\mu_4)} \lhd^{1+(2+\mu_3)^h} G.$$

The proof is completed by defining $\lambda_1(h, r) = \mu_4 + 2 + (2 + \mu_3)^h$. A tedious check reveals that $\lambda_1(h, r) \geqq r$, a necessary condition to cover the case $h_i = 0$, for some i, as we pointed out at the beginning of the proof. However, it is clearly quicker and quite in order to modify our definition of $\lambda_1(h, r)$ so that r is an obvious lower bound, for example by adding $r - 1$. □

§3.2 Applications to the join problem and coalescence

Roseblade's theorem (Theorem 3.1.1) enabled him to improve the results we have established in Chapter 1 on coalescence and the join problem. In particular we recall Theorem 1.6.12, which implies that if $\mathfrak{X}$ is a subjunctive class and if H and K are $\mathfrak{X}$-subnormals of a group G with H/H' and K/K' finitely generated, then the join J of H and K is an $\mathfrak{X}$-subnormal in G. Now H/H' and K/K' finitely generated implies that J/J' is finitely generated (but not of course conversely). Roseblade's results on derived series are precisely what is needed to show that finite generation of the derived quotient of J is enough to guarantee that J is subnormal in G. In fact we have

Theorem 3.2.1 [128]. *Suppose that $\mathfrak{X}$ is subjunctive and J is the join of finitely many subnormal $\mathfrak{X}$-subgroups of the group G. If J/J' is finitely generated, then J is a subnormal $\mathfrak{X}$-subgroup of G.*

As a special case we note at once that if J is a join of finitely many subnormal subgroups of a group G and if J/J' is finitely generated, then J is subnormal in G—a significant generalization of Wielandt's original join theorem. However, in Theorem 3.4.1 we shall show that we can weaken the hypothesis even further—we need only assume that $J'/\gamma_3(J)$ has finite rank. This, the widest generalization of Wielandt's theorem to date, requires new techniques which we shall develop in the appropriate place. We pause here merely to point out that they will imply the existence of various bounds, for example for the defect of J in G in terms of the defects of H and K in G and the rank of $J'/\gamma_3(J)$. Even under the stronger hypothesis that J/J' is finitely generated, this fact will not be obtainable from the proof of Theorem 3.2.1.

The proof of Theorem 1.6.12 relied heavily on properties of the permutizer. So also does the proof of Theorem 3.2.1. Indeed, we shall deduce it from the following consequence of Theorem 3.1.1.

Theorem 3.2.2 [128]. *Suppose that $J = \langle H_1, H_2, \ldots, H_n \rangle$ with $H_i \lhd^{h_i} G$ for $1 \leqq i \leqq n$. Then there exists $\lambda_8 = \lambda_8(h)$ and subnormal subgroups $P_1, P_2, \ldots, P_n$ of $H_1, H_2, \ldots, H_n$, respectively, such that*

$$J^{(\lambda_8)} \lhd^{\lambda_8} G,$$

$$J^{(\lambda_8)} \leqq P_1 P_2 \cdots P_n$$

and

$$P_{i+1} P_1 P_2 \cdots P_i = P_1 P_2 \cdots P_i P_{i+1} \qquad (1 \leqq i < n),$$

whenever $h_1 + h_2 + \cdots + h_n \leqq h$.

Proof. Assume the hypotheses and suppose that $h \geqq h_1 + \cdots + h_n$. We define subgroups P_i, Q_i of J, for $1 \leqq i \leqq n$, and integers p_i, q_i, for $1 \leqq i \leqq h$, inductively by

$$P_1 = Q_1 = H_1, \tag{1}$$

$$P_{i+1} = P_{H_{i+1}}(Q_i) \qquad (1 \leqq i < n), \tag{2}$$

$$Q_{i+1} = Q_i P_{i+1} \qquad (1 \leqq i < n), \tag{3}$$

$$p_1 = q_1 = h, \tag{4}$$

$$p_{i+1} = \mu_6(h + q_i) \qquad (1 \leqq i < n), \tag{5}$$

where $\mu_6(r) = \max(d(m,n))$ for $m + n \leqq r$, as in Theorem 1.6.9, and

$$q_{i+1} = (p_{i+1} + q_i)! \qquad (1 \leqq i < n). \tag{6}$$

A routine induction using Theorem 1.6.9 and §3.1 (3) shows that

$$P_i \lhd^{p_i} G \tag{7}$$

and

$$Q_i \lhd^{q_i} G \tag{8}$$

for $1 \leqq i \leqq n$.

We now set $r_1 = 0$ and $r_{i+1} = \lambda_4(h + q_i)$ for $1 \leqq i < n$, where λ_4 comes from p. 91. From Theorem 1.6.8 applied to the pair H_{i+1}, Q_i we obtain via (8) that

$$H_{i+1}^{(r_{i+1})} \leqq P_{H_{i+1}}(Q_i) \qquad (1 \leqq i < n). \tag{9}$$

Now (2) and (9) show that $H_{i+1}^{(r_{i+1})} \leqq P_{i+1}$ for $1 \leqq i < n$. Hence

$$H_i^{(r_i)} \leqq P_i \qquad (1 \leqq i \leqq n).$$

The result follows from Theorem 3.1.1 by defining
$\lambda_8(h) = \lambda_1(h, r_1 + \cdots + r_n)$. □

We may now move towards a proof of Theorem 3.2.1. Suppose in Theorem 3.2.2 that $H_1, H_2, \ldots, H_n$ are all in the subjunctive class $\mathfrak{X}$. Clearly, so also are $P_1, \ldots, P_n$. Since $Q_i = P_1 P_2 \cdots P_i$ for $1 \leqq i \leqq n$, we have $P_{i+1} Q_i = Q_i P_{i+1}$ for $1 \leqq i < n$, by Theorem 3.2.2, and a simple induction argument (via Proposition 1.6.3) shows that each Q_i lies in $\mathfrak{X}$. Thus $Q_n \in \mathfrak{X}$ and so we have as a consequence of Theorem 3.2.2

Corollary 3.2.3 [128]. *Suppose that $J = \langle H_1, H_2, \ldots, H_n \rangle$ with $H_i \lhd^{h_i} G$, for $1 \leqq i \leqq n$. If $\mathfrak{X}$ is any subjunctive class containing all H_i, then*

$$J^{(\lambda_8(h))} \in \mathfrak{X}$$

whenever $h_1 + h_2 + \cdots + h_n \leqq h$.

In order to prove Theorem 3.2.1 we also need a result from [124].

Lemma 3.2.4. *If $G = N'Y$ with $N \lhd G$ and $Y \operatorname{sn} G$, then $G = N^{(r)}Y$ for any positive integer r.*

Proof. As a characteristic subgroup of N, $N^{(r)}$ is normal in G. Therefore $T = N^{(r)}Y$ is subnormal in G, G/T^G is soluble and perfect, so $G = T^G$. Hence $G = T$ as required. □

Proof of Theorem 3.2.1. Suppose that $\mathfrak{X}$ is a subjunctive class and $J = \langle H_1, H_2, \ldots, H_n \rangle$ where the H_i are $\mathfrak{X}$-subnormals of G. Suppose further that J/J' is finitely generated. Then there exists a finitely generated subgroup F of J such that $J = J'F$.

By Theorem 1.6.14, there is a subnormal $\mathfrak{X}$-subgroup X of G such that $F \leqq X \leqq J$. Since $J = J'X$, it follows by Lemma 3.2.4 that $J = J^{(\lambda_8)}X$, where λ_8 is given by Theorem 3.2.2. Now $J^{(\lambda_8)}$ *sn* G by Theorem 3.2.2 and from Corollary 3.2.3 it is in $\mathfrak{X}$. Therefore J, as the join of a pair of permutable subnormal $\mathfrak{X}$-subgroups of G, is itself an $\mathfrak{X}$-subnormal of G by Proposition 1.6.3 and Theorem 1.2.1. □

To conclude this section we mention a further corollary of Theorem 3.2.2 which should be compared with Corollary 3.2.3.

Corollary 3.2.5. *Suppose that $\mathfrak{X} = \mathrm{QS_n}\mathfrak{X}$ and that $J = \langle H_1, \ldots, H_n \rangle$ with $H_i \in \mathfrak{X}$, $H_i \lhd^{h_i} G$ for $1 \leqq i \leqq n$. Then there exists $\lambda_9 = \lambda_9(h)$ such that*

$$J^{(\lambda_8)} \in \mathfrak{X}^{\lambda_9}$$

whenever $h_1 + h_2 + \cdots + h_n \leqq h$.

This corollary improves upon results by Stonehewer [139] and we note that it follows at once (either from Corollary 3.2.5 or from Corollary 3.2.3) that *if*

$\mathfrak{X} = \mathfrak{X}^2 = \mathrm{Qs_n}\mathfrak{X}$ contains all abelian groups, then in any group the join of finitely many subnormal $\mathfrak{X}$-subgroups is in $\mathfrak{X}$.

In order to prove Corollary 3.2.5, we need the following version of it in the special case of a pair of permuting subgroups.

Lemma 3.2.6. *Suppose that $J = H_1 H_2$ with $H_1 \lhd^{h_1} J$ and $H_2 \lhd^{h_2} J$. Then there exists $\mu_7 = \mu_7(h)$ such that whenever $\mathfrak{X} = \mathrm{Qs_n}\mathfrak{X}$ is a class containing H_1 and H_2 and $h_1 + h_2 \leqq h$, J is in $\mathfrak{X}^{\mu_7}$.*

We omit the proof which may be modelled on that of Proposition 1.6.3.

Proof of Corollary 3.2.5. We define a sequence of positive integers t_i depending only upon h by putting $t_0 = t_1 = 1$ and

$$t_{i+1} = t_i \mu_7 (q_i + p_{i+1}) \qquad (1 \leqq i < h),$$

where the p_i and q_i are as in (4), (5) and (6). With the notation as in the proof of Theorem 3.2.2, suppose that $\mathfrak{X} = \mathrm{Qs_n}\mathfrak{X}$ is a class containing $H_1, H_2, \ldots, H_n$. Clearly $P_1, P_2, \ldots, P_n$ are also in $\mathfrak{X}$ and, by (1), so also is Q_1.

Suppose that for some i, $1 \leqq i < n$, the subgroup Q_i is in $\mathfrak{X}^{t_i}$. Since $\mathfrak{X} \leqq \mathfrak{X}^{t_i}$, we may apply Lemma 3.2.6 to the pair P_{i+1}, Q_i to give, using (7), (8) and (3), that Q_{i+1} is in $\mathfrak{X}^{t_{i+1}}$. It follows that Q_n is in $\mathfrak{X}^{t_n}$, and therefore in $\mathfrak{X}^{t_h}$. But $\mathfrak{X}^{t_h}$ is $\mathrm{s_n}$-closed and so it follows from Theorem 3.2.2 that $J^{(\lambda_8)}$ is in $\mathfrak{X}^{t_h}$. On defining $\lambda_9(h) = t_h$ the proof is complete. □

§3.3 The lower central series of a join of subnormal subgroups

In §3.2 we have seen that properties of the derived series of a join of subnormal subgroups enable us to make considerable progress with the join problem. It turns out that the behaviour of the lower central series of a join of subnormal subgroups is the key to deeper results.

It is of course an immediate consequence of Fitting's theorem that if $H_1, H_2, \ldots, H_n$ are normal subgroups of their join J, then, given any positive integers $a_1, \ldots, a_n$, we have, for any $r \geqq a_1 + \cdots + a_n - (n-1)$,

$$\gamma_r(J) \leqq \gamma_{a_1}(H_1)\gamma_{a_2}(H_2) \cdots \gamma_{a_n}(H_n). \tag{1}$$

If the subgroups $H_1, \ldots, H_n$ are allowed to be subnormal rather than normal, it is not hard to see that, in general, even in the case $n = 2$, things go badly wrong. For example, if we take for J the group G of Theorem 1.5.2 and for H_1 and H_2 the subnormal subgroups VH and VK respectively, then we obtain

$$\gamma_3(H_1) = \gamma_3(H_2) = 1,$$

but $\gamma_4(J) = \gamma_3(J) = V \neq 1$, so that no analogue of (1) holds in J. (For the special case of two permuting subgroups, see Proposition 3.3.12.)

In this example $J'/\gamma_3(J)$ has infinite rank and it turns out that in general if we require $J'/\gamma_3(J)$ to have finite rank, then we obtain not only an analogue of (1) but also that J is subnormal in G. The main objective in §3.3 and §3.4 is to prove two theorems of John P. Williams which we have sharpened to include relevant bounds. The second will be Theorem 3.4.3. The first is

Theorem 3.3.1 [166]. *Let $H_1, H_2, \ldots, H_n$ be subnormal subgroups, of defects $h_1, h_2, \ldots, h_n$ respectively, of a group G and let $J = \langle H_1, \ldots, H_n \rangle$. Suppose further that $J'/\gamma_3(J)$ has finite rank r. Then given positive integers $a_1, \ldots, a_n$, there exists a function $c = c(a_1, \ldots, a_n, h_1, \ldots, h_n, r)$ such that*

$$\gamma_c(J) \leqq \gamma_{a_1}(H_1) \cdots \gamma_{a_n}(H_n).$$

We note that we shall frequently abbreviate expressions like $c(a_1, \ldots, a_n, h_1, \ldots, h_n, r)$ to $c(\boldsymbol{a}, \boldsymbol{h}, r)$.

This theorem was proved in the case $n = 2$ for J/J' finitely generated in [79] and, again in the case $n = 2$, was established in [80] for J/J' of finite rank.

We note that, for any group X, X/X' of finite rank r implies $X/\gamma_n(X)$ of finite rank bounded by a function of r and n, for any n (see §5.3, p. 162). Hence in particular $X'/\gamma_3(X)$ has finite rank.

Suppose then that H ($= H_1$) and K ($= H_2$) are subnormal subgroups of their join J and that we are trying to prove that something like (1) holds. Roseblade's Theorem 3.1.2 will allow us to assume that J is soluble. Induction on the derived length of J followed by a further induction will give a reduction to the case where J is abelian-by-nilpotent. Thus J has an abelian normal subgroup A such that $X = J/A$ is nilpotent. We are really only concerned with the action of J on A by conjugation, so that we are effectively working in the split extension $G = A]X$. Now X is of course the join of the subgroups $Y = HA/A$ and $Z = KA/A$ and so we are led to consider the way in which the action of X on the module A is built up out of the actions of Y and Z on A in light of the fact H and K are subnormal in J. In this way we are led to examine a question which is parallel to the original one, but framed in terms of the integral group ring $\mathbb{Z}X$ of the nilpotent group $X = \langle Y, Z \rangle$. Of course $\mathbb{Z}X$ is the set of all finite formal sums $\Sigma n_x x$, $n_x \in \mathbb{Z}$, $x \in X$, with the obvious addition and multiplication.

We are concerned with the *integral augmentation ideal* $\mathfrak{x}$ which is the kernel of the ring homomorphism $\Sigma n_x x \mapsto \Sigma n_x$ from $\mathbb{Z}X$ to $\mathbb{Z}$. (We use this notation for the augmentation ideal consistently.) Under certain conditions, for example X/X' of finite rank, given positive integers a, b, there exists a positive integer c such that

$$\mathfrak{x}^c \leqq \mathbb{Z}X\mathfrak{y}^a + \mathbb{Z}X\mathfrak{z}^b.$$

(Here $\mathfrak{x}^c$ is the ideal of $\mathbb{Z}X$ generated by all products of c elements of $\mathfrak{x}$.)

More generally we shall be concerned with a group J generated by subnormal subgroups $H_1, \ldots, H_n$ and, following Williams, we shall say that J satisfies the *augmentation condition* if, given positive integers $b_1, \ldots, b_n$, there exists a positive integer e such that

$$\mathfrak{j}^e \leqq \mathbb{Z}J\mathfrak{h}_1^{b_1} + \cdots + \mathbb{Z}J\mathfrak{h}_n^{b_n}.$$

In the strategy behind the proof of Theorem 3.3.1, the augmentation condition plays a central role.

The first step is to establish the augmentation condition for nilpotent groups of finite rank.

Theorem 3.3.2. *Let X be a nilpotent group of class c and finite rank r and suppose that X is generated by subgroups $Y_1, \ldots, Y_n$. Then given positive integers $b_1, \ldots, b_n$, there exists $e = e(\mathbf{b}, c, r)$ such that*

$$\mathfrak{x}^e \leqq \mathbb{Z}X\mathfrak{y}_1^{b_1} + \cdots + \mathbb{Z}X\mathfrak{y}_n^{b_n}.$$

This result is then used to obtain

Theorem 3.3.3. *Suppose that $J = \langle H_1, \ldots, H_n \rangle$, H_i sn J and $J'/\gamma_3(J)$ has finite rank. Then nilpotent homomorphic images of J satisfy the augmentation condition. More precisely, suppose that $X = J/N$ is a nilpotent (of class c) image of J and $Y_i = H_iN/N$. Then if the rank of $J'/\gamma_3(J)$ is r, given $a_1, \ldots, a_n$, there exists $e = e(\mathbf{a}, c, r)$ such that*

$$\mathfrak{x}^e \leqq \mathbb{Z}X\mathfrak{y}_1^{a_1} + \cdots + \mathbb{Z}X\mathfrak{y}_n^{a_n}.$$

The proof of Theorem 3.3.1 is then completed by means of

Theorem 3.3.4. *If the nilpotent homomorphic images of a join J of subnormal subgroups $H_1, \ldots, H_n$ satisfy the augmentation condition, then given $a_1, \ldots, a_n$, there exists c such that*

$$\gamma_c(J) \leqq \gamma_{a_1}(H_1) \cdots \gamma_{a_n}(H_n).$$

Moreover if $J'/\gamma_3(J)$ has finite rank r, then $c = c(\mathbf{a}, \mathbf{h}, r)$, where $H_i \lhd^{h_i} J$.

We first prove Theorem 3.3.2. The main ingredient is

Theorem 3.3.5. *Suppose that X is a finitely generated nilpotent group generated by subgroups Y and Z. Then X satisfies the augmentation condition.*

We remark here that there is no hope for a result like Theorem 3.3.1 if the corresponding ring-theoretic version does not hold. To illustrate this in the case of two subgroups, take $X = \langle Y, Z \rangle$ to be nilpotent of class c and form the natural split extension $G = \mathbb{Z}X]X$, where X acts on $\mathbb{Z}X$ by right multiplication. Set $H = \langle \mathbb{Z}X, Y \rangle$ and $K = \langle \mathbb{Z}X, Z \rangle$. Since X is nilpotent, H

and K are subnormal in G. If $d \geqq 1$, then

$$\gamma_{c+d+1}(G) \leqq \mathbb{Z}X(X-1)^d = \mathfrak{x}^d$$

(see Lemma 3.3.7) $= [\mathbb{Z}X, {}_dX] \leqq \gamma_{d+1}(G)$. Similarly, for $a, b \geqq 1$, $\gamma_{c+a+1}(H) \leqq \mathbb{Z}X\mathfrak{y}^a \leqq \gamma_{a+1}(H)$ and $\gamma_{c+b+1}(K) \leqq \mathbb{Z}X\mathfrak{z}^b \leqq \gamma_{b+1}(K)$. Now consider the following two properties:

(P1) for all $u, v \geqq 1$, there exists $w \geqq 1$ such that $\gamma_w(G) \leqq \gamma_u(H)\gamma_v(K)$;
(P2) for all $a, b \geqq 1$, there exists $d \geqq 1$ such that $\mathfrak{x}^d \leqq \mathbb{Z}X\mathfrak{y}^a + \mathbb{Z}X\mathfrak{z}^b$.

Then it follows from above that (P1) and (P2) are equivalent.

Theorem 3.3.5 is proved using techniques developed by Philip Hall, but before we give the proof, we show how Theorem 3.3.2 is deduced from it. We proceed by induction on the number of subgroups n and thus we consider first the case $n = 2$. Suppose then that X is a nilpotent group of finite rank r, generated by subgroups which we shall denote by Y and Z instead of Y_1 and Y_2, for notational convenience. Let Y_1, Z_1 be finitely generated subgroups of Y, Z, respectively, and let $X_1 = \langle Y_1, Z_1 \rangle$. Then X_1 is finitely generated and so by Theorem 3.3.5, given a, b, there is an integer e such that

$$\mathfrak{x}_1^e \leqq \mathbb{Z}X_1\mathfrak{y}_1^a + \mathbb{Z}X_1\mathfrak{z}_1^b. \tag{2}$$

Now e can be bounded by a function of a, b, c and r (where c is the nilpotency class of X) using the following argument due Brian Hartley. Let H and K be free nilpotent groups of class c on r generators and form the free product $F = H * K$ of H and K. Since the rank of X is r, Y_1 and Z_1 can each be generated by at most r elements. Hence if

$$\bar{F} = F/\gamma_{c+1}(F),$$

there is an epimorphism from F to X_1 in which H and K map onto Y_1 and Z_1, respectively, and this epimorphism factors through $\bar{F}$. Thus there is an epimorphism $\theta: \bar{F} \to X_1$ in which $\bar{H}^\theta = Y_1$, $\bar{K}^\theta = Z_1$. On applying Theorem 3.3.5 to $\bar{F} = \langle \bar{H}, \bar{K} \rangle$, we see that there is an integer e such that

$$\bar{\mathfrak{f}}^e \leqq \mathbb{Z}\bar{F}\bar{\mathfrak{h}}^a + \mathbb{Z}\bar{F}\bar{\mathfrak{k}}^b.$$

Clearly e depends only on a, b, c and r, because $\bar{F}$, $\bar{H}$ and $\bar{K}$ are defined by c and r. Applying the extension of θ to the group rings, we obtain (2).

Finally, since every finitely generated subgroup of X is contained in one like X_1, we have

$$\mathfrak{x}^e \leqq \mathbb{Z}X\mathfrak{y}^a + \mathbb{Z}X\mathfrak{z}^b,$$

where $e = e(a, b, c, r)$ as required. The case $n = 2$ is complete and an easy induction gives the case for general n. Thus Theorem 3.3.2 will follow from Theorem 3.3.5.

The key result used in the proof of Theorem 3.3.5 is a slight generalization of the fact, due to Hall [48] (Lemma 9), that if M is a noetherian $\mathbb{Z}X$-module, where X is a finitely generated nilpotent group, and if some central element of $\mathbb{Z}X$ annihilates every simple image of M, then some power of that element annihilates M.

Lemma 3.3.6. *Let X be a finitely generated nilpotent group and M a finitely generated left $\mathbb{Z}X$-module. Suppose that $\mathfrak{x}M \leqq N$ for every maximal submodule N of M. Then $\mathfrak{x}^e M = 0$, for some positive integer e.*

Proof. Hall shows in [44] that if G is a polycyclic-by-finite group, then finitely generated $\mathbb{Z}G$-modules are noetherian. Thus M is noetherian and we may therefore assume that the result is true with any quotient M/L in place of M. We may in addition clearly suppose that the action of X on M is faithful.

Choose $z \neq 1$ in the centre of X. Then it follows from the hypothesis that $z-1$ annihilates all simple quotient modules of M. Therefore by the first of Hall's results above, there is an integer a such that

$$(z-1)^a M = 0.$$

Furthermore, since the lemma is true for the module $M/(z-1)M$, there is an integer b such that

$$\mathfrak{x}^b M \leqq (z-1)M.$$

Hence $\mathfrak{x}^{ab} M = 0$. □

In what follows we shall need

Lemma 3.3.7. *Suppose that a group G is generated by a subset S. Then*

$$(S-1)\mathbb{Z}G = \mathbb{Z}G(S-1) = \mathfrak{g}.$$

Proof. Clearly $(S-1)\mathbb{Z}G \leqq \mathfrak{g}$. However, for $s_1, s_2 \in S$, we have

$$s_1(s_2-1) = (s_1-1)(s_2-1) + (s_2-1) \in (S-1)\mathbb{Z}G,$$

and
$$\begin{aligned} s_1^{-1}(s_2-1) &= (s_1^{-1}-1)(s_2-1) + (s_2-1) \\ &= (1-s_1)s_1^{-1}(s_2-1) + (s_2-1) \in (S-1)\mathbb{Z}G. \end{aligned}$$

Hence $(S-1)\mathbb{Z}G$ is an ideal of $\mathbb{Z}G$. Since G acts trivially by left multiplication on $\mathbb{Z}G/(S-1)\mathbb{Z}G$, we must have $(S-1)\mathbb{Z}G = \mathfrak{g}$. Similarly $\mathbb{Z}G(S-1) = \mathfrak{g}$. □

All is now ready for the proof of Theorem 3.3.5. Recall that X is a finitely generated nilpotent group generated by subgroups Y and Z. We must show that X satisfies the augmentation condition relative to Y and Z. Thus let $a, b \geqq 1$. Take M to be the left $\mathbb{Z}X$-module

$$\mathbb{Z}X/(\mathbb{Z}X\mathfrak{y}^a + \mathbb{Z}X\mathfrak{z}^b)$$

and denote by μ the coset represented by 1 in M. Then $M = \mathbb{Z}X\mu$. We note that M is noetherian [44]. Suppose that N is a maximal submodule of M. Then

M/N ($= \bar{M}$, say) is a simple left $\mathbb{Z}X$-module. Since X is polycyclic, it follows that $\bar{M}$ is a finite elementary abelian p-group, for some prime p ([47]; see also [120], Theorem 9.55 (ii)). Thus $\bar{M}$ is an irreducible $\mathbb{Z}_p X$-module and since Y is subnormal in X, Clifford's theorem (see [26]) shows that $\bar{M}$ (as $\mathbb{Z}_p Y$-module) is a direct sum of irreducible $\mathbb{Z}_p Y$-submodules $\bar{M}_\lambda$, $\lambda \in \Lambda$, say. Writing $\bar{\mu} = \mu + N$, we have

$$\bar{\mu} = \bar{\mu}_1 + \cdots + \bar{\mu}_k,$$

where $\bar{\mu}_i \in \bar{M}_{\lambda_i}$ say. Then $\mathfrak{y}^a \mu = 0$ implies that $\mathfrak{y}^a \bar{\mu} = 0$ and so $\mathfrak{y}^a \bar{\mu}_i = 0$, all i. But $\mathfrak{y}\bar{\mu}_i$ is a $\mathbb{Z}_p Y$-submodule of $\bar{M}_{\lambda_i}$ and thus $\mathfrak{y}\bar{\mu}_i = 0$. Therefore $\mathfrak{y}\bar{\mu} = 0$, that is $\mathfrak{y}\mu \leqq N$. Similarly $\mathfrak{z}\mu \leqq N$. However, by Lemma 3.3.7,

$$\mathbb{Z}X\mathfrak{y} + \mathbb{Z}X\mathfrak{z} = \mathfrak{x}$$

and hence $\mathfrak{x}\mu \leqq N$. Then $\mathfrak{x}M \leqq N$.

Lemma 3.3.6 guarantees the existence of an integer e such that $\mathfrak{x}^e M = 0$ and so we have

$$\mathfrak{x}^e \leqq \mathbb{Z}X\mathfrak{y}^a + \mathbb{Z}X\mathfrak{z}^b. \quad \square$$

We want to use Theorem 3.3.2 in order to prove Theorem 3.3.3, that is if $J = \langle H_1, \ldots, H_n \rangle$ is a join of subnormal subgroups $H_1, \ldots, H_n$ and $J'/\gamma_3(J)$ has finite rank, then nilpotent images of J satisfy the augmentation condition. This is already clear in the case when J/J' has finite rank, for then $J/\gamma_n(J)$ has finite rank and so we can apply Theorem 3.3.2. To deal with the situation where $J'/\gamma_3(J)$ has finite rank we need the following result.

Lemma 3.3.8. *Let $X = \langle Y_1, \ldots, Y_n \rangle$ be a nilpotent group of class c and suppose that $X'/\gamma_3(X)$ has rank r. Then there exists an integer t, depending only on c and r, such that if F is a finitely generated subgroup of X', then there exist t-generator subgroups A_i of Y_i with $F \leqq \langle A_1, A_2, \ldots, A_n \rangle$.*

The proof of this result is rather long and so we split off the first part into two lemmas. In what follows, a *simple* commutator of *weight* c will mean an element of the form $[x_1, x_2, \ldots, x_c]$.

Lemma 3.3.9. *Let X be a nilpotent group of class 2 and suppose that X' has finite rank r. Let F be a finitely generated subgroup of X'. Then there exists a subgroup F_1 generated by at most $2r$ simple commutators of weight 2 such that $F \leqq F_1 \leqq X'$.*

Proof. Since F is finitely generated, there exists a finitely generated subgroup X_1 of X such that $F \leqq X_1'$. Hence we may assume that X is finitely generated and $F = X'$.

Consider first the case when X is finite. Then X is a direct product of finite p-groups and it is easy to see that we may assume that X is a p-group. The simple commutators of weight 2 generate X' and by Burnside's basis theorem, r of them suffice.

Now suppose that X is infinite but X' is finite. Since X is finitely generated, it follows that $X/\zeta_1(X)$ is finite. Let N be a maximal torsion-free subgroup of $\zeta_1(X)$. Then X/N is finite and $N \cap X' = 1$. By the first case X' can be generated, modulo N, by at most r simple commutators. In fact since $N \cap X' = 1$, they generate X'.

Suppose finally that X' is infinite. Any set of generators of a free abelian group A of rank r contains a subset T with $|T| = r$ and $|A : \langle T \rangle|$ finite. (For, we choose T so that $\langle T \rangle A^2 = A$.) Thus, by considering X modulo the torsion subgroup of X', we see that there is a subgroup M, generated by at most r commutators, which has finite index in X'. Now $(X/M)' = X'/M$ is finite and so we may apply the previous case to X/M giving a further r commutators which generate X' modulo M. Hence $2r$ commutators suffice for X'. □

Lemma 3.3.10. *Suppose that X is a nilpotent group of class $c \geqq 2$ and $X'/\gamma_3(X)$ has finite rank r. Then there exists a positive integer $m = m(c, r)$ such that if F is a finitely generated subgroup of $\gamma_c(X)$, there is a subgroup F_1 generated by at most m simple commutators of weight c such that*

$$F \leqq F_1 \leqq \gamma_c(X).$$

Proof. The case $c = 2$ is Lemma 3.3.9 and we proceed by induction on c. Since F is finitely generated, there exist finitely generated subgroups A of $\gamma_{c-1}(X)$ and B of X such that $F \leqq [A, B]$. We may assume inductively that A is generated by $m_1 = m(c-1, r)$ simple commutators of weight $c-1$. Thus a typical generator of A has the form $[u, v]$, where u is a simple commutator of weight $c-2$ and v is an element of X.

Let U be the subgroup of $\gamma_{c-2}(X)$ generated by all such u and V be the subgroup generated by all such v. Thus U, V can each be generated by m_1 elements and $A \leqq [U, V]$, so

$$F \leqq [U, V, B] \leqq \gamma_c(X).$$

Each of the subgroups $[V, B, U]$ and $[B, U, V]$ is in the centre of X (since $\gamma_c(X)$ is; see [119] (Lemma 2.21 (ii))) and so we have by the 3-subgroup lemma (see 4.3.4) $F \leqq [V, B, U][B, U, V]$. Now $[V, B]$ is a finitely generated subgroup of X'. Hence by Lemma 3.3.9, there are subgroups L and M of X, each generated by at most $2r$ elements, such that $[V, B] \leqq [L, M]\gamma_3(X)$. Now $[\gamma_3(X), U] \leqq \gamma_{c+1}(X) = 1$ and so $[V, B, U] \leqq [L, M, U]$. Using the 3-subgroup lemma once more, we obtain

$$[L, M, U] \leqq [M, U, L][U, L, M].$$

Now $[M, U]$ is generated (modulo $\gamma_c(X)$) by $2m_1 r$ simple commutators of weight $c-1$ and hence $[M, U, L]$ is generated by $4m_1r^2$ commutators of weight c. The same argument applies to $[U, L, M]$.

Turning to $[B, U]$, which is a finitely generated subgroup of $\gamma_{c-1}(X)$, we see from the induction hypothesis that there are m_1 simple commutators of weight $c-1$ which generate a subgroup H of $\gamma_{c-1}(X)$ such that $[B, U] \leqq H\gamma_c(X)$. Since $[\gamma_c(X), V] = 1$, we obtain $[B, U, V] \leqq [H, V]$.

Now V can be generated by m_1 elements of X and so $[H, V]$ can be generated by m_1^2 simple commutators of weight c. Putting these facts together, we obtain

$$F \leqq [V, B, U][B, U, V] \leqq [L, M, U][H, V] \leqq [M, U, L][U, L, M][H, V]$$

and $F_1 = [M, U, L][U, L, M][H, V]$ can be generated by

$$(4m_1r^2)(4m_1r^2)(m_1^2) = 16m_1^4r^4$$

simple commutators of weight c. Setting $m(c, r) = 16m(c-1, r)^4r^4$ completes the proof. □

All is now ready for the proof of Lemma 3.3.8. So suppose that $X = \langle Y_1, \ldots, Y_n \rangle$ is a nilpotent group of class c and that $X'/\gamma_3(X)$ has rank r. Let F be a finitely generated subgroup of X'. Then we may assume that $c > 1$ and by induction on c that there are t_1-generator subgroups W_i of Y_i, $t_1 = t(c-1, r)$, such that

$$F \leqq \langle W_1, \ldots, W_n \rangle \gamma_c(X).$$

Let f be an element of F. Then $f = we$ for some $w \in \langle W_1, \ldots, W_n \rangle$, $e \in \gamma_c(X)$. Let E be the subgroup of $\gamma_c(X)$ generated by all such e as f runs through a finite set of generators of F. Then E is a finitely generated subgroup of $\gamma_c(X)$ and $F \leqq \langle W_1, \ldots, W_n, E \rangle$.

By Lemma 3.3.10 there is a subgroup E_1 of $\gamma_c(X)$ containing E and which can be generated by $m = m(c, r)$ simple commutators of weight c. Let $[x_1, \ldots, x_c]$ be such a commutator. We can write

$$x_j = y_{1,j}y_{2,j} \cdots y_{r,j}w_j,$$

where $y_{i,j} \in Y_i$ and $w_j \in X'$, $1 \leqq i \leqq n$, $1 \leqq j \leqq c$. Let V_i be the (at most mc-generator) subgroup of Y_i generated by all the $y_{i,j}$ appearing in some such expressions of all the x_j occurring in the m generators of E_1. Since $\gamma_{c+1}(X) = 1$, commutators of length c are homomorphic in each argument and hence

$$[x_1, x_2, \ldots, x_c] = \Pi\,[y_{i_1,1}, y_{i_2,2}, \ldots, y_{i_c,c}],$$

where the product is over all $1 \leqq i_1 \leqq r, 1 \leqq i_2 \leqq r, \ldots, 1 \leqq i_c \leqq r$. It follows that $E_1 \leqq \langle V_1, \ldots, V_n \rangle$. Hence

$$F \leqq \langle W_1, \ldots, W_n, V_1, \ldots, V_n \rangle,$$

so the result follows on setting $A_i = \langle V_i, W_i \rangle$ which is a subgroup of Y_i, generated by at most $t_1 + mc$ elements. □

The final ingredient we need for the proof of Theorem 3.3.3 is a special situation in which the augmentation condition is invariably satisfied.

Proposition 3.3.11 [79]. *Suppose that Y and Z are subnormal subgroups of the group $X = YZ$. Then given any positive integers a, b, there exists c, depending only on a, b and the defects of Y and Z in X, such that*

$$\mathfrak{x}^c \leqq \mathbb{Z}X\mathfrak{y}^a + \mathbb{Z}X\mathfrak{z}^b.$$

Proof. Suppose that $Y \lhd^m X$ and $Z \lhd^n X$. We argue by induction on $m+n$. If $m = 0$ or $n = 0$, the result is clear, so we may assume that $m > 0, n > 0$. Let Y_1, Z_1 be the normal closures of Y, Z in X, respectively. Then $Y_1 = Y(Y_1 \cap Z)$ since $X = YZ$, and so by induction there is an integer $c_1 = c_1(a, b, m, n-1)$ such that

$$\mathfrak{y}_1^{c_1} \leqq \mathbb{Z}Y_1\mathfrak{y}^a + \mathbb{Z}Y_1\mathfrak{z}^b.$$

Similarly there exists $c_2 = c_2(a, b, m-1, n)$ such that

$$\mathfrak{z}_1^{c_2} \leqq \mathbb{Z}Z_1\mathfrak{y}^a + \mathbb{Z}Z_1\mathfrak{z}^b.$$

Moreover $Y_1 \lhd X, Z_1 \lhd X$ and $\mathfrak{x} = \mathbb{Z}X\mathfrak{y}_1 + \mathbb{Z}X\mathfrak{z}_1$ (by Lemma 3.3.7) and thus we conclude that

$$\mathfrak{x}^{c_1+c_2} \leqq \mathbb{Z}X\mathfrak{y}^a + \mathbb{Z}X\mathfrak{z}^b. \quad \square$$

We may now prove Theorem 3.3.3. Suppose that J is the join of subnormal subgroups $H_1, \ldots, H_n$ and that $J'/\gamma_3(J)$ has finite rank r. Our aim is to show that each nilpotent image $X = J/N$ of J satisfies the augmentation condition.

Suppose that X has class c and set $Y_i = H_i N/N$. Let F be any finitely generated subgroup of X'. Then, by Lemma 3.3.8,

$$F \leqq C = \langle A_1, \ldots, A_n \rangle,$$

where $A_i \leqq Y_i$ and the number of generators of each A_i depends only on r and c. Since C is nilpotent of class at most c, its rank is also bounded (see §5.3, p. 162). Let $b_1, \ldots, b_n$ be any positive integers. We now apply Theorem 3.3.2 to obtain an integer s such that

$$\mathfrak{f}^s \leqq \mathfrak{c}^s \leqq \mathbb{Z}C\mathfrak{a}_1^{b_1} + \cdots + \mathbb{Z}C\mathfrak{a}_n^{b_n} \leqq \mathbb{Z}X\mathfrak{y}_1^{b_1} + \cdots + \mathbb{Z}X\mathfrak{y}_n^{b_n},$$

where s is independent of F. This is the crucial point. Indeed $s = s(b, r, c)$. Hence on setting $Q = X'$ we obtain

$$\mathfrak{q}^s \leqq \mathbb{Z}X\mathfrak{y}_1^{b_1} + \cdots + \mathbb{Z}X\mathfrak{y}_n^{b_n}.$$

Furthermore $X = Y_1 Y_2 \cdots Y_r Q$ and so, by repeated application of Proposition

3.3.11, there is an integer $e = e(\boldsymbol{b}, r, c)$ such that

$$\mathfrak{x}^e \leqq \mathbb{Z}X\mathfrak{y}_1^{b_1} + \cdots + \mathbb{Z}X\mathfrak{y}_n^{b_n} + \mathbb{Z}X\mathfrak{q}^s$$

and the proof is now complete. □

In order to attack the proof of Theorem 3.3.4 we need one more preliminary result. It is the group-theoretic version of Proposition 3.3.11.

Proposition 3.3.12. *Suppose that H and K are subnormal subgroups of defects m and n, respectively, in their join J and that $J = HK$. Then there is a function $c = c(a, b, m, n)$, where a, b are any positive integers, such that*

$$\gamma_c(J) \leqq \gamma_a(H)\gamma_b(K).$$

This result was in essence proved independently by Stonehewer [142] and Brewster [15]. We give Brewster's version of the proof showing that there is a function $c = c(h, s)$ such that $\gamma_c(J) \leqq \gamma_a(H)\gamma_b(K)$ whenever $h \geqq m+n$ and $s \geqq a+b$. This is clearly equivalent to what we require.

The function $c(h, s)$ is defined inductively as follows. Put $c(0, s) = c(1, s) = c(2, s) = s-1$ for $s \geqq 2$. We then define

$$c(h, s) = c(h-1, c(h-1, s)+s),$$

$h \geqq 3$. The proof goes by induction on h. If $m = n = 1$, then Fitting's theorem gives the result at once, as was observed in the introduction to this section. So we may assume that $m > 1$. Suppose that H_{m-1} is the $(m-1)$th normal closure of H in J (so that $J = H_{m-1}K$). Then we obtain from the induction hypothesis that

$$\gamma_{c(h-1,\, c(h-1,\, s)+s)}(J) \leqq \gamma_{c(h-1,\, s)}(H_{m-1})\gamma_s(K). \tag{3}$$

Also by the induction hypothesis applied to $H_{m-1} = H(K \cap H_{m-1})$, we obtain

$$\gamma_{c(h-1,\, b)}(H_{m-1}) \leqq \gamma_a(H)\gamma_b(K \cap H_{m-1}), \tag{4}$$

so that we may now deduce from (3) and (4) that

$$\gamma_{c(h,\, s)}(J) \leqq \gamma_a(H)\gamma_b(K). \quad \square$$

Proof of Theorem 3.3.4. Suppose that J is the join of subnormal subgroups $H_1, \ldots, H_n$ of defects $h_1, \ldots, h_n$, respectively, and that nilpotent images of J satisfy the augmentation condition. Given integers $a_1, \ldots, a_n$, we wish to prove that there exists c such that

$$\gamma_c(J) \leqq \gamma_{a_1}(H_1) \cdots \gamma_{a_n}(H_n).$$

We split the proof into four steps: the cases where J is

(i) abelian-by-nilpotent,
(ii) nilpotent-by-nilpotent,
(iii) polynilpotent (that is soluble),

(iv) arbitrary.
Case (iv) will be reduced to case (iii) by applying Theorem 3.1.1.

(i) *J abelian-by-nilpotent*

For some positive integer r_1, we have $\gamma_{r_1}(J) = A$, say, is abelian. By Proposition 3.3.12, there exist positive integers $b_i = b_i(a_i, h_i)$ such that

$$\gamma_{b_i}(H_i A) \leqq \gamma_{a_i}(H_i). \tag{5}$$

We now put $X = J/A$ and $Y_i = H_i A/A$, so that X is a nilpotent image of J and satisfies the augmentation condition. Hence there exists r_2 such that

$$\mathfrak{x}^{r_2} \leqq \mathbb{Z}X\mathfrak{y}_1^{b_1} + \cdots + \mathbb{Z}X\mathfrak{y}_n^{b_n}.$$

If further $J'/\gamma_3(J)$ has finite rank r, then we have, by Theorem 3.3.3, $r_2 = r_2(\boldsymbol{b}, r_1, r)$. For the purpose of keeping track of bounds, we assume in this special case that $r_1 = r_1(\boldsymbol{a}, \boldsymbol{h}, r)$. The reader who is not interested in bounds can simply ignore them at each stage.

It follows at once that

$$A\mathfrak{x}^{r_2} \leqq A\mathfrak{y}_1^{b_1} + \cdots + A\mathfrak{y}_n^{b_n},$$

which, when translated into multiplicative notation, becomes

$$[A, {}_{r_2}J] \leqq [A, {}_{b_1}H_1] \cdots [A, {}_{b_n}H_n].$$

But $A = \gamma_{r_1}(J)$ and hence, on setting $c = r_1 + r_2$ and using (5), we obtain $\gamma_c(J) \leqq \gamma_{a_1}(H_1) \cdots \gamma_{a_n}(H_n)$ as required.

(ii) *J metanilpotent*

Now there is a positive integer $r_3 = r_3(\boldsymbol{a}, \boldsymbol{h}, r)$ such that $N = \gamma_{r_3}(J)$ is nilpotent of class at most r_3. Then by Proposition 3.3.12, there exists $c_i = c_i(r_3, h_i)$ such that

$$C_i = \gamma_{c_i}(H_i N) \leqq \gamma_{a_i}(H_i) \cap N. \tag{6}$$

Clearly $C_i \lhd N$ and hence the C_i normalize each other. By induction on the class of N we may assume that there exists $r_4 = r_4(\boldsymbol{c}, \boldsymbol{h}, r, r_3)$ such that if Z denotes the centre of N, then

$$\gamma_{r_4}(J) \leqq \gamma_{c_1}(H_1 Z) \cdots \gamma_{c_n}(H_n Z)Z.$$

The induction is started by the result in step (i). Therefore on setting $A = \gamma_{r_4}(J)$, we have

$$A \leqq \gamma_{c_1}(H_1 N) \cdots \gamma_{c_n}(H_n N)Z.$$

It follows that

$$A' \leqq C_1 \cdots C_n = C. \tag{7}$$

Moreover J/A' is abelian-by-nilpotent and so by step (i) there exists $r_5 = r_5(c, h, r)$ such that

$$\gamma_{r_5}(J) \leqq \gamma_{c_1}(H_1 A') \cdots \gamma_{c_n}(H_n A')A' \leqq C_1 \cdots C_n A'. \tag{8}$$

It follows at once from (6), (7) and (8) that

$$\gamma_{r_5}(J) \leqq C_1 \cdots C_n \leqq \gamma_{a_1}(H_1) \cdots \gamma_{a_n}(H_n).$$

Remark. If the subgroups H_i $(i = 1, 2, \ldots, n)$ are all subnormal in some group G and J is metanilpotent, then by repeated application of Theorem 1.2.5, C is subnormal of bounded defect in G. Hence so is its normal subgroup $\gamma_5(J)$.

(iii) *J polynilpotent*

Now J has a series $1 = N_{p+1} < N_p < \cdots < N_1 < N_0 = J$ with $\gamma_{q_j}(N_j) = N_{j+1}$, for some integers q_j depending only on a, h and r. Note that $p = p(a, h, r)$ in the special case where $J'/\gamma_3(J)$ has rank r. Without loss of generality we can replace each q_j by their maximum q. The case $p = 1$ is of course step (ii) and we proceed by induction on p. Assume therefore that $p \geqq 2$ and that the result is true for $p-1$.

By Proposition 3.3.12 there exist integers $s_i = s_i(q, h_i, a_i)$, which we may assume to be not less than q, such that

$$\gamma_{s_i}(H_i N_1) \leqq \gamma_{a_i}(H_i) N_2.$$

Then $\gamma_{s_i}^2(H_i N_1) = \gamma_{s_i}(\gamma_{s_i}(H_i N_1)) \leqq \gamma_{a_i}(H_i) N_3$ and so on until we obtain

$$D_i = \gamma_{s_i}^p(H_i N_1) \leqq \gamma_{a_i}(H_i). \tag{9}$$

(Here $\gamma_i^1(G) = \gamma_i(G)$ and, for $j \geqq 2$, $\gamma_i^j(G)$ is defined inductively to be $\gamma_i(\gamma_i^{j-1}(G))$.) Note that $D_i \lhd H_i$, $H_i \lhd^{h_i} J$, $N_1 \lhd J$. It follows that $H_i N_1$ and therefore D_i have bounded defect in J. Our strategy now is to find $A \lhd J$ such that $A \leqq D_1 \cdots D_n$ and J/A is polynilpotent with a series of length p, so that we may apply the induction hypothesis to J/A.

Applying the induction hypothesis to J/N_p, we obtain the existence of $t_1 = t_1(a, h, r)$ such that

$$\begin{aligned}\gamma_{t_1}(J) &\leqq \gamma_{s_1}(H_1 N_p) \cdots \gamma_{s_n}(H_n N_p) N_p \\ &\leqq \gamma_{s_1}(H_1 N_1) \cdots \gamma_{s_n}(H_n N_1) N_p.\end{aligned}$$

Now $\gamma_{s_i}(H_i N_1) \leqq N_1$, since $s_i \geqq q$, so $\gamma_{s_i}(H_i N_1) \lhd N_1$, for $i = 1, \ldots, n$. Also $N_p \lhd N_1$. Thus

$$G_1 = \gamma_{s_1}(H_1 N_1) \cdots \gamma_{s_n}(H_n N_1) N_p$$

is a group, as it is a product of normal subgroups of N_1.

We now apply Fitting's theorem repeatedly to G_1 and obtain an integer

$t_2 = t_2(\boldsymbol{a}, \boldsymbol{h}, r)$ such that

$$\gamma_{t_2}\gamma_{t_1}(J) \leqq \gamma_{t_2}(G_1) \leqq \gamma^2_{s_1}(H_1 N_1) \cdots \gamma^2_{s_n}(H_n N_1).$$

Again, each $\gamma^2_{s_i}(H_i N_1)$ is contained in N_2 as a normal subgroup and so their product is a group G_2. As before, we obtain t_3 such that

$$\gamma_{t_3}\gamma_{t_2}\gamma_{t_1}(J) \leqq \gamma_{t_3}(G_2) \leqq \gamma^3_{s_1}(H_1 N_1) \cdots \gamma^3_{s_n}(H_n N_1),$$

and so on until we have

$$A = \gamma_{t_p}\gamma_{t_{p-1}} \cdots \gamma_{t_1}(J) \leqq \gamma^p_{s_1}(H_1 N_1) \cdots \gamma^p_{s_n}(H_n N_1). \tag{10}$$

By Proposition 3.3.12 we can choose $e_i = e_i(h_i, a_i, q)$ such that

$$\gamma_{e_i}(H_i N_p) \leqq \gamma_{a_i}(H_i).$$

We may assume that $e_i \geqq q$, by increasing e_i if necessary.

Observe that J/A is polynilpotent with a series of length p, so our induction hypothesis produces an integer $c = c(\boldsymbol{e}, \boldsymbol{h}, r)$ with

$$\begin{aligned}\gamma_c(J) &\leqq \gamma_{e_1}(H_1 A) \cdots \gamma_{e_n}(H_n A)A \\ &\leqq \gamma_{e_1}(H_1 N_p) \cdots \gamma_{e_n}(H_n N_p)A.\end{aligned} \tag{11}$$

Further $E_i = \gamma_{e_i}(H_i N_p) \leqq N_1$ since $e_i \geqq q$, so E_i normalizes each D_j, since $D_j \lhd N_1$. From (9), (10) and (11) we obtain

$$\gamma_c(J) \leqq E_1 \cdots E_n D_1 \cdots D_n = (E_1 D_1) \cdots (E_n D_n) \leqq \gamma_{a_1}(H_1) \cdots \gamma_{a_n}(H_n),$$

as required.

(iv) *J arbitrary*

Set $A_i = \gamma_{a_i}(H_i)$, $i = 1, \ldots, n$. For each pair i, j, $1 \leqq i \leqq n$, $1 \leqq j \leqq i$, define subgroups $Q_{i,j}$ as follows: $Q_{i,i} = A_i$ and inductively (with j decreasing)

$$Q_{i,j-1} = P_{Q_{i,j}}(A_{j-1}).$$

Let $P_i = Q_{i,1}$, $i \geqq 1$. Then

$$\begin{aligned}P_i A_1 A_2 \cdots A_i &= Q_{i,1} A_1 A_2 \cdots A_i = A_1 Q_{i,1} A_2 \cdots A_i \\ &\leqq A_1 Q_{i,2} A_2 \cdots A_i = A_1 A_2 Q_{i,2} A_3 \cdots A_i \\ &\leqq A_1 A_2 Q_{i,3} A_3 \cdots A_i \\ &= \cdots \leqq A_1 \cdots A_{i-1} Q_{i,i} A_i = A_1 A_2 \cdots A_i.\end{aligned}$$

Hence $P_i A_1 A_2 \cdots A_i = A_1 A_2 \cdots A_i$ for $i = 1, \ldots, n$, and therefore

$$P_1 P_2 \cdots P_n A_1 A_2 \cdots A_n = A_1 A_2 \cdots A_n. \tag{12}$$

By Theorem 1.6.8 it is easy to see that there exists α, depending only on the defects of $H_1, \ldots, H_n$ in J, such that $A_i^{(\alpha)} \leqq P_i$. Moreover, by Theorem 3.1.1

there exists $m = m(\boldsymbol{a}, \boldsymbol{h}, \alpha)$ such that

$$\begin{aligned} J^{(m)} &\leqq H_1^{(a_1+\alpha)} \cdots H_n^{(a_n+\alpha)} \\ &= (H_1^{(a_1)})^{(\alpha)} \cdots (H_n^{(a_n)})^{(\alpha)} \\ &\leqq \gamma_{a_1}(H_1)^{(\alpha)} \cdots \gamma_{a_n}(H_n)^{(\alpha)} \\ &\leqq P_1 \cdots P_n. \end{aligned}$$

Finally $J/J^{(m)}$ is polynilpotent ($p = m-1$, $q = 2$ in the notation of step (iii)) and so by step (iii) there exists an integer $c = c(\boldsymbol{a}, \boldsymbol{h}, r)$ such that

$$\begin{aligned} \gamma_c(J) &\leqq \gamma_{a_1}(H_1 J^{(m)}) \cdots \gamma_{a_n}(H_n J^{(m)}) J^{(m)} \\ &\leqq \gamma_{a_1}(H_1) \cdots \gamma_{a_n}(H_n) J^{(m)} \\ &= J^{(m)} \gamma_{a_1}(H_1) \cdots \gamma_{a_n}(H_n) \\ &\leqq P_1 \cdots P_n A_1 \cdots A_n \\ &= A_1 \cdots A_n, \end{aligned}$$

by (12). □

Before pressing on, we pause here to record the fact (pointed out originally by Dan Segal for the case $n = 2$ and X/X' of finite rank) that Theorem 3.3.1 enables us to strengthen Theorem 3.3.3 to

Theorem 3.3.13. *Let J be the join of subnormal subgroups $H_i \lhd^{h_i} J$, $i = 1, 2, \ldots, n$, with $J'/\gamma_3(J)$ of finite rank r. Then, given any positive integers $a_1, \ldots, a_n$, there exists a function $f = f(\boldsymbol{a}, \boldsymbol{h}, r)$ such that*

$$\mathfrak{j}^f \leqq \mathbb{Z}J\mathfrak{h}_1^{a_1} + \cdots + \mathbb{Z}J\mathfrak{h}_n^{a_n}.$$

The proof requires

Lemma 3.3.14. *Suppose that G is any group and that $Y_n = \gamma_n(G)$. Then $\mathfrak{y}_n \leqq \mathfrak{g}^n$, for $n = 1, 2, \ldots$.*

Proof. The result is clear for $n = 1$. Suppose that $\mathfrak{y}_n \leqq \mathfrak{g}^n$ for some $n \geqq 1$. Then, for $y \in Y_n$, $g \in G$, we have

$$(y-1)(g-1) \in \mathfrak{y}_n\mathfrak{g} \leqq \mathfrak{g}^{n+1}$$

and

$$(g-1)(y-1) \in \mathfrak{g}\mathfrak{y}_n \leqq \mathfrak{g}^{n+1}.$$

Therefore $yg - gy \in \mathfrak{g}^{n+1}$, so that $[y, g] - 1 \in \mathfrak{g}^{n+1}$. But $Y_{n+1} = [Y_n, G]$ and Lemma 3.3.7 shows that $\mathfrak{y}_{n+1}$ is generated as a right $\mathbb{Z}Y_{n+1}$-module by

$$\{[y, g] - 1 \mid y \in Y_n, g \in G\}.$$

Hence $\mathfrak{y}_{n+1} \leqq \mathfrak{g}^{n+1}$ and the result follows by induction on n. □

Proof of Theorem 3.3.13. By Theorem 3.3.1, there exists $c = c(\boldsymbol{a}, \boldsymbol{h}, r)$ such that

$$\gamma_c(J) \leqq \gamma_{a_1}(H_1) \cdots \gamma_{a_n}(H_n).$$

If $u \in \gamma_c(J)$, then $u = v_1 \cdots v_n$, where $v_i \in \gamma_{a_i}(H_i)$. Hence

$$u - 1 = (v_1 - 1) + v_1(w - 1),$$

where $w = v_2 \cdots v_n$. Further

$$w - 1 = (v_2 - 1) + v_2(v_3 \cdots v_n - 1),$$

etc., so that it readily follows, on setting $U = \gamma_c(J)$ and $V_i = \gamma_{a_i}(H_i)$, $i = 1, 2, \ldots, n$, that

$$\mathbb{Z}J\mathfrak{u} \leqq \mathbb{Z}J\mathfrak{v}_1 + \cdots + \mathbb{Z}J\mathfrak{v}_n.$$

By Lemma 3.3.14, $\mathfrak{v}_i \leqq \mathfrak{h}_i^{a_i}$, $i = 1, 2, \ldots, n$, so that

$$\mathbb{Z}J\mathfrak{u} \leqq \mathbb{Z}J\mathfrak{h}_1^{a_1} + \cdots + \mathbb{Z}J\mathfrak{h}_n^{a_n}. \tag{13}$$

Now applying Theorem 3.3.3 to $J/U = \langle H_1 U/U, \ldots, H_n U/U \rangle$ yields an $f = f(\boldsymbol{a}, \boldsymbol{h}, r)$ such that

$$\mathfrak{j}^f \leqq \mathbb{Z}J\mathfrak{u} + \mathbb{Z}J\mathfrak{h}_1^{a_1} + \cdots + \mathbb{Z}J\mathfrak{h}_n^{a_n}.$$

The result follows at once from (13). □

We narrow our attention now to the situation where $J = \langle H, K \rangle$ is the join of a *pair* of subnormal subgroups of a group G. In light of the work of Williams which will be described in Chapter 5, it is reasonable to conjecture that if $[H, K]/[H, K, J]$ is the direct product of a group of finite rank and a periodic divisible group, then, given integers a, b, there exists an integer c such that

$$\gamma_c(J) \leqq \gamma_a(H)\gamma_b(K)$$

and, furthermore, that $J \operatorname{sn} G$. In this direction, Howard Smith has proved

Theorem 3.3.15. *Let $J = \langle H, K \rangle$, where H, K are subnormal subgroups of a group G, and suppose that*

$$[H, K]\gamma_n(J)/\gamma_n(J)$$

has finite rank for all $n = 1, 2, \ldots$. Then
(i) *given a, b, there exists c such that $\gamma_c(J) \leqq \gamma_a(H)\gamma_b(K)$, and*
(ii) *$J \operatorname{sn} G$.*

We defer the proof of the second part to the next section. The first part requires

Proposition 3.3.16. *Suppose that X is a finite nilpotent group, of class d, generated by subgroups Y and Z such that $[Y, Z]$ has rank r. Then*

(i) *$[Y, Z]$ can be generated by at most r commutators of the form $[y, z]$, with $y \in Y$, $z \in Z$; and*
(ii) *given positive integers a, b, there exists $c = c(a, b, r, d)$ such that*

$$\mathfrak{x}^c \leqq \mathbb{Z}X\, \mathfrak{y}^a + \mathbb{Z}X\mathfrak{z}^b.$$

Proof. Of course $[Y, Z]$ is generated by commutators $[y, z]$, $y \in Y$, $z \in Z$. Write $y = y_{p_1} \cdots y_{p_l}$, $z = z_{p_1} \cdots z_{p_l}$, where the p_i are the primes dividing the order of X and y_{p_i}, z_{p_i} are the components of y, z, respectively, in the Sylow p_i-subgroup S_i of X. Since X is the direct product of the S_i, we have

$$[y, z] = \prod_{i=1}^{l} [y_{p_i}, z_{p_i}].$$

Clearly the p_i-component of $[Y, Z]$ is generated by elements of the form $[y_{p_i}, z_{p_i}]$ and hence by at most r such elements, by Burnside's basis theorem. Hence $[Y, Z]$ can be generated by at most r commutators of the form $[y, z]$, with $y \in Y$, $z \in Z$. This gives (i). Furthermore, if $Y_1 = \langle y_1, \ldots, y_r \rangle$ and $Z_1 = \langle z_1, \ldots, z_r \rangle$, we have

$$W = [Y, Z] = [Y_1, Z_1]$$

and W is contained in the $2r$-generator subgroup $X_1 = \langle Y_1, Z_1 \rangle$ of X.

By Theorem 3.3.2, there exists $e_1 = e_1\,(a, b, d, 2r)$ such that

$$\mathfrak{w}^{e_1} \leqq \mathfrak{x}_1^{e_1} \leqq \mathbb{Z}X_1\mathfrak{y}_1^a + \mathbb{Z}X_1\mathfrak{z}_1^b \leqq \mathbb{Z}X\mathfrak{y}^a + \mathbb{Z}X\mathfrak{z}^b.$$

Let $YW = V$. Now $X = VZ$, $V \lhd X$ and $W \lhd V$, so by Proposition 3.3.11 there exists $e_2 = e_2\,(a, e_1, d)$ such that

$$\mathfrak{v}^{e_2} \leqq \mathbb{Z}V\mathfrak{y}^a + \mathbb{Z}V\mathfrak{w}^{e_1}$$

and there exists $e_3 = e_3\,(e_2, b, d)$ such that $\mathfrak{x}^{e_3} \leqq \mathbb{Z}X\mathfrak{v}^{e_2} + \mathbb{Z}X\mathfrak{z}^b$. Therefore

$$\mathfrak{x}^c \leqq \mathbb{Z}X\mathfrak{y}^a + \mathbb{Z}X\mathfrak{z}^b$$

for $c = e_3$. □

Proceeding now to the proof of Theorem 3.3.15, we suppose that J is the join of subnormal subgroups H and K and that $[H, K]\gamma_n(J)/\gamma_n(J)$ has finite rank for all n. Using Theorem 3.1.2 and Proposition 3.3.12 in a routine way, we can reduce to the case where J is abelian-by-nilpotent. Assume then that $A = \gamma_{d+1}(J)$ is abelian. It is clearly enough to show that, given α, β, there exists c such that

$$[A, {}_c J] \leqq [A, {}_\alpha H]\,[A, {}_\beta K].$$

Now J is the set-theoretic union of finitely generated subgroups F of the form $F = \langle L, M \rangle$, where L, M are finitely generated subgroups of H, K, respectively. By applying Theorem 3.3.3 to the nilpotent image $F/F \cap A$ of F, we deduce that there is an integer e such that

$$[A,_e F] \leqq [A,_\alpha L]\ [A,_\beta M] \leqq [A,_\alpha H]\ [A,_\beta K].$$

It suffices to show that e can be bounded independently of F. Working modulo $[A,_e F]$, we may assume that AF is nilpotent. Let U be a finitely generated subgroup of A, so that $\langle U,F\rangle = X$, say, is finitely generated and nilpotent. Suppose that $[H,K]A/A$ has rank r and let X/N be a finite quotient of X. Then from Proposition 3.3.16 applied to $FN/(A \cap F)N$ (which is finite and nilpotent of class at most d), we see that there is an integer $c = c(\alpha, \beta, r, d)$ such that

$$[U^F,_c F] \leqq [U^F,_\alpha L]\ [U^F,_\beta M]N.$$

The integer c is independent of N. Also a subgroup of a polycyclic group is the intersection of those subgroups of finite index which contain it [99]. Therefore

$$[U^F,_c F] \leqq [U^F,_\alpha L]\ [U^F,_\beta M] \leqq [A,_\alpha L]\ [A,_\beta M].$$

Similarly c is independent of F and hence

$$[A,_c F] \leqq [A,_\alpha L]\ [A,_\beta M]$$

as required. □

§3.4 Further applications to the join problem

We come now to the promised application, due to Williams [166], of the results on lower central series to the task of proving that a join J of subnormal subgroups $H_1, \ldots, H_n$ of a group G is subnormal in G, provided that $J'/\gamma_3(J)$ has finite rank. Since the case $n = 2$ is considerably easier from a technical point of view, we split that case off first in order to discuss the conceptual points arising. So suppose that J is the join of subnormal subgroups H and K of a group G and that $J'/\gamma_3(J)$ has finite rank. It follows from Theorem 3.3.1 that, for some integer c, $\gamma_c(J) \leqq HK$. Thus if we set $L = H\gamma_c(J)$, we obtain $L = H(K \cap L)$. Now $\gamma_c(K) \leqq K \cap L$, so $K \cap L \lhd^{c-1} K \operatorname{sn} G$ and hence L is subnormal in G, by Theorem 1.2.5. But $\gamma_c(J) \lhd L$, so that $\gamma_c(J)$ is subnormal in G. The subnormality of $\gamma_c(J)$ will play a central role in what follows, though it must be pointed out that in itself it is not sufficient to ensure the subnormality of J. This fact is demonstrated by the example given in Theorem 1.5.1 of a join J of a pair of subnormal subgroups of a group G, where J is nilpotent but not subnormal in G. However, in the case where $J'/\gamma_3(J)$ has finite rank, the subnormality of $\gamma_c(J)$ in G can be lifted to that of J by using techniques already developed in Chapter 1.

In order to express the first main result concisely, we introduce some further notation. If $\mathfrak{X}$ is a class of groups, $\mathbf{U}\mathfrak{X}$ is the class of all groups which are the union of an ascending chain, of length at most ω, of $\mathfrak{X}$-subgroups. We recall that $\mathrm{N}_2\mathfrak{X}$ is the class of all groups which are expressible as the product of two normal $\mathfrak{X}$-subgroups.

Our first objective is to establish the following portmanteau theorem, which was first proved in [80] for the case J/J' of finite rank.

Theorem 3.4.1. *Let G be a group with subnormal subgroups H and K of defects m, n, respectively, and let J be their join. Suppose that $J'/\gamma_3(J)$ has finite rank r. Then*

(i) *there is a function $d = d(m,n,r)$ such that J is subnormal in G of defect at most d. Moreover, if $\mathfrak{X}$ is the class $(H) \cup (K)$, then*

(ii) *there is a function $f = f(m,n,r)$ such that*

$$J' \in \mathrm{U}(\mathrm{N}_2\mathrm{S}_\mathrm{n})^f\mathfrak{X},$$

and

(iii) *there are functions $g = g(m,n), c = c(m,n,r)$ and a normal subgroup N of J such that*

$$N \in (\mathrm{N}_2\mathrm{S}_\mathrm{n})^g\mathfrak{X}$$

and J/N is nilpotent of class at most c.

When J/J' has finite rank, it will be apparent from the proof that J belongs to the class described in part (ii). Moreover, when J/J' is finitely generated, then the operation U is redundant. (See Theorem 3.2.1.) Then all the known coalescence results appear as special cases of this theorem. For convenience, we collect together the more important of them, along with a note, if relevant, of where they appear in Chapter 1.

(a) Finite groups (Theorem 1.3.3).
(b) Finitely generated nilpotent groups (a special case of Theorem 1.6.2).
(c) Any class $\mathfrak{X}$ of groups satisfying the maximal condition for subgroups for which $\mathfrak{X} = \mathrm{N}_0\mathfrak{X} = \mathrm{S}\mathfrak{X}$ (P. Hall, unpublished). We note that $\mathfrak{X} = \mathrm{N}_0\mathfrak{X}$ is equivalent to $\mathfrak{X} = \mathrm{N}_2\mathfrak{X}$.
(d) Any class $\mathfrak{X} = \mathrm{N}_0\mathfrak{X} \leqq \mathfrak{Min}$-*sn* (Theorem 1.7.4).
(e) Any class $\mathfrak{X} = \mathrm{N}_0\mathfrak{X} \leqq \mathfrak{Max}$-*sn* (Theorem 1.7.1).
(f) Finitely generated $\mathfrak{X}$-groups, for any class $\mathfrak{X} = \mathrm{N}_0\mathfrak{X} = \mathrm{S}_\mathrm{n}\mathfrak{X}$ (Theorem 1.6.2).
(g) The class of all $\mathfrak{X}$-groups of finite rank, where $\mathfrak{X}$ is the class of all groups, nilpotent, soluble, locally nilpotent or locally soluble groups (Corollary 1.7.13).

Clearly (a), (b) and (c) are special cases of (e) and there is no difficulty in deducing (e) from Theorem 3.4.1. For, if $\mathfrak{X} = \mathrm{N}_0\mathfrak{X} \leqq \mathfrak{Max}$-*sn* and if H is a subnormal $\mathfrak{X}$-subgroup of a group $J \in \mathfrak{Max}$-*sn*, then (as we saw in the proof of Theorem 1.7.1) $H^J \in \mathfrak{X}$. Similarly (d) is a special case of Theorem 3.4.1(i) and (iii). For, with the above notation, but $\mathfrak{X} = \mathrm{N}_0\mathfrak{X} \leqq \mathfrak{Min}$-*sn*, $H^J \in \mathfrak{X}$ follows from Theorem 1.7.10 and a routine induction.

Observe next that (f) is a special case of Theorem 3.2.1 in the situation where J is the join of just two subnormal subgroups and in this case Theorem 3.2.1 is an obvious consequence of Theorem 3.4.1. The coalescence of the five classes listed under (g) is immediate from Theorem 3.4.1. (The N_0-closure of the class of groups of finite rank has already been noted on p. 39 as a consequence of Lemma 1.7.11. Also the N_0-closure of the class of locally soluble groups of finite rank has already been recorded on pp. 40 and 41.)

We note, too, that Robinson's result (Theorem 1.3.13) that if $J = \langle H, K \rangle$, where H, K are subnormal in G, then $J\,sn\,G$, provided $J' \in \mathfrak{Max}\text{-}sn$, is an immediate consequence of Theorem 3.4.1.

The result which enables us to lift the subnormality of $\gamma_c(J)$ to that of J, as mentioned on p. 116, is obtained by combining Theorem 1.2.5 and Corollary 1.4.4.

Lemma 3.4.2. *Let H and K be subnormal subgroups of a group G, of defects m and n respectively, and let J be their join. Suppose that there is a normal subgroup N of J with $N \leqq H \cap K$ and each of $H/N, K/N$ generated by at most r elements, where r is finite. Then J is subnormal in G of defect bounded by a function of m, n and r. Also there is a function $f(m, n, r)$ such that $J \in (\mathrm{N}_2\mathrm{S}_\mathrm{n})^f((H) \cup (K))$.*

Proof. Let $H \lhd^{m_0} J$ and proceed by induction on m_0. If $m_0 = 1$, then the result follows by Theorem 1.2.1. We suppose therefore that $m_0 \geqq 2$ and assume the natural induction hypothesis. For all $k \in K$, $N \leqq H^k$ and so by Corollary 1.4.4 (assuming that $n \geqq 1$, which is permissible)

$$H^K = L[H, {}_nK],$$

where L is generated by at most $s = 1 + r + r^2 + \cdots + r^{n-1}$ conjugates of H under K. By induction on m_0 and a second induction on s, we see that L is subnormal in G with defect bounded by a function of m, n and r. Also

$$[H, {}_nK] \lhd K \lhd^n G$$

and therefore, by Theorem 1.2.5, H^K is subnormal in G with defect bounded by a function of m, n and r. A final application of Theorem 1.2.5 shows that $J = H^K K$ is also subnormal in G with defect similarly bounded.

The final statement of the lemma follows easily from the induction argument and Proposition 1.6.3. □

We may now proceed with the proof of Theorem 3.4.1. Suppose that J is the join of subnormal subgroups H and K of defects m and n, respectively, in a group G. Suppose further that $J'/\gamma_3(J)$ has rank r. By Theorem 3.3.1 there is a function c of m, n and r such that

$$N = \gamma_{c+1}(J) \leqq HK.$$

Thus

$$HN = H(HN \cap K). \tag{1}$$

Since $HN \lhd^m J$, we see that $(HN \cap K) \lhd^{m+n} G$. Therefore, by Theorem 1.2.5, HN is subnormal in G of defect bounded by a function of m and n. Set $L = HN$. Similarly $M = KN$ is subnormal in G with defect bounded in terms of m and n.

Let F be any subgroup of J' containing N such that F/N is finitely generated. (Clearly we may suppose that $c \geqq 1$.) By Proposition 3.3.8 there is an integer t, depending only on c and r, such that there exist subgroups A of L and B of M with $F \leqq \langle A, B \rangle$ and $A/N, B/N$ have at most t generators. Clearly A and B are subnormal in G of defect depending only on c, m and n. It follows from Lemma 3.4.2 that $\langle A, B \rangle$ is subnormal in G of defect bounded by a function of m, n, c and r, and so of m, n and r. Now $N \leqq F \leqq \langle A, B \rangle \leqq J$, so F is subnormal in G of defect bounded by a function of m, n and r. Thus J' is subnormal in G with defect similarly bounded, by Lemma 1.7.5. Finally $J = H(KJ')$ and two further applications of Theorem 1.2.5 give (i) of the theorem.

We recall that $\mathfrak{X} = (H) \cup (K)$. Then by Proposition 1.6.3 it follows that L, as given by (1), lies in $(\mathrm{N}_2\,\mathrm{S_n})^s\,\mathfrak{X}$, for some $s = s(m, n)$. By Lemma 3.3.8, if F/N is a finitely generated subgroup of J'/N, then there are t-generator subgroups A/N, B/N of L/N, M/N respectively, with $F \leqq \langle A, B \rangle$ and $t = t(c, r)$. Now by Lemma 3.4.2 there is a function $f = f(m, n, r)$ such that

$$F \in (\mathrm{N}_2\,\mathrm{S_n})^f\,\mathfrak{X}.$$

Abelian groups of finite rank are countable and hence nilpotent groups of finite rank are countable. Therefore since J'/N has finite rank, it is countable and

$$J' \in \mathrm{U}(\mathrm{N}_2\,\mathrm{S_n})^f\,\mathfrak{X}.$$

To prove (iii) we can take $g(m, n) = s + 1$ and since $c = c(m, n, r)$ has already been found, the proof of Theorem 3.4.1 is complete. □

We now state and prove the generalization, due to Williams, of the foregoing theorem to n subgroups, $n \geqq 2$.

Theorem 3.4.3 [166]. *Suppose that $H_1, \ldots, H_n$ are subnormal subgroups of a group G and let $J = \langle H_1, \ldots, H_n \rangle$. If $J'/\gamma_3(J)$ has finite rank, then J is subnormal in G.*

Remark. The subnormal defect of J in G can be checked from the proof to be bounded in terms of n, the defects of the H_i in G and the rank of $J'/\gamma_3(J)$.

Proof. The main difficulty to be overcome in showing that J is subnormal in G is proving that some term of the lower central series of J is subnormal in G. In the case of two subgroups, this was a simple consequence of the pair case of Theorem 3.3.1. For more than two subgroups, it is by no means an immediate deduction from Theorem 3.3.1 and in fact one has to go back to examining certain facets of the proof of that theorem in order to circumvent the difficulty. The argument proceeds in two steps.

Step (i). Assume that J is polynilpotent, so that there is a series

$$1 = N_{p+1} < N_p < N_{p-1} < \cdots < N_0 = J,$$

where $\gamma_{q_j}(N_j) = N_{j+1}$ for some q_j. That $\gamma_c(J)\, sn\, G$, for some c, in the case $p=1$ was indicated at the end of part (ii) of the proof of Theorem 3.3.4. Our objective is to extend that argument by induction on p. We show by induction on p that, given integers $a_1, \ldots, a_n$, there exist an integer h and subgroups $C_{p,j,i}$ and $L_{p,j}$ of J ($1 \leqq j \leqq p$, $1 \leqq i \leqq n$) such that

(a) $1 = L_{p,p} \leqq L_{p,p-1} \leqq \cdots \leqq L_{p,1}$ and $L_{p,j} \lhd J$, all j;
(b) $C_{p,j,i}\, sn\, J$, for all i,j;
(c) $C_{p,j,i}$ and $C_{p,j,k}$ permute, for all j, i, k;
(d) $L_{p,j} \leqq C_{p,j+1,1} C_{p,j+1,2} \cdots C_{p,j+1,n}$, for all $j < p$;
(e) $C_{p,j,i} \leqq \gamma_{a_i}(H_i) L_{p,j}$, for all j, i;
(f) $\gamma_h(J) \leqq C_{p,1,1} C_{p,1,2} \cdots C_{p,1,n}$.

We consider first of all the case $p = 1$. Then J is metanilpotent. Set $L_{1,1} = 1$, $C_{1,1,i} = C_i$ and $h = r_5$, where C_i and r_5 are defined in step (ii) of the proof of Theorem 3.3.4 (pp. 110 and 111). The above conditions are clearly satisfied in this case.

Now let $p > 1$ and assume the natural induction hypothesis. We then put $C_{p,p,i} = D_i$, where D_i is the subgroup in (9) on p. 111. Let

$$A = \gamma_{t_p} \gamma_{t_{p-1}} \cdots \gamma_{t_1}(J)$$

as in step (iii) of the proof of Theorem 3.3.4. Then

$$A \leqq C_{p,p,1} C_{p,p,2} \cdots C_{p,p,r}.$$

Each $C_{p,p,i} \leqq N_p$ and so e_i exists such that $\gamma_{e_i}(H_i A) \leqq \gamma_{a_i}(H_i)$, by Proposition 3.3.12. We now apply the induction hypothesis to the factor group J/A with the integers e_i replacing the a_i. Thus we obtain an integer h and subgroups $C_{p-1,j,i}$ and $L_{p-1,j}$ of J/A, for $1 \leqq j \leqq p-1$, $1 \leqq i \leqq n$, satisfying the conditions corresponding to (a)–(f).

For $1 \leqq j \leqq p-1$, define $C_{p,j,i}$ to be the full pre-image in J of $C_{p-1,j,i}$ and $L_{p,j}$ the full pre-image in J of $L_{p-1,j}$. Thus in particular $L_{p,p-1} = A$. Finally set $L_{p,p} = 1$. It is routine to check that conditions (a) to (f) hold for this choice of $h, C_{p,j,i}$ and $L_{p,j}$.

The next step is to prove that $C_{p,j,i}\, sn\, G$. This is certainly the case for $j = p$, as $C_{p,p,i}\, sn\, J$ and $C_{p,p,i} \leqq \gamma_{a_i}(H_i)$ together imply $C_{p,p,i}\, sn\, \gamma_{a_i}(H_i)$, and of course $\gamma_{a_i}(H_i) \lhd H_i\, sn\, G$. So we may proceed by reverse induction on j and assume that each $C_{p,j+1,i}\, sn\, G$. As these subgroups permute, their product is subnormal in G. It now follows from (a) and (d) that $L_{p,j}$ is subnormal in G.

Furthermore, since $L_{p,j} \lhd J$, we have that $L_{p,j}$ permutes with $\gamma_{a_i}(H_i)$ and hence $\gamma_{a_i}(H_i)L_{p,j} sn\, G$. Combining (b) and (e) gives

$$C_{p,j,i}\, sn\, \gamma_{a_i}(H_i)L_{p,j},$$

so that $C_{p,j,i}\, sn\, G$, as claimed.

Therefore each $C_{p,1,i}$ is subnormal in G and, by repeated application of Theorem 1.2.5, their product is subnormal in G. The fact that $\gamma_h(J) \lhd J$, together with (f), now shows that $\gamma_h(J)\, sn\, G$.

Step (ii). We are now in a position to deal with the general case. By Theorem 3.1.1 there is an integer s such that $J^{(s)}\, sn\, G$. Of course $J/J^{(s)}$ is soluble and hence polynilpotent, so that the above argument, with $p = s-1$ and each $a_i = 1$, yields an integer h and subgroups $C_{p,j,i}$ and $L_{p,j}$ of $J/J^{(s)}$ satisfying the conditions (a) to (f) stated above.

For $1 \leqq j \leqq p = s-1$ and $1 \leqq i \leqq n$, define $F_{j,i}$ to be the full pre-image in J of $C_{p,j,i}$ and K_j to be the full pre-image in J of $L_{p,j}$, so that $K_p = J^{(s)}$ in particular. The $F_{j,i}$, K_j and $N = \gamma_h(J)$ satisfy

(a′) $1 \leqq K_p \leqq \cdots \leqq K_1$ and $K_j \lhd J$, for all j;
(b′) $F_{j,i}\, sn\, J$, for all i, j;
(c′) $F_{j,i}$ and $F_{j,k}$ permute, for all j, i, k;
(d′) $K_j \leqq F_{j+1,1}F_{j+1,2}\cdots F_{j+1,n}$, for all $j < s-1$;
(e′) $F_{j,i} \leqq H_iK_j$, for all i, j;
(f′) $N = \gamma_h(J) \leqq F_{1,1}F_{1,2}\cdots F_{1,n}$.

The argument is now as in step (i). We claim that $F_{j,i}\, sn\, G$, for all i, j. For, from (b′) and (e′) we have $F_{p,i}\, sn\, H_i J^{(s)}$, but $H_i\, sn\, G$ and $J^{(s)}\, sn\, G$, so $H_i J^{(s)}\, sn\, G$ and hence $F_{p,i}\, sn\, G$. By induction on j decreasing, we may assume that each $F_{j+1,i}$ is subnormal in G and hence, by repeated application of Theorem 1.2.5, so is their product. By a familiar argument it now follows from (d′) that K_j is subnormal in G and then that $H_iK_j\, sn\, G$. From (e′) it readily follows that $F_{j,i}\, sn\, G$, as required.

Hence $F_{1,i}\, sn\, G$, for each i, and, again by Theorem 1.2.5, so is their product. But this product contains N and is contained in J. It follows at once that $N\, sn\, G$. We have now established that $\gamma_h(J)\, sn\, G$ for some h. We finally deduce that J is subnormal in G as in the pair case Theorem 3.4.1, except this time by using repeated applications of Lemma 3.4.2 at the appropriate juncture. □

Our final task in this section is to prove Theorem 3.3.15(ii), which Smith [135] established while seeking a common origin to the version of Theorem 3.4.1 with J/J' of finite rank and Robinson's result (Corollary 1.3.13, p. 9). We suppose that H and K are subnormal subgroups of a group G, $J = \langle H, K \rangle$ and that

$$[H,K]\gamma_n(J)/\gamma_n(J)$$

has finite rank, for all $n \geqq 2$. We prove that J is subnormal in G. Recall that we have already established in Theorem 3.3.15(i) that, under these hypotheses, $\gamma_c(J) \leqq HK$, for some c, and hence, by the usual argument (see page 000), $\gamma_c(J)\, sn\, G$. We require

Proposition 3.4.4. *Suppose that* $H \lhd^m G$, $K \lhd^n G$, $J = \langle H, K \rangle$ *and that there exist a positive integer* c *and a normal subgroup* N *of* J *such that*

$$\gamma_c(J) \leqq N \leqq H \cap K,$$

with J/N *finitely generated and* $[H, K]N/N$ *of rank* r. *Then* J *is subnormal in* G *with defect bounded in terms of only* m, n, c *and* r.

Proof. For $i = 0, 1, \ldots, c-3$, the factor group

$$L_i = [H, K, {}_iJ]N/[H, K, {}_{i+1}J]N$$

can be generated by commutators of the form $[h, k, s_1, \ldots, s_i]$ modulo $[H, K, {}_{i+1}J]N$, where $h \in H$, $k \in K$ and $s_j \in H \cup K$ for $j = 1, \ldots, i$. Since the torsion-free rank of L_i is at most r, there exists a subgroup R_i generated by at most r commutators of the above form, each commutator involving $i + 2\,(\leqq c-1)$ elements of $H \cup K$, such that

$$|[H, K, {}_iJ]N : R_i[H, K, {}_{i+1}J]N| < \infty.$$

We now set $R = \langle R_0, \ldots, R_{c-3} \rangle$, so that R is generated by a set of at most $r(c-2)$ commutators with entries in $H \cup K$. It is easy to see that $|[H, K]N : RN|$ is finite. Therefore, if H^* and K^* are the subgroups generated by the H-entries and K-entries, respectively, in these commutators and we put $J^* = \langle H^*, K^* \rangle$, then J^* can be generated by $2r(c-1)(c-2)$ elements and $|[H, K]N : J^* \cap [H, K]N|$ is finite. Note also that

$$H^*N \lhd^c H \lhd^m G \quad \text{and} \quad K^*N \lhd^c K \lhd^n G.$$

We may now apply Lemma 3.4.2 to conclude that J^*N is subnormal in G with defect in G bounded (by a function of m, n, c and r). Hence RN has bounded defect, since $RN \lhd^c J^*N$. Since RN has finite index in $[H, K]N$ and $[H, K]N/N$ is finitely generated, $M = \text{core}_J(RN)$ (that is the intersection of the conjugates of RN in J) has finite index in $[H, K]N$ and contains N. (Finitely generated groups have only finitely many subgroups of a given finite index [42].) Of course $M \lhd RN$ and so M has bounded defect in G. Hence MH has bounded defect in G, by Theorem 1.2.1, and MH has finite index in $H[H, K]N = H^K N = H^K$.

Now let $Q = \text{core}_J(MH)$. Then Q has finite index in H^K, $Q \lhd J$ and $Q \leqq MH$. Hence Q has bounded defect in G so that QK has bounded defect in G. Moreover, QK has finite index in J and therefore so has $T = \text{core}_J(QK)$. Clearly T has bounded defect in G. Observe that J/T is a finite nilpotent group and, modulo T, $[H, K]$ has rank at most r. By Proposition 3.3.16(i), we now see

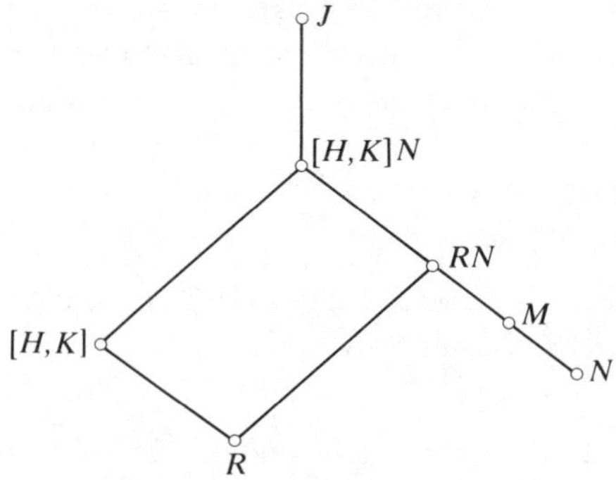

that $[H,K]$ can be generated, modulo T, by at most r commutators $[h_i, k_i]$, where $h_i \in H$, $k_i \in K$. Let $X = \langle h_1, \ldots, h_r \rangle$ and $Y = \langle k_1, \ldots, k_r \rangle$. Then TX and TY have bounded defect in G, since $TX \lhd^c TH$ and $TY \lhd^c TK$. We can now apply Lemma 3.4.2 once more to obtain that $T\langle X, Y \rangle$ is subnormal of bounded defect in G. But $[H,K] \lhd T\langle X, Y \rangle$ and hence $[H,K]$ is subnormal in G and has bounded defect. Finally, a double application of Theorem 1.2.1 yields that $J = ([H,K]H)K$ is subnormal of bounded defect in G. □

Continuing with the proof of Theorem 3.3.15(ii), as we have already pointed out on p. 122, $N = \gamma_c(J)\, sn\, G$. Then $[H,K]N/N$ has finite rank r, say, by hypothesis. Let F be a finitely generated subgroup of J. Then $F \leqq V = \langle L, M \rangle$, where L, M are finitely generated subgroups of H, K, respectively. Now $LN \lhd^c HN\, sn\, G$ and so LN, and similarly MN, are subnormal in G of defect independent of F. Moreover, since VN/N is finitely generated and $[LN, MN]N/N = [L, M]N/N$ has rank at most r, we deduce from Proposition 3.4.4 that VN is subnormal in G of defect d, say, independent of F. By Lemma 1.7.5 we have $J \lhd^d G$. □

§3.5 More on the classes $\mathfrak{S}$ and $\mathfrak{S}^\infty$

In this section we give applications, due to Smith [137], of the foregoing work in this chapter to the classes $\mathfrak{S}$ and $\mathfrak{S}^\infty$. We are interested first of all in extending Robinson's result (Theorem 1.4.6) that

$$\mathrm{P}(\mathfrak{Max}\text{-}sn \cup \mathfrak{Min}\text{-}sn)\mathfrak{S} = \mathfrak{S}$$

to

Theorem 3.5.1. *An extension of a group of finite rank by an $\mathfrak{S}$-group is an $\mathfrak{S}$-group.*

At first sight this result does not appear to be an extension of Robinson's theorem. However, we notice that soluble groups in the class

$\mathfrak{M} = \mathrm{P}(\mathfrak{Max}\text{-}sn \cup \mathfrak{Min}\text{-}sn)$ have finite rank. We note further that the class $\mathfrak{M}\mathfrak{S}$ is closed under Q and $\mathrm{s_n}$. Robinson's result follows from Theorem 3.5.1 on using the following very helpful reduction argument which Smith has established. It is dependent upon Roseblade's work on derived series of joins of subnormal subgroups and we record it as

Proposition 3.5.2. *Suppose that $\mathfrak{X}$ is a* Q- *and* $\mathrm{s_n}$*-closed class of groups and that soluble $\mathfrak{X}$-groups belong to $\mathfrak{S}$. Then $\mathfrak{X} \leqq \mathfrak{S}$.*

Proof. Let $G \in \mathfrak{X}$ and suppose that H and K are subnormal subgroups of defects m and n in G. We prove that $J = \langle H, K \rangle$ is subnormal in G, by induction on m. If $m = 1$, then $J = HK \lhd^n G$, by Proposition 1.2.1. Suppose therefore that $m \geqq 2$ and that the usual induction hypothesis holds. Denote by H_i the ith term of the normal closure series of H in G and set $L = H_{m-1}$. By the induction hypothesis

$$J_0 = \langle L, K \rangle \, sn \, G.$$

Hence $J_0 \in \mathrm{s_n}\mathfrak{X} = \mathfrak{X}$. We now set $P = P_L(K)$, so that

$$H^K = H^{PK} \lhd \langle H, PK \rangle = J_1,$$

say, and therefore $J = H^K K \, sn \, J_1$. By Theorem 1.6.8 there exists a positive integer λ with $L^{(\lambda)} \leqq P$ and, by Theorem 3.1.2, there exists μ such that $J_0^{(\mu)} \leqq L^{(\lambda)}K \subseteq PK \leqq J_1$. Now PK is subnormal in J_0, by Theorems 1.6.9 and 1.2.5. Furthermore $J_0/J_0^{(\mu)}$ is a soluble Q$\mathfrak{X}$ ($= \mathfrak{X}$)-group and hence lies in $\mathfrak{S}$ by hypothesis. It follows at once that $J_1 \, sn \, J_0$ and so $J_1 \, sn \, G$. But $J \, sn \, J_1$ and the result is proved. □

Proof of Theorem 3.5.1. Suppose that G is an extension of a group of finite rank by an $\mathfrak{S}$-group. By Proposition 3.5.2 we may assume that G is soluble. Let N be a normal subgroup of G such that N has finite rank and $G/N \in \mathfrak{S}$. We proceed by induction on the derived length of N. Suppose that A is the last non-trivial term of the derived series of N. Then $A \lhd G$ and we may assume that $G/A \in \mathfrak{S}$.

Let H and K be subnormal subgroups of G and set $J = \langle H, K \rangle$. Since $G/A \in \mathfrak{S}$, we have $JA \, sn \, G$ and so we may replace G by JA. Then, without loss of generality, $J \cap A = 1$, since $J \cap A \lhd G$, so that $G = JA$ splits over A. Also, since $C_J(A) \lhd G$, we may assume that $C_J(A) = 1$. It follows at once from Theorem 2.5.4 that H and K are nilpotent and hence that $G = \langle A, H, K \rangle$ is a Baer group. Then G is locally nilpotent. Let P be the periodic subgroup of A. Thus $P \lhd G$ and A/P is torsion-free. It now follows from [120] (Lemma 6.37) that if r is the rank of A, then $A/P \leqq \zeta_r(G/P)$, so that $[A, {}_rG] \leqq P$. Therefore we certainly have

$$[A, {}_r JP] \leqq JP,$$

from which it follows that $JP \lhd^r G$.

Clearly G can be replaced by JP, that is we may now assume that A is periodic and, as before, that $C_J(A) = 1$. Let F be any finitely generated subgroup of J. Then $F \leqq \langle H_0, K_0 \rangle = J_0$, where H_0, K_0 are finitely generated subgroups of H, K, respectively. It is enough to show that there is an integer d, independent of F, such that $[A, {}_d J_0] = 1$, for then $[A, {}_d J] = 1$ and $J \lhd^d G$. Let A_p denote the p-component of A for each relevant prime p. Then $A_p \lhd G$ and so it is sufficient to show that $[A_p, {}_d J_0] = 1$, for some fixed d independent of p. Let $a \in A_p$ and set $A_0 = \langle a \rangle^J$. Then $A_0 \leqq \langle a, J_0 \rangle$, a finitely generated and therefore nilpotent subgroup of G. So A_0 is finitely generated and hence finite, since A is periodic.

It is clearly sufficient to prove that, for some suitably bounded d, $[A_0, {}_{\bar{d}} J_0] = 1$. Let $C = C_{J_0}(A_0)$. Then $\bar{J}_0 = J_0/C$ is a finite p-group which acts as a group of automorphisms of the finite p-group A_0. By a result of P. Hall (which we find convenient to prove later as Lemma 6.1.10), $\bar{J}_0$ has rank at most $r(5r-1)/2$. Now H and K are nilpotent of classes c_m and c_n depending only on m and n, the defects of H and K, respectively, in G. (In fact $c_m \leqq m-1$, $c_n \leqq n-1$.) It follows that $H_0 \lhd^{c_m+m} J_0$, $K_0 \lhd^{c_n+n} J_0$. By Theorem 3.3.1 we deduce that $J_0 = \langle \bar{H}_0, \bar{K}_0 \rangle$ has nilpotency class depending only on m, n and r. Regarding A_0 in the usual way as a $\mathbb{Z}\bar{J}_0$-module, we may apply Theorem 3.3.2 to give an integer $d = d(m, n, r)$ such that

$$[A_0, {}_d \bar{J}_0] \leqq [A_0, {}_m \bar{H}_0][A_0, {}_n \bar{K}_0] = 1.$$

Hence $[A_0, {}_d J_0] = 1$, as required. □

The main result on the class $\mathfrak{S}^\infty$ was $\mathfrak{M}\mathfrak{S}^\infty = \mathfrak{S}^\infty$ (Theorem 2.6.1) and it was proved (Theorem 2.4.5) that $\mathfrak{S}^\infty(\mathfrak{Max}\text{-}sn) = \mathfrak{S}^\infty$, but that $\mathfrak{Max}\text{-}sn$ cannot be replaced by $\mathfrak{Min}\text{-}sn$ here; the most that can be said is that $\mathfrak{S}^\infty(\mathfrak{Min}\text{-}sn) \leqq \mathfrak{S}$ (Corollary 2.4.8). Our remaining objective in this section is to extend these results to show

Theorem 3.5.3 [137]. (i) *An extension of an* $\mathfrak{S}^\infty$*-group by a group of finite rank is in* $\mathfrak{S}$. (ii) $\mathfrak{S}^\infty\mathfrak{M} \leqq \mathfrak{S}$.

(Recall that $\mathfrak{M} = \mathrm{P}(\mathfrak{Max}\text{-}sn \cup \mathfrak{Min}\text{-}sn)$. Also (ii) follows from (i) and Proposition 3.5.2.)

In fact Smith has extended this result to show that if $\mathfrak{X}$ is the class of finite rank-by-abelian-by-finite rank groups, then $\mathfrak{S}^\infty\mathfrak{X} \leqq \mathfrak{S}$, but we content ourselves here with the special case. As a preliminary, we need

Lemma 3.5.4. *Suppose that the group G has a normal nilpotent subgroup N of class c such that G/N has finite rank r. Then if H, K are subnormal subgroups of*

G of defects m, n, respectively, their join J is subnormal in G with defect d depending only on m, n, c and r. If further H and K are nilpotent, then so is J.

Proof. We prove the first part by induction on m. The case $m = 0$ is trivial and the case $m = 1$ is dealt with by applying Proposition 1.1.4. Assume that $m \geqq 2$ and that the natural induction hypothesis holds. Let F be a finitely generated subgroup of H^K. Then there exist elements $k_1, \ldots, k_s$ of K such that

$$F \leqq \langle H^{k_1}, \ldots, H^{k_s} \rangle \leqq H^{\langle k_1, \ldots, k_s \rangle} = H^{K^*},$$

where $K^* = (K \cap N)\langle x_1, \ldots, x_r \rangle$, for some suitably chosen elements $x_1, \ldots, x_r$ in K. We may now apply Corollary 1.4.4 to obtain, for $n \geqq 1$,

$$H^{K^*} = \langle H^{y_1}, \ldots, H^{y_t} \rangle^{K \cap N}[H, {}_n K^*], \tag{1}$$

where $y_1, \ldots, y_t \in K$ and $t = t(r, n)$.

By the induction hypothesis, together with a second induction on t, $M = \langle H^{y_1}, \ldots, H^{y_t} \rangle$ is subnormal in H^G with defect at most

$$d_1 = d_1(m-1, c, r, t) = d_1(m-1, c, r, n).$$

Moreover $R = M^{K \cap N}$ is normal in $(K \cap N)^M M$, which, by Theorem 1.2.1, is subnormal in G with defect at most $d_2 = d_2(d_1, c+1)$, since $(K \cap N)^M \leqq N \lhd G$. Furthermore, it follows from (1) that

$$[H, {}_n K^*] \leqq P = P_K(R),$$

so that

$$F \leqq H^{K^*} = R[H, {}_n K^*] \leqq (RP) \cap H^K \lhd RP.$$

Combining Theorems 1.6.9 and 1.2.5 we obtain that RP is subnormal in G of defect d_3, bounded in terms of n and d_2. Hence $H^K \lhd^{d_3+1} G$, by Lemma 1.7.5, and therefore $J = H^K K \lhd^d G$, $d = d(m, n, r, c)$, by Theorem 1.2.1. The proof of the first part of the result is complete.

In order to prove the second part, we may assume that $G = \langle H, K \rangle$ and that H, K are nilpotent of classes a, b, respectively. The proof proceeds exactly as in the first part, by induction on the defect of H in G, to show that G is nilpotent of class bounded in terms of a, b, m, n, c, r. All that is needed is to keep careful track of the classes involved and to apply Proposition 1.6.3 at suitable junctures. □

All is now ready for the proof of Theorem 3.5.3. Suppose that G has a normal $\mathfrak{S}^\infty$-subgroup N with G/N of finite rank r. Let H and K be subnormal subgroups of G with join J. By Proposition 3.5.2 we may assume that G is soluble and by a familiar argument we may also assume that $G = JA$, where $A\,(\leqq N)$ is a non-trivial normal abelian subgroup of G. Without loss of generality we may suppose that $\mathrm{core}_J(A) = 1$, so that $A \cap J = 1$ and Theorem 2.5.4 applies to give H and K nilpotent. Hence $G = \langle H, A, K \rangle$ is a Baer group. By Corollary 2.5.2 all subgroups of N are subnormal in N and hence in G. In particular, $J \cap N \text{ sn } G$ and $J \cap N$ is nilpotent. But $J/J \cap N$ has finite rank and so J is nilpotent, by the second part of Lemma 3.5.4. Hence $G/A \cong J$ is nilpotent, of class c, say.

Suppose now that F is a finitely generated subgroup of J. Then there exists $F_0 = \langle H_0, K_0 \rangle$, where H_0, K_0 are r-generator subgroups of H, K, respectively, and $F \leq F_0(J \cap N)$. Furthermore HA, KA and $(J \cap N)A$ are nilpotent, by Proposition 1.6.3, and so there exist s, t, u with

$$[A, {}_sH] = [A, {}_tK] = [A, {}_u(J \cap N)] = 1.$$

It now follows from Theorem 3.3.2 that there exists $d = d(s, t, r, c)$ such that

$$[A, {}_dF_0] = 1.$$

Write $F_0 = X$ and $J \cap N = Y$. Then, in additive notation, we have $A\mathfrak{x}^d = 0$ and $A\mathfrak{y}^u = 0$. Since $Y \lhd \langle X, Y \rangle = Z$, say, Proposition 3.3.11 shows that there is an integer $v = v(u, d)$ (in fact $v = ud$ will do) such that $A\mathfrak{z}^v = 0$. Retranslating this into multiplicative notation we obtain $[A, {}_vF] = 1$. It now follows that $[A, {}_vJ] = 1$. Therefore $J \text{ sn } AJ = G$, as required. □

To complete the section we give an easy example (due to Smith) of a soluble group of finite rank which is not in the class $\mathfrak{S}^\infty$.

Theorem 3.5.5. *There exists a metabelian Baer group of rank 2 which does not belong to the class* $\mathfrak{S}^\infty$.

Proof. For each positive integer n, let p_n denote the nth prime. Define

$$H_n = \langle a, b \mid a^{p_n^n} = b^{p_n^{n-1}} = 1, a^b = a^{1+p_n} \rangle.$$

Then H_n is metacyclic and hence of rank 2. Also $[a, {}_ib] = a^{p_n^i}$ and so $[a, {}_nb] = 1$, that is H_n has nilpotency class n. Let G be the direct product of the H_n, for all $n \geq 1$, so that G is metabelian and is a Baer group. If G were in $\mathfrak{S}^\infty$, then, by Corollary 2.5.2, we would have every subgroup of G subnormal in G. By considering the direct product of the subgroups $\langle b \rangle$ (for all n), it is easy to see that this is false. □

Observe, for example by Corollary 1.2.4, that $G \in \mathfrak{S}$.

Smith has also studied the class $\mathfrak{L}$ of all groups G such that, given subnormal subgroups H, K of G and positive integers a, b, there exists c with $\gamma_c(J) \leq \gamma_a(H)\gamma_b(K)$, where $J = \langle H, K \rangle$. He proves that $\mathfrak{L} < \mathfrak{S}$ (that is the inclusion is strict). The reader is referred to [137] for the details.

4
THE PERMUTABILITY OF SUBNORMAL SUBGROUPS

It is abundantly clear from the theory we have developed so far in this book that subnormal subgroups which permute with each other play a very important part. It is therefore appropriate that a chapter should be devoted to this situation in its own right and we have elected to place the section at this point in view of the fact that the most efficient way of obtaining the best results is to use certain properties of the lower central series of subnormal subgroups.

§4.1 Wielandt's theory of operators

As our starting point we take the following seminal result of Wielandt (later generalized by Roseblade, see Theorem 1.6.6).

Theorem 4.1.1 [152] (Satz 23). *A perfect subnormal subgroup of a group with a composition series permutes with every subnormal subgroup of that group.*

Wielandt's method of proof here was first of all to show that, in a group G with a composition series, a perfect subnormal subgroup P can be written as the join of a finite number of perfect subnormal subgroups P_i, $i = 1, \ldots, s$, say, of a special type, namely, each P_i has a unique maximal normal subgroup. He then showed that if K is a subnormal subgroup which does not contain P_i, then K normalizes P_i and thus certainly permutes with P_i. Therefore P_i and hence P itself permute with all subnormal subgroups of G and the result follows. (See Wielandt [152], (20) and (23) or [159], 4.1 and 4.3.)

However, as we have already recorded a proof of the generalization of Theorem 4.1.1 to all groups in Theorem 1.6.6, we do not pursue the original proof any further here.

The first main result that we shall prove on permutability was conjectured by Wielandt on the basis of Theorem 4.1.1 and was eventually proved by him in [153].

Theorem 4.1.2. *Suppose that G satisfies* Min-*sn and that H, K are subnormal subgroups of G. Then $HK = KH$ provided that $\pi(H/H') \cap \pi(K/K')$ is empty.*

Here, as usual, for a group X, $\pi(X)$ denotes the set of primes dividing the orders of elements of finite order in X.

Once again this result has been superseded in time by a more general theorem due to Roseblade which depends on different arguments (see Theorem 4.3.1). However, Wielandt's proof is of considerable interest as it involves a very elegant theory of operators which he developed as a tool to obtain his result and which we reproduce here. The theory originally concerned a group G with a composition series and its lattice Σ of subnormal subgroups. (Recall that Σ is a lattice by Proposition 1.1.2(ii) and Theorem 1.3.2.)

For an arbitrary group G and any lattice Σ of subgroups of G, a map $\omega\colon \Sigma \to \Sigma$ is called an *operator* on Σ if, for all H, K in Σ,

(i) $\langle H, K\rangle^{\omega} = \langle H^{\omega}, K^{\omega}\rangle$, and

(ii) $H \lhd K \Rightarrow H^{\omega} \lhd K$.

Here, of course, H^{ω} denotes the image of H under the map ω. Note that (ii) implies that $H^{\omega} \lhd H$.

We denote by Ω the set of all operators on Σ. In showing that various maps are operators, our interest lies in the fact that they then satisfy condition (i) above.

The proof of Theorem 4.1.2 is carried out in two major stages. In the first we construct certain maps and demonstrate that they give rise to operators and in the second (in §4.2) we develop the quite remarkable permutability properties of operators which allow us to read off the result.

Constructing operators

Wielandt was concerned, as mentioned above, with groups that have a composition series, that is groups which satisfy both Max-*sn* and Min-*sn* (see p. 56). We shall, however, present this part of the discussion in the somewhat wider context of groups which are required to satisfy only Min-*sn*, as developed by Roseblade in Cambridge lectures.

To develop the theory of operators we need first some notation. If $\mathfrak{X}$ is a class of groups and G is any group, then we shall say that a normal subgroup K of G is an $\mathfrak{X}$-*kernel* if G/K is in $\mathfrak{X}$. The intersection of all $\mathfrak{X}$-kernels of G will be denoted by $G^{\mathfrak{X}}$, the $\mathfrak{X}$-*residual* of G. Clearly $G/G^{\mathfrak{X}}$ is residually an $\mathfrak{X}$-group.

We remark that, since the class $\mathfrak{Min}$-*sn* is coalescent by Theorem 1.7.4, we may speak of the lattice Σ of subnormal subgroups of a group in this class. Also when $\mathfrak{X}$ is the class of soluble groups, the map $H \mapsto H^{\mathfrak{X}}$ is an operator on Σ, by Theorem 3.1.2, and the same is true when $\mathfrak{X}$ is the class of nilpotent groups, by Theorem 3.3.1. Our first main theorem includes the former of these results.

Theorem 4.1.3. *Suppose that* $\mathfrak{X} = \text{PQS}_{\text{n}}\mathfrak{X}$ *and that G is a group satisfying* Min-*sn. Then the map* $H \mapsto H^{\mathfrak{X}}$ *is an operator on the lattice* Σ *of subnormal subgroups of* G.

As a consequence of this result we shall prove

Corollary 4.1.4. *Suppose that $\mathfrak{X}_1, \ldots, \mathfrak{X}_n$ are pairwise disjoint* $\mathbf{PQS}_n$*-closed classes of groups and G is any group satisfying* Min-*sn*. *Then the map*

$$H \mapsto H^{\mathfrak{X}_1} \cap H^{\mathfrak{X}_2} \cap \cdots \cap H^{\mathfrak{X}_n}$$

is an operator on Σ, the lattice of subnormal subgroups of G.

Here, by *pairwise disjoint* we mean, of course, that $\mathfrak{X}_i \cap \mathfrak{X}_j = (1)$ for $i \neq j$.

Preparatory to the proof of Theorem 4.1.3 we need some properties of $\mathfrak{X}$-residuals. We recall from p. 45 that a class $\mathfrak{X}$ is said to be $\mathbf{R}$-closed, that is $\mathfrak{X} = \mathbf{R}\mathfrak{X}$, if $G^{\mathfrak{X}}$ is an $\mathfrak{X}$-kernel for all groups G. A class $\mathfrak{X}$ is said to be $\mathbf{R}_0$*-closed* if the intersection of two (and hence finitely many) $\mathfrak{X}$-kernels of any group G is an $\mathfrak{X}$-kernel of G. Thus $\mathbf{R}_0$ is the 'finite' version of $\mathbf{R}$ so that if, for example, G satisfies Min-*sn* and $\mathfrak{X}$ is $\mathbf{R}_0$-closed, then $G^{\mathfrak{X}}/(G^{\mathfrak{X}})^{\mathfrak{X}} \in \mathfrak{X}$ and, since $G^{\mathfrak{X}}$ is a characteristic subgroup of G, we obtain $(G^{\mathfrak{X}})^{\mathfrak{X}} \lhd G$. Hence $G/(G^{\mathfrak{X}})^{\mathfrak{X}} \in \mathfrak{X}^2$. Thus if in addition $\mathfrak{X} = \mathbf{P}\mathfrak{X}$, we must have

$$G^{\mathfrak{X}} = (G^{\mathfrak{X}})^{\mathfrak{X}}.$$

This means that $G^{\mathfrak{X}}$ has no non-trivial $\mathfrak{X}$-quotients. A group with no $\mathfrak{X}$-quotients other than 1 is called $\mathfrak{X}$*-perfect*. This means that $G^{\mathfrak{X}}$ is $\mathfrak{X}$-perfect.

Summing up we have

Lemma 4.1.5. *Suppose that G satisfies* Min-*sn and that $\mathfrak{X} = \mathbf{R}_0\mathfrak{X} = \mathbf{P}\mathfrak{X}$. Then $G/G^{\mathfrak{X}} \in \mathfrak{X}$ and $G^{\mathfrak{X}}$ is $\mathfrak{X}$-perfect.*

We now record a useful sufficient condition for $\mathfrak{X} = \mathbf{R}_0\mathfrak{X}$, where $\mathfrak{X}$ is a class of groups. To express this concisely we write $\mathbf{A} \leqq \mathbf{B}$, for operations $\mathbf{A}$ and $\mathbf{B}$ on group classes, to denote that $\mathbf{A}\mathfrak{X} \leqq \mathbf{B}\mathfrak{X}$ for *every* class $\mathfrak{X}$ of groups.

Lemma 4.1.6. $\mathbf{R}_0 \leqq \mathbf{PS}_n$.

Proof. Suppose that $\mathfrak{X} = \mathbf{PS}_n\mathfrak{X}$ and that H and K are $\mathfrak{X}$-kernels of G. Then we have

$$H/H \cap K \cong KH/K \lhd G/K \in \mathfrak{X} = \mathbf{S}_n\mathfrak{X},$$

so that $H/H \cap K \in \mathfrak{X}$. It follows at once that $G/H \cap K \in \mathfrak{X}^2 = \mathfrak{X}$. □

We also need

Lemma 4.1.7. *Suppose that $\mathfrak{X} = \mathbf{QS}_n\mathfrak{X}$ and G is any group with $H \lhd G$ and $K \operatorname{sn} G$. Then if $H \in \mathfrak{X}$ and K is $\mathfrak{X}$-perfect, we obtain $K^H = K$.*

Proof. Let $J = HK$ and $K_1 = K^J$, $K_2 = K^{K_1}$. Since $K \operatorname{sn} G$, it will suffice to prove that $K_1 = K_2$. Now

$$K_1 = K(H \cap K_1) = K_2(H \cap K_1),$$

so that

$$K_1/K_2 \cong H \cap K_1/H \cap K_2 \in \text{QS}_n\mathfrak{X} = \mathfrak{X}.$$

On the other hand K_1 is $\mathfrak{X}$-perfect since K is and, as is easily seen, the class of $\mathfrak{X}$-perfect groups is N-closed. It now follows that $K_1 = K_2$, as required. □

As an immediate consequence of this result, we see that the condition $H \lhd G$ can be replaced by $H \, sn \, G$ provided we assume in addition that $\mathfrak{X} = \text{N}\mathfrak{X}$. However, we need something rather stronger than this in order to be able to prove Theorem 4.1.3. It turns out that the condition $\mathfrak{X} = \text{N}_0\mathfrak{X}$ is enough provided that the group G in question satisfies Min-*sn*. Now it is easy to see that

$$\text{N}_0 \leqq \text{PQ} \leqq \text{PQS}_n$$

and we present the result in a form shaped to our needs as

Lemma 4.1.8. *Suppose that* $\mathfrak{X} = \text{PQS}_n\mathfrak{X}$, $G \in \mathfrak{Min}$-*sn*, *and* H *sn* G, K *sn* G. *Then if* $H \in \mathfrak{X}$ *and* K *is* $\mathfrak{X}$-*perfect, we have* $K^H = K$.

It is clear that Lemma 4.1.8 will follow from Lemma 4.1.7 together with

Lemma 4.1.9. *Suppose that* $G \in \mathfrak{Min}$-*sn and* H *is a subnormal* $\mathfrak{X}$-*subgroup of* G, *where* $\mathfrak{X}$ *is any class of groups. Then* $H^G \in \text{N}_0\mathfrak{X}$.

Proof. By induction on the defect of H in G we may assume that $H \lhd^2 G$. By Theorem 1.7.10 H has only a finite number of conjugates in G. Thus H^G, as the product of these conjugates, is in $\text{N}_0\mathfrak{X}$, as required. □

For further results on relative normalizing properties of subnormal subgroups see §4.5 and §4.6 and in particular Theorem 4.5.10.

Proof of Theorem 4.1.3. Suppose that $\mathfrak{X} = \text{PQS}_n\mathfrak{X}$ and that G is a group satisfying Min-*sn*. Let Σ be the lattice of subnormal subgroups of G and let $H, K \in \Sigma$. We require first of all to prove

(i) $\langle H,K \rangle^{\mathfrak{X}} = \langle H^{\mathfrak{X}}, K^{\mathfrak{X}} \rangle$.

Set $J = \langle H, K \rangle$. Then $H/H \cap J^{\mathfrak{X}} \cong HJ^{\mathfrak{X}}/J^{\mathfrak{X}} \in \mathfrak{X}$ so that $H^{\mathfrak{X}} \leqq J^{\mathfrak{X}}$. Similarly $K^{\mathfrak{X}} \leqq J^{\mathfrak{X}}$ and thus

$$\langle H^{\mathfrak{X}}, K^{\mathfrak{X}} \rangle \leqq J^{\mathfrak{X}}.$$

We now treat the special case where H is $\mathfrak{X}$-perfect, that is $H = H^{\mathfrak{X}}$. Then

$$J^{\mathfrak{X}} \geqq \langle H, K^{\mathfrak{X}} \rangle.$$

Denoting factors modulo the normal closure of $K^{\mathfrak{X}}$ in J by bars, we see that $\bar{K}$ is an $\mathfrak{X}$-group, since $K/K^{\mathfrak{X}} \in \mathfrak{X}$, and $\bar{H}$ is $\mathfrak{X}$-perfect. Therefore $\bar{K}$ normalizes $\bar{H}$ by Lemma 4.1.8. Hence $J = KH(K^{\mathfrak{X}})^J$ and it follows that

$$(K^{\mathfrak{X}})^J = ((K^{\mathfrak{X}})^H)^{(K^{\mathfrak{X}})^J}.$$

Therefore

$$(K^{\mathfrak{X}})^J = (K^{\mathfrak{X}})^H$$

since the latter is subnormal in J, by Theorems 1.7.4 and 1.7.10. Hence $J = KQ$, where $Q = \langle K^{\mathfrak{X}}, H \rangle$, and $Q \leqq J^{\mathfrak{X}}$ from above. Let Q_1 be the normal closure of Q in $J^{\mathfrak{X}}$. Then

$$J^{\mathfrak{X}} = Q_1 (K \cap J^{\mathfrak{X}})$$

so that

$$J^{\mathfrak{X}}/Q_1 \in \text{QS}_n(K/K^{\mathfrak{X}}) \leqq \mathfrak{X}.$$

But $J^{\mathfrak{X}}$ is $\mathfrak{X}$-perfect so that we now must have $J^{\mathfrak{X}} = Q_1$ and therefore $J^{\mathfrak{X}} = Q$. This completes the first step.

For the general case we consider $U = \langle H^{\mathfrak{X}}, K \rangle^{\mathfrak{X}}$. Since $H^{\mathfrak{X}}$ is $\mathfrak{X}$-perfect, it follows from the first case that $U = \langle H^{\mathfrak{X}}, K^{\mathfrak{X}} \rangle$. But U is normalized by K, by definition, and by symmetry it is normalized by H as well. Hence $U \lhd J$. But J/U is the join of two subnormal $\mathfrak{X}$-subgroups HU/U and KU/U and we can now use the fact that $\mathfrak{X} = \text{N}_0 \mathfrak{X}$ together with Lemma 4.1.9 to deduce that J/U is in $\mathfrak{X}$, so that $J^{\mathfrak{X}} \leqq U$. But $U \leqq J^{\mathfrak{X}}$ so that we obtain $U = J^{\mathfrak{X}}$. This completes the proof of (i).

For (ii) we simply note that if $H \lhd K$, then $H^{\mathfrak{X}} \lhd K$ since $H^{\mathfrak{X}}$ is characteristic in H. The proof that

$$H \mapsto H^{\mathfrak{X}}$$

is an operator is now complete. □

Examples. Obvious examples of classes fulfilling the hypotheses of Theorem 4.1.3 are $\mathfrak{X} = \text{P}\mathfrak{A}$ (as we have already pointed out on p. 129), $\mathfrak{F}$ and $\mathfrak{F}_p$, the class of finite p-groups, p a prime.

Proof of Corollary 4.1.4. We proceed by induction on the number n of classes involved, the case $n = 1$ being Theorem 4.1.3.

We suppose then that $n > 1$ and that the natural induction hypothesis holds. We let $\lambda\colon H \mapsto H^{\mathfrak{X}_1} \cap \cdots \cap H^{\mathfrak{X}_n}$, $\mu_i\colon H \mapsto \bigcap_{j \neq i} H^{\mathfrak{X}_j}$ and

$$\lambda_i\colon H \mapsto H^{\mathfrak{X}_i}, \qquad i = 1, \ldots, n.$$

By induction we know that if H and K are subnormal in G with join J, then

$$J^{\mu_i} = \langle H^{\mu_i}, K^{\mu_i} \rangle.$$

Now

$$\langle H^{\lambda}, K^{\lambda} \rangle \leqq J^{\lambda} \leqq J^{\mu_i} = \langle H^{\mu_i}, K^{\mu_i} \rangle.$$

By Theorem 4.1.3 we have

$$\langle H^{\mu_i}, K^{\mu_i} \rangle^{\lambda_i} = \langle H^{\mu_i \lambda_i}, K^{\mu_i \lambda_i} \rangle.$$

Now

$$H^{\mu_i}/H^{\lambda} \cong H^{\lambda_i}H^{\mu_i}/H^{\lambda_i} \in \mathfrak{X}_i.$$

Hence

$$H^{\mu_i\lambda_i} \leqq H^{\lambda} \quad \text{and similarly} \quad K^{\mu_i\lambda_i} \leqq K^{\lambda}.$$

Therefore we have

$$\langle H^{\mu_i}, K^{\mu_i} \rangle^{\lambda_i} \leqq \langle H^{\lambda}, K^{\lambda} \rangle,$$

so that

$$J^{\mu_i\lambda_i} \leqq \langle H^{\lambda}, K^{\lambda} \rangle$$

for $i = 1, 2, \ldots, n$.

Now

$$J^{\mu_i\lambda_i} \leqq J^{\lambda}$$

and

$$J^{\lambda}/J^{\mu_1\lambda_1}J^{\mu_2\lambda_2} \in \mathfrak{X}_1 \cap \mathfrak{X}_2 = (1).$$

Hence $J^{\lambda} = \langle H^{\lambda}, K^{\lambda} \rangle$. The fact that λ is an operator is now clear. □

Suppose that H is subnormal in G and H_i $(i \geqq 0)$ are the terms of the normal closure series. If $H_i/H_{i+1} \in \mathfrak{X}$ for all i, then we call H an *$\mathfrak{X}$-subkernel* of G.

In order to gear in now to Wielandt's original formulation of his results, we specialize to the class $\mathfrak{C}$ of groups with a finite composition series, that is groups which satisfy both Max-*sn* and Min-*sn*. Suppose, then, that $G \in \mathfrak{C}$. Let Λ be a set consisting of abstract simple groups and suppose that H is a subnormal subgroup of G with the property that all the composition factors of G above H lie in Λ. Thus if $H \leqq K \lhd L$ *sn* G, L/K simple, then $L/K \in \Lambda$ (see Remark 1 on p. 57).

We write $\mathfrak{X}$ for the PQS$_n$-closure of Λ, that is the smallest P, Q and S$_n$-closed class of groups containing Λ. Moreover, since Λ is a set of simple groups, $\mathfrak{X}$ will in fact be the P-closure of $\Lambda \cup (1)$. With this notation it is clear that H is an $\mathfrak{X}$-subkernel of G. Conversely, if M is an $\mathfrak{X}$-subkernel of G and Δ is a composition factor of G above M, then we clearly have $\Delta \in \Lambda$. Thus the $\mathfrak{X}$-subkernels of G are precisely those subnormal subgroups M of G with $\mathscr{C}(G:M) \subseteq \Lambda$, that is such that $G^{\mathfrak{X}} \leqq M$. We note that Wielandt in his original formulation used the notation H^{Λ}, where H *sn* G, to denote $H^{\mathfrak{X}}$, but since Λ is a class of groups, the notation is now ambiguous and we do not use it.

§4.2 The permutability properties of operators

In the interests of maximal generality we set the initial discussion in a wider context than that of groups with Min-*sn*. Suppose that G is an arbitrary, but fixed, group. Following Wielandt we shall say that a subset $\mathscr{L}$ of the set of subgroups of G is a *subnormality lattice* if and only if

(i) $1 \in \mathscr{L}$,

(ii) $H, K \in \mathscr{L} \Rightarrow H \cap K, \langle H, K \rangle \in \mathscr{L}$,

(iii) $H, K \in \mathscr{L} \Rightarrow$ there is at least one series $H = L_r \lhd L_{r-1} \lhd \cdots \lhd L_0 = \langle H, K \rangle$, where $L_i \in \mathscr{L}$, $i = 0, 1, \ldots, r$.

Thus, for example, if $G \in \mathfrak{Min}\text{-}sn$, then the set of all subnormal subgroups of G is a subnormality lattice.

Let $\mathscr{L}$ be a subnormality lattice and let $\mho$ be the set of all maps of $\mathscr{L}$ into itself. Then $\mho$ is an associative semigroup under the usual law of product of maps. The identity map is the unit of $\mho$ and the map $0: H \mapsto 1, H \in \mathscr{L}$, is a right zero. We write Ω for the set of all operators on $\mathscr{L}$ (as defined on p. 129), so that $\Omega \subseteq \mho$. We now write down some of the basic properties of Ω, of which the first three are self-evident.

(1) $\alpha, \beta \in \Omega \Rightarrow \alpha\beta \in \Omega$,

(2) $H^{\omega} \lhd H$ for all $H \in \mathscr{L}, \omega \in \Omega$,

(3) $H \leqq K,\ H, K \in \mathscr{L} \Rightarrow H^{\omega} \leqq K^{\omega}$ for all $\omega \in \Omega$.

The properties of Ω in which we are most interested are those relating to permutability and in order to write them down concisely we shall use the notation $H\,v\,K$ (the v stands for 'vertauschbar') to denote $HK = KH$, for H, K in $\mathscr{L}$.

(4) $H\,v\,K \Rightarrow H^{\omega}\,v\,K^{\omega}$, for all $\omega \in \Omega$,

(5) $H\,v\,K \Rightarrow H^{\omega}\,v\,K$, for all $\omega \in \Omega$,

(6) if $\omega \in \Omega$, then $H^{\omega}\,v\,K^{\omega} \Rightarrow H^{\omega}\,v\,K$,

(7) if $\omega \in \Omega$, $H\,v\,K^{\omega} \Rightarrow H^{\omega}\,v\,K$,

and finally the most important of all:

(8) Suppose that α, β are in Ω. Then $H^{\alpha}\,v\,K^{\beta} \Leftrightarrow H^{\beta}\,v\,K^{\alpha}$.

We prove (4)–(8). Suppose that $H\,v\,K$. According to (iii) on p. 133 there is a series

$$H = L_r \lhd L_{r-1} \lhd \cdots \lhd L_0 = HK = J,$$

where $L_i \in \mathscr{L}, i = 0, 1, \ldots, r$. In order to see (4) we proceed by induction on r. If $r = 0$, then $H = J$ so that $K \leqq H$ and hence, since $K^{\omega} \leqq H^{\omega}$ by (3), we have $K^{\omega}\,v\,H^{\omega}$ trivially. So suppose that $r > 0$ and assume the natural induction hypothesis. Then $L_1 = H(K \cap L_1)$ and from the induction hypothesis we have $L_1^{\omega} = H^{\omega}(K \cap L_1)^{\omega}$. But $L_1^{\omega} \lhd J$ so that $L_1^{\omega}\,v\,K^{\omega}$. Moreover $(K \cap L_1)^{\omega} \leqq K^{\omega}$ by (3), so that we obtain $L_1^{\omega}K^{\omega} = H^{\omega}K^{\omega}$, a subgroup, as required. Furthermore, since $H^{\omega}K^{\omega} = (HK)^{\omega} \lhd HK$, we have $H^{\omega}K^{\omega}\,v\,K$. Hence $H^{\omega}\,v\,K$, and thus (5) is established. The same argument goes for (6). To see (7) we apply (5) with K^{ω} replacing K and then (7) follows from (6). Finally, in order to get (8) we note that $H^{\alpha}\,v\,K^{\beta}$ yields $H\,v\,K^{\beta\alpha}$ by (7), from which it follows, again by (7), that $H^{\beta\alpha}\,v\,K$ and then a third application of (7) gives $H^{\beta}\,v\,K^{\alpha}$, as required.

We have now reached a point in the theory where we have enough information to give the shorter version of the proof of Theorem 4.1.2 due to Hainzl [41]. We shall reproduce his proof below (see p. 137). However, Wielandt's original method involved some further properties of operators which are of sufficient interest to be included and so we shall give these first. The fundamental result is

Proposition 4.2.1. *Let $\omega_1, \ldots, \omega_k$ be elements of Ω. Let α, β, α', β' be products of positive powers of $\omega_1, \ldots, \omega_k$ such that in both $\alpha\beta$ and $\alpha'\beta'$ each ω_i occurs at least once. Suppose that H, K are in $\mathscr{L}$. Then $H^\alpha v K^\beta$ if and only if $H^{\alpha'} v K^{\beta'}$. In particular, $H^\alpha v K^\beta$ if and only if*

$$H^{\omega_1 \cdots \omega_k} v K.$$

In order to see this we define an equivalence relation $\sim$ on Ω as follows:

$\alpha \sim \beta$ if and only if, for H, K in $\mathscr{L}$, $H^\alpha v K$ implies $H^\beta v K$ and conversely.

If this is so, we shall say that α and β are *permutationally equivalent*. We then have that the associative product in Ω induces an associative, commutative, idempotent product in $\Omega/\sim$. This is the content of

Lemma 4.2.2.

(i) *If $\alpha, \beta, \gamma \in \Omega$ and $\alpha \sim \beta$, then $\alpha\gamma \sim \beta\gamma$;*
(ii) *if $\alpha, \beta \in \Omega$, then $\alpha\beta \sim \beta\alpha$;*
(iii) *if $\omega \in \Omega$, then $\omega^2 \sim \omega$.*

Proof. (i) By hypothesis $H^\alpha v K \Leftrightarrow H^\beta v K$. If $H^{\alpha\gamma} v K$, then by (7) $H^\alpha v K^\gamma$. Hence $H^\beta v K^\gamma$. Again by (7), $H^{\beta\gamma} v K$. It now follows that $\alpha\gamma \sim \beta\gamma$.

(ii) Suppose that $H^{\alpha\beta} v K$. Then $H^\alpha v K^\beta$ by (7) and hence $H^\beta v K^\alpha$ by (8). A final application of (7) yields $H^{\beta\alpha} v K$.

(iii) $H^{\omega^2} v K$ implies $H^\omega v K^\omega$ by (7) from which $H^\omega v K$ by (6). Conversely, if $H^\omega v K$ then $H^{\omega^2} v K$ by (5). □

Proof of Proposition 4.2.1. By Lemma 4.2.2 and the hypothesis, it follows that both $\alpha\beta$ and $\alpha'\beta'$ are permutationally equivalent to $\omega_1\omega_2 \cdots \omega_k$. Therefore

$$\begin{aligned} H^\alpha v K^\beta &\Leftrightarrow H^{\alpha\beta} v K \qquad \text{(by (7))} \\ &\Leftrightarrow H^{\omega_1 \cdots \omega_k} v K \\ &\Leftrightarrow H^{\alpha'} v K^{\beta'}. \quad \square \end{aligned}$$

If, in addition, $\mathscr{L}$ satisfies the minimal condition, as is the case, for example, when $\mathscr{L}$ is the lattice of subnormal subgroups of a group satisfying Min-*sn*, then we obtain a particularly useful corollary to Proposition 4.2.1.

Proposition 4.2.3. *Suppose that $\mathscr{L}$ satisfies the minimal condition. Let α, β be elements of Ω such that for H, K in $\mathscr{L}$, $H v K$ always follows from $H = H^\alpha = H^\beta$. Then, for all H, K in $\mathscr{L}$, $H^\alpha v K^\beta$.*

Proof. The sequence $H^{(\alpha\beta)^n}$, $n = 1, 2, \ldots$ must break off. The limit L of the sequence clearly satisfies $L = L^\alpha = L^\beta$ and so permutes with K by hypothesis.

It now follows from Proposition 4.2.1 that $H^{\alpha}\, v\, K^{\beta}$, as required. □

We are now in a position to give the long awaited

Proof of Theorem 4.1.2. Suppose then that G is a group satisfying Min-*sn* and that H, K are subnormal subgroups of G such that $\pi(H/H') \cap \pi(K/K')$ is empty. We have to show that $H\, v\, K$.

We set $J = \langle H, K \rangle$ and note first of all that the result is essentially a theorem about *finite* groups. To see this, observe that $\mathfrak{F} = \mathrm{PQS_n}\mathfrak{F}$, so that by Theorem 4.1.3 the map $H \mapsto H^{\mathfrak{F}}$ is an operator on the subnormal subgroups of G. This means

$$J^{\mathfrak{F}} = \langle H^{\mathfrak{F}}, K^{\mathfrak{F}} \rangle.$$

Moreover, since G is a group satisfying Min-*sn*, $H^{\mathfrak{F}}$ normalizes every subnormal subgroup in G by Theorem 1.7.10. Hence

$$J^{\mathfrak{F}} = H^{\mathfrak{F}} K^{\mathfrak{F}}.$$

We can then work in $J/J^{\mathfrak{F}}$, which inherits the hypotheses, since $J = HKJ^{\mathfrak{F}} = HJ^{\mathfrak{F}}K$ implies $J = HK$ directly. Thus we may assume that J is finite.

Let $p_1, \ldots, p_n$ be the primes dividing $|J|$ which do not appear in $\pi(K/K')$ and let λ be the map on the subnormal subgroups of J given by

$$\lambda\colon L \mapsto \bigcap_{i=1}^{n} L^{\mathfrak{F}_{p_i}}.$$

Similarly let $q_1, \ldots, q_m$ be the primes dividing $|J|$ which do not appear in $\pi(H/H')$ and let μ be the map

$$\mu\colon L \mapsto \bigcap_{i=1}^{m} L^{\mathfrak{F}_{q_i}}.$$

By Corollary 4.1.4 both λ and μ are operators. Suppose that $L\, sn\, J$ and $L^{\lambda} = L^{\mu} = L$. Then $\pi(L/L')$ is empty since $p \in \pi(L/L')$ means that p is different from $p_1, \ldots, p_n, q_1, \ldots, q_m$. Hence $L = L'$ and L permutes with all subnormal subgroups of J by Theorem 4.1.1. It now follows from Proposition 4.2.3 that $H^{\mu}\, v\, K^{\lambda}$. But $H^{\mu} = H$ and $K^{\lambda} = K$, so that $H\, v\, K$, as required. □

As a further consequence of Proposition 4.2.3, we have the following result.

Theorem 4.2.4. *Suppose that G is a group with* Min-*sn and H, K are subnormal subgroups of G. Then*

(i) $\langle H, K \rangle^{\mathfrak{N}} = \langle H^{\mathfrak{N}}, K^{\mathfrak{N}} \rangle$;

(ii) $H^{\mathfrak{N}}\, v\, K$.

Proof. (i) follows from Theorem 3.3.1 and then (ii) is a consequence of Proposition 4.2.3 (defining $\alpha\colon H \mapsto H^{\mathfrak{N}}$ and β as the identity) and Corollary 1.6.7. However, Theorem 3.3.1 is far too powerful a result to use here and so we offer the following alternative simple proof of (i).

We may assume that $G = \langle H, K \rangle$. Clearly

$$B = \langle H^{\mathfrak{N}}, K^{\mathfrak{N}} \rangle \leqq G^{\mathfrak{N}} = A,$$

say. Let p_i denote the ith prime and let $\mathfrak{X}_i$ be the class of soluble p_i-groups. If

$$\gamma\colon H \mapsto \bigcap_{i \geqq 1} H^{\mathfrak{X}_i},$$

then Corollary 4.1.4 gives $G^\gamma = \langle H^\gamma, K^\gamma \rangle \leqq B$. Since nilpotent residuals are preserved by homomorphisms (for groups in the class $\mathfrak{Min}$-*sn*), we may assume that $G^\gamma = 1$ and hence G is a *Černikov* group (that is a finite extension of an abelian group satisfying Min) and a direct product of p_i-groups, $i = 1, 2, \ldots$.

Let G^* be the minimal subgroup of finite index in G. Then G^* is abelian, $A \leqq G^*$ and $HG^*/H^{\mathfrak{N}}$ is nilpotent, by Proposition 1.6.3. Since the divisible subgroups of a nilpotent *Černikov* group are central (see Corollary 5.3.4), we have

$$[H, G^*] \leqq H^{\mathfrak{N}} \leqq B.$$

Similarly $[K, G^*] \leqq B$. Therefore, since $B \leqq A \leqq G^*$, $B \lhd G$ and A/B is central in G. Thus $A = B$, as required. □

We have a more detailed look at similar permutability properties of residuals in §4.4.

Hainzl's proof of Theorem 4.1.2

We now give the proof of Theorem 4.1.2 due to Hainzl [41] which avoids the appeal to 4.2.1–4.2.3.

Suppose then that $G \in \mathfrak{Min}$-*sn* and H, K are subnormal in G with $\pi(H/H') \cap \pi(K/K')$ empty. As before we can reduce to the case where $G = \langle H, K \rangle$ is finite. Let Λ be the set of all cyclic groups of prime order which appear as composition factors of G. Let Λ_1 be the set of groups of prime order p with $p \in \pi(H/H')$ and $\Lambda_2 = \Lambda \backslash \Lambda_1$. Write $\mathfrak{X}_1$ and $\mathfrak{X}_2$ for the $\mathrm{PQS_n}$-closure of Λ_1 and Λ_2, respectively. Then the maps

$$L \mapsto L^{\mathfrak{X}_1}, \qquad L \mapsto L^{\mathfrak{X}_2}$$

are operators on the lattice of subnormal subgroups of G (Theorem 4.1.3).

We proceed by induction on $|H|$. It is clear that $H^{\mathfrak{X}_2} = H$ and $K^{\mathfrak{X}_1} = K$ from the definition of $\mathfrak{X}_1$ and $\mathfrak{X}_2$, and if $H^{\mathfrak{X}_1} = H$, we have $H = H'$ and the result then follows from Theorem 4.1.1. Thus we may assume that $H^{\mathfrak{X}_1} < H$. Now it is easy to see that

$$\pi(H^{\mathfrak{X}_1}/(H^{\mathfrak{X}_1})') \cap \pi(K^{\mathfrak{X}_2}/(K^{\mathfrak{X}_2})')$$

is empty, from the definition of Λ_1, Λ_2. By the induction hypothesis we conclude that $H^{\mathfrak{X}_1} v K^{\mathfrak{X}_2}$. But then $H^{\mathfrak{X}_2} v K^{\mathfrak{X}_1}$ by (8) and so $H v K$, as required. □

Let G be a group with a composition series and Ω be the set of operators on the subnormal subgroups of G. If $\alpha \in \Omega$, following Wielandt [160], we set $G_\alpha = \langle S | S \, sn \, G, S^\alpha = 1 \rangle$. It is clear that $(G_\alpha)^\alpha = 1$ and we have

Proposition 4.2.5 [160] (Satz 12.1). *Suppose that G is a group with a composition series, A and B are subnormal subgroups of G and $\alpha \in \Omega$. If $A^\alpha \leqq B^\alpha$, then A normalizes B^α. In particular G_α normalizes B^α.*

Proof. Since $\alpha \in \Omega$, we have $\langle A, B \rangle^\alpha = \langle A^\alpha, B^\alpha \rangle \lhd \langle A, B \rangle$. But $\langle A^\alpha, B^\alpha \rangle = B^\alpha$, by hypothesis, and the result follows. □

Thus, for example, $G_{\mathfrak{N}}$ normalizes $A^{\mathfrak{N}}$ for all $A \, sn \, G$. Also $G_{\mathfrak{F}_\pi}$ normalizes $A^{\mathfrak{F}_\pi}$ for all sets π of primes. ($\mathfrak{F}_\pi$ denotes the class of finite π-groups.)

§4.3 Roseblade's Permutability Theorem

Suppose that H and K are subnormal subgroups of a group G. Then, as we saw in Wielandt's Theorem 4.1.2, H and K permute provided that G satisfies Min-*sn* and $\pi(H/H') \cap \pi(K/K')$ is empty. The latter condition, for groups with Min-*sn*, implies that the tensor product

$$H/H' \otimes K/K'$$

is trivial. (For elementary results on tensor products we refer the reader to Fuchs [34].) In such a case we say that the subgroups H and K are *orthogonal* and write

$$H \perp K.$$

We note that the condition depends only on the groups themselves in their abstract structure and not on the way in which they are embedded in some larger group. Also the condition $H \perp K$ does not necessarily imply $\pi(H/H') \cap \pi(K/K') = \varnothing$, as may be seen, for example, by taking $H \cong C_{p^\infty} \cong K$. However, $H \perp K$ does imply $\tau(H) \cap \tau(K) = \varnothing$, where $\tau(H) = \{p | C_p \in \mathrm{Q}(H)\}$. One example of a result about orthogonal subnormal subgroups is afforded by Lemma 1.7.8, where it was proved that if $H \cong C_{p^\infty}$, $K \cong C_{q^\infty}$, then H and K commute.

In the light of Wielandt's result mentioned above, P. Hall conjectured that any pair of orthogonal subnormal subgroups of an arbitrary group permute. Roseblade established the truth of this conjecture in 1965 in what will form the main result of this section.

Theorem 4.3.1 [124]. *If H and K are orthogonal subnormal subgroups of a group, then $HK = KH$.*

Suppose that H and K are subnormal subgroups of a group G and write $J = \langle H, K \rangle$. Since $J = H[H, K]K$ and since $HK = KH$ is equivalent to $J = HK$, it is clear that H and K are permutable if and only if $[H, K] \leqq HK$.

Now since H, K are subnormal in G, there exist m and n with

$$[K, {}_m H] \leqq H \qquad \text{and} \qquad [H, {}_n K] \leqq K.$$

Thus Theorem 4.3.1 follows from the apparently stronger, though, as we shall see on p. 141, equivalent

Theorem 4.3.2 [124] (Theorem 2). *If H and K are orthogonal subgroups of a group, then*

$$[K, H] = [K, {}_m H][H, {}_n K]$$

for any positive integers m and n.

Roseblade's theorem is, of course, a significant generalization of Wielandt's Theorem 4.1.2, and his proof is completely different from that of Wielandt in that the theory of operators has to be replaced by commutator calculations and induction on the sum of the defects of H and K. We shall not, in fact, give Roseblade's original proof here, but rather the following simplified version due to Brewster [15] who reduces the problem to one about nilpotent groups by means of his

Theorem 4.3.3. *If H and K are subnormal subgroups of their join J, then H and K permute if and only if their images in any nilpotent factor of J permute.*

Remarks

1. We note that this result is a generalization of that of Wielandt [160] (Satz 10.12) proving that, for J finite, $J = HK$ provided $J = HKJ^{\mathfrak{N}}$, whenever J is generated by subnormal subgroups H, K.
2. If J/J' happens to be of finite rank, then Theorem 4.3.3 follows at once from Theorem 3.3.1. However, we do not need to invoke Theorem 3.3.1 at any stage of the proof.
3. Granted Theorem 4.3.3, we note how easy it is to deduce Theorem 4.1.2 for finite groups (and hence for groups with Min-*sn* by the usual arguments): by Theorem 4.3.3 we may assume that J is nilpotent. Let $\pi = \pi(H/H')$. Then H is a π-group since H is nilpotent. The hypothesis $\pi(H/H') \cap \pi(K/K')$ is empty then yields that K is a π'-group. Thus H and K, being subgroups of coprime order in a nilpotent group, commute.
4. In [81] Lennox and Wilson use Brewster's result to prove another permutability criterion. *Suppose that H and K are subnormal subgroups of their join J and that J/J' is finitely generated. If $H^\phi K^\phi = K^\phi H^\phi$ for all homomorphisms ϕ from J onto finite nilpotent groups, then $HK = KH$.*

From 4 it follows by a routine argument, similar to that given in [99] (Section 7), that there is an algorithm for deciding whether or not two subnormal subgroups of a group J, with J/J' finitely generated, permute.

Before we give the proof of Brewster's result, we show how it facilitates the proof of Roseblade's theorem (Theorem 4.3.1). First of all we need two lemmas.

Lemma 4.3.4 (Hall's 3-subgroup lemma). *Suppose that H, K and L are subgroups of a group G. Then any normal subgroup of G containing two of the subgroups $[H, K, L]$, $[K, L, H]$ and $[L, H, K]$ contains the third.*

Proof. $[H, K, L]$ can be generated by conjugates of elements of the form $[h, k, l]$, where $h \in H$, $k \in K$, $l \in L$. Thus Lemma 2.5.6 gives the result. □

Lemma 4.3.5 [124] (Lemma 1). *If H and K are groups, then $H/H' \otimes K/K'$ is isomorphic to the factor*

$$[H, K]/[H, K, K][K, H, H]$$

in the free product $H * K$.

Proof. Suppose that $G = H * K$, where H and K are embedded in G in the natural way. Both $[H, K, K]$ and $[K, H, H]$ are normal subgroups of $[H, K] = X$, say. Let $N = [H, K, K][K, H, H]$. Clearly, $N \lhd G$ and by Lemma 4.3.4 it follows that both $[H', K]$ and $[K', H]$ are contained in N. The elements $N[h, k], h \in H, k \in K$, of X/N therefore depend only on the cosets $H'h$ and $K'k$. A simple verification, using the fact that X is free on the commutators $[h, k]$, $h \in H$, $k \in K$, $h \neq 1 \neq k$, shows that the map

$$N[h, k] \mapsto H'h \otimes K'k$$

is an isomorphism of X/N onto $H/H' \otimes K/K'$. This completes the proof. □

Thus if $H \perp K$, we must have

$$[H, K] = [H, K, K][K, H, H]$$

in $H * K$. However, if this equation holds in $G = H * K$, it must also hold in any group L which contains H and K as subgroups. This is because commutator subgroups are preserved under homomorphisms and there is a homomorphism of G onto the subgroup $\langle H, K \rangle$ of L which maps H and K (as subgroups of G) isomorphically onto H and K (as subgroups of L). This particular case of Theorem 4.3.2 we record as

Proposition 4.3.6. *If H and K are orthogonal subgroups of a group, then*

$$[H, K] = [K, H, H][H, K, K].$$

A special case of this result is

Corollary 4.3.7. *If H and K are orthogonal subgroups of a residually nilpotent group, then* $[H, K] = 1$.

Before moving on we prove that a second consequence of Lemma 4.3.5 is that if H and K are *not* orthogonal, then they can always be subnormally embedded in a group such that $HK \neq KH$.

Theorem 4.3.8 [124] (Theorem 3). *If H and K are groups, then there exists a group $G = \langle H, K \rangle$ such that*

(i) $[K, H, H] = [H, K, K] = 1$,
(ii) $H \cap (K[H, K]) = K \cap (H[H, K]) = 1$ *and*
(iii) $[H, K] \cong H/H' \otimes K/K'$.

Proof. Let $F = H * K$ and $N = [H, K, K][K, H, H]$. Let bars denote factor groups modulo N. Clearly $\bar{F} = \langle \bar{H}, \bar{K} \rangle$ and $[\bar{K}, \bar{H}, \bar{H}] = [\bar{H}, \bar{K}, \bar{K}] = 1$. Now $H \cap (K^H) = 1$ and $N \leqq K^H$. Therefore

$$(NH) \cap K^H = N$$

so that $\bar{H} \cap (\bar{K}^{\bar{H}}) = 1$. Similarly $\bar{K} \cap (\bar{H}^{\bar{K}}) = 1$. Lemma 4.3.5 shows that

$$[\bar{H}, \bar{K}] \cong \bar{H}/\bar{H}' \otimes \bar{K}/\bar{K}',$$

since $\bar{H} \cong H$, $\bar{K} \cong K$ both follow from $H \cap N = K \cap N = 1$. Let $G = \bar{F}$. We identify H with $\bar{H}$ and K with $\bar{K}$ in the natural way and note that if $HK = KH$, then $H \lhd G$ and $K \lhd G$ so that $[H, K] = 1$. Therefore if $H/H' \otimes K/K'$ is non-trivial, then $HK \neq KH$. □

In order to see that Theorem 4.3.2 follows from Theorem 4.3.1, and thus to complete our justification of the statement on p. 139 that they are equivalent, we proceed as follows. It suffices to show that Theorem 4.3.2 holds when H and K are embedded in $H * K$. Denote the terms of the normal closure series of H, K in G by H_m, K_n, respectively. Since H_m is generated by conjugates of H and therefore by subgroups isomorphic with H, it is not hard to show that $H_m \perp K_n$. By Theorem 4.3.1

$$G = H_m K_n, \qquad m \geqq 0, n \geqq 0.$$

We then have

$$H_1 \cap K_1 = (H_m \cap K_1)(K_n \cap H_1),$$

for all $m, n \geqq 1$. But $H \cap K_1 = 1$ and $H_m = H[K, {}_m H]$ so that $H_m \cap K_1 = [K, {}_m H]$. Similarly $K_n \cap H_1 = [H, {}_n K]$ and the result follows. □

We return to the deduction of Roseblade's theorem (Theorem 4.3.1) from Brewster's. Suppose that H and K are orthogonal subnormal subgroups of their join G. Then the images of H and K remain orthogonal in any factor group of G. Therefore by Theorem 4.3.3 we may assume that G is nilpotent. Hence $[H, K] = 1$, by Corollary 4.3.7. □

We observe at this point that it is an immediate consequence of Theorem 4.3.2 that if H and K are orthogonal subnormal subgroups of a group and H is nilpotent, then K is normalized by H. (See §4.5 for further normalizing properties.)

Proof of Brewster's Theorem 4.3.3. Suppose that $J = \langle H, K \rangle$ where H and K are subnormal subgroups of J and that

$$J = HK\gamma_c(J)$$

for all $c \geqq 1$. In order to prove that $J = HK$, we may assume by Theorem 1.6.4 that J is a soluble group. We proceed by induction on the derived length of J. Let A be the last non-trivial term of the derived series of J. Then by the natural induction hypothesis we have $J = HAK$. By Proposition 3.3.12, there exists r such that

$$\gamma_r(HA) \leqq H\gamma_2(A) = H.$$

Again by Proposition 3.3.12 there exists s such that

$$\gamma_s(J) \leqq \gamma_r(HA)K \leqq HK.$$

By assumption $J = H\gamma_s(J)K$ so that $J = HK$ as required. □

Another consequence of Theorem 4.3.3 is the following result which settles a conjecture of Lennox [74].

Brewster [15] (Corollary 2): *Suppose that H and K are subnormal subgroups of their join J and that*

$$H/(H \cap K)H' \otimes K/(H \cap K)K' = 0.$$

Then $J = HK$.

This result had already been established in [74] for the case where H and K satisfy Min-*sn*. Brewster also proves:

Suppose that H and K are subgroups of a group and that

$$H/(H \cap K)H' \otimes K/(H \cap K)K' = 0.$$

Then

$$[H, K] = [H, {}_mK][H \cap K, K][H \cap K, H][K, {}_nH],$$

for $m, n \geqq 1$.

§4.4 Permutability properties of soluble and nilpotent residuals of subnormal subgroups

The starting point for this section is again Wielandt's theory as expounded in §4.1. Let $\mathfrak{S}$ denote the class of soluble groups. It is an immediate consequence of Theorems 1.6.6, 4.1.3 and 4.2.4 that we have

Theorem 4.4.1. *Suppose that H and K are subnormal subgroups of a group G satisfying* Min-*sn*. *Then*

(W1) (i) $H^{\mathfrak{S}}K = KH^{\mathfrak{S}}$ *and* (ii) $\langle H, K\rangle^{\mathfrak{S}} = H^{\mathfrak{S}}K^{\mathfrak{S}}$,
(W2) (i) $H^{\mathfrak{N}}K = KH^{\mathfrak{N}}$ *and* (ii) $\langle H, K\rangle^{\mathfrak{N}} = H^{\mathfrak{N}}K^{\mathfrak{N}}$.

Of course (i) follows from (ii) in both cases. The hypothesis that G satisfies Min-*sn* cannot be omitted in either result. For, according to Theorem 2.1.3, any group L can be embedded in the product G of two normal free subgroups, say H and K. By Iwasawa's theorem (already quoted on p. 45) showing that free groups are residually nilpotent,

$$H^{\mathfrak{N}} = H^{\mathfrak{S}} = K^{\mathfrak{N}} = K^{\mathfrak{S}} = 1.$$

However, $G^{\mathfrak{S}} = 1$ (or $G^{\mathfrak{N}} = 1$) would imply $L^{\mathfrak{S}} = 1$ (or $L^{\mathfrak{N}} = 1$). (Cf. Roseblade's construction in Theorem 3.1.3.) In spite of this Stonehewer [139], [142] has shown that (W1) and (W2) can be substantially generalized.

Theorem 4.4.2. *Suppose that H and K are subnormal subgroups of their join J and that $H/H^{\mathfrak{S}}$ and $K/K^{\mathfrak{S}}$ are soluble. Then* (W1) *holds for H and K and $J/J^{\mathfrak{S}}$ is soluble.*

It is obvious that this result (and also Theorem 4.4.1 (W1) of course) is an immediate corollary of Roseblade's theorem (Theorem 3.1.2). For, by hypothesis, we have $H^{\mathfrak{S}} = H^{(m)}$ and $K^{\mathfrak{S}} = K^{(n)}$ for some integers m and n. Then by Theorem 3.1.2 there exists a positive integer l such that

$$J^{(l)} \leqq H^{(m)}K^{(n)}.$$

Therefore

$$J^{\mathfrak{S}} \leqq J^{(l)} \leqq H^{\mathfrak{S}}K^{\mathfrak{S}} \leqq J^{\mathfrak{S}}$$

and it follows that

$$J^{\mathfrak{S}} = H^{\mathfrak{S}}K^{\mathfrak{S}}$$

and $J/J^{\mathfrak{S}}$ is soluble. Moreover $J^{\mathfrak{S}}K = H^{\mathfrak{S}}K$ so that the latter is a group. Then we get

$$H^{\mathfrak{S}}K = KH^{\mathfrak{S}}$$

and so J satisfies (W1).

By an analogous argument we obtain from Theorem 3.3.1

Theorem 4.4.3. *Suppose that H and K are subnormal subgroups of their join J and that $J'/\gamma_3(J)$ has finite rank. Then if $H/H^{\mathfrak{N}}$ and $K/K^{\mathfrak{N}}$ are nilpotent,* (W2) *holds for H and K and $J/J^{\mathfrak{N}}$ is nilpotent.*

Stonehewer in [142] had earlier proved weaker versions of Theorem 4.4.3, firstly where H and K had finite rank and secondly in the case where they were groups possessing series each factor of which satisfied either Max-*sn* or Min-*sn*. The hypothesis of Theorem 4.4.3 is clearly satisfied in both cases. We mention, however, part of Stonehewer's method here, as it raises an interesting question. What he did was to show that the hypotheses of the following result were satisfied in each of the cases he considered.

Proposition 4.4.4 [142]. *Suppose that H and K are subnormal subgroups of their join J and that $J^{\mathfrak{N}} = \langle H^{\mathfrak{N}}, K^{\mathfrak{N}} \rangle$. Then $(W2)$ holds for H and K provided only that $H/H^{\mathfrak{N}}$ is nilpotent.*

Proof. Let $A = J^{\mathfrak{N}}$ and, for some $c \geqq 1$, let $N = \gamma_c(A)$. Then $N \lhd J$. If we apply Proposition 3.3.12 to the product $HA/N = (HN/N)(A/N)$, we deduce that

$$(HA/N)^{\mathfrak{N}} = (HN/N)^{\mathfrak{N}}$$

since A/N is nilpotent. However, it is easy to see that $(HN/N)^{\mathfrak{N}} = H^{\mathfrak{N}}N/N$, since $H/H^{\mathfrak{N}}$ is nilpotent, and it follows that $H^{\mathfrak{N}}N/N \lhd HA/N$. In particular $H^{\mathfrak{N}}N \lhd A$ and hence from $A = \langle H^{\mathfrak{N}}, K^{\mathfrak{N}} \rangle$ we see that

$$A = H^{\mathfrak{N}}K^{\mathfrak{N}}\gamma_c(A).$$

This is true for all $c \geqq 1$ and therefore $A = H^{\mathfrak{N}}K^{\mathfrak{N}}$, by Theorem 4.3.3. □

The results which we have considered having a conclusion of type $H^{\mathfrak{S}}K = KH^{\mathfrak{S}}$ (or $H^{\mathfrak{N}}K = KH^{\mathfrak{N}}$) all involve the hypothesis that $H/H^{\mathfrak{S}}$ is soluble (or $H/H^{\mathfrak{N}}$ is nilpotent) so that in fact some term of the derived (or lower central) series of H actually permutes with K. Thus we are led to the question of whether it is generally true for subnormal subgroups H and K of a group that, for some finite i, the ith term of the derived (lower central) series of H permutes with K. For derived series the question is open, whereas for lower central series the answer is in the negative. To see this let L be an infinite elementary abelian 2-group and $K = \langle \xi \rangle$ be a cyclic group of order 2. Form $W = L \wr K$. Let F be the field of two elements, $B = FW$, the group ring, and form the split extension $B] W = G$. (In other words G is $C_2 \wr (L \wr C_2)$.) Over F we have $(\xi - 1)^2 = 0$, so that

$$K \lhd^2 BK \lhd^2 G.$$

Put $H = BL \lhd^2 G$. We show that $\gamma_n(H)$ does not permute with K, for any n.

Clearly $\langle H, K \rangle = G \neq HK$, because $W \neq LK$. For $n \geqq 1$, $\gamma_{n+1}(H) = [B, {}_nL] = B_n$, say. Suppose B_n permutes with K. Then

$$\langle B_n, B_n^{\xi} \rangle \leqq \langle B_n, \xi \rangle \cap B = B_n \langle \xi \rangle \cap B = B_n,$$

from which it follows that $B_n = B_n^{\xi}$. Then if $C = F(L \times L^{\xi})$, it is easy to deduce that $[C, {}_nL] = [C, {}_nL^{\xi}]$, which in additive notation reads $C\mathfrak{l}^n = C\mathfrak{l}^{\xi n}$. Applying the epimorphism $C \to FL$ defined by $L^{\xi} \to 1$ then yields $(FL)\mathfrak{l}^n = 0$, that is $\mathfrak{l}^n = 0$, a contradiction, since L is infinite.

In this example it is also not hard to check that if $M = L^B$, then

$$P_M(K) = M \cap M^\xi = [B, L] \cap [B, L^\xi] = [B, L, L^\xi],$$

from which it follows that $B_n \nleqq P_M(K)$, for all n. Hence $\gamma_n(M)$ is not contained in $P_M(K)$, for any n.

What we can say in general is that some term of the derived series of H is always contained in $P_H(K)$ (assuming of course that H and K are subnormal in their join; this is Theorem 1.6.8) and by using a similar argument, based on Theorem 3.3.1, it is easy to establish the following analogue of Theorem 1.6.8.

Theorem 4.4.5. *Suppose that H and K are subnormal subgroups of their join J and that $J'/\gamma_3(J)$ has finite rank. Then $\gamma_r(H) \leqq P_H(K)$, for some $r \geqq 1$.*

Concluding remarks. Jennifer Whitehead [150] considers locally nilpotent and hypercentral residuals and proves

(i) *Suppose that H and K are subnormal subgroups of their join J and that $J/J^{\mathrm{L}\mathfrak{N}}$ is locally finite. Then*

$$J^{\mathrm{L}\mathfrak{N}} = \langle H^{\mathrm{L}\mathfrak{N}}, K^{\mathrm{L}\mathfrak{N}} \rangle = H^{\mathrm{L}\mathfrak{N}} K^{\mathrm{L}\mathfrak{N}} \qquad \textit{and} \qquad H^{\mathrm{L}\mathfrak{N}} K = K H^{\mathrm{L}\mathfrak{N}}.$$

(ii) *Suppose that J is a group generated by subnormal subgroups H and K, each of finite rank. If H/H^* and K/K^* are hypercentral, then*

$$J^* = \langle H^*, K^* \rangle = H^* K^*$$

and J/J^* is hypercentral of finite rank.

Here, for a group G, G^* denotes the hypercentral residual of G, which in fact coincides with the locally nilpotent residual for a group of finite rank (Mal'cev [98]; see also, for example, [120] page 38). Of course H having finite rank does not guarantee that H/H^* is hypercentral in general, as can be seen from the infinite dihedral group. Also there is no precise analogue of Theorem 3.3.1, as the following example shows. Let

$$A = \langle a_i \mid a_0 = 1,\ a_{i+1}^2 = a_i,\ i \geqq 0 \rangle,$$
$$B = \langle b_i \mid b_0 = 1,\ b_{i+1}^2 = b_i,\ i \geqq 0 \rangle$$

be C_{2^∞}-groups. Put $K = A \times B$ and define

$$G = \langle K, c, t \mid a_i^c = a_i,\ b_i^c = a_i b_i,\ a_i^t = a_i^{-1},\ b_i^t = b_i,\ c^t = c^{-1},\ t^2 = 1,\ i \geqq 1 \rangle.$$

Let $H = \langle A, c, t \rangle$. Then $H, K \lhd G$ and $G = HK$. If $\bar{H} = \langle c \rangle$ and $\bar{K} = 1$, then $H/\bar{H}$ and $K/\bar{K}$ are hypercentral. However, $\bar{H}$ is core-free in G and G is not hypercentral, since it contains the infinite dihedral subgroup $\langle c, t \rangle$. Thus there is no normal subgroup N of G contained in $\langle \bar{H}, \bar{K} \rangle$ such that G/N is hypercentral. Also G has rank 4.

§4.5 Relative normalizing and centralizing properties

The starting point of our discussion of the join problem for subnormal subgroups was Theorem 1.2.1, where we proved that if H, K are subnormal in G, $J = \langle H, K \rangle$ and $H^K = H$, then $J \, sn \, G$ and we promised to consider situations where $H^K = H$. We devote this section to the topic, including also a discussion of the related question of when subnormal subgroups centralize each other. Indeed if H and K are trivially intersecting subnormal subgroups of a group, they centralize each other if and only if they normalize each other and we note incidentally that it was Speiser who in 1927 first pointed out the elementary but useful fact that normal subgroups of a group centralize each other if they intersect trivially.

Subnormal subgroups intersecting trivially do not of course necessarily centralize each other—consider the dihedral group of order 8. However, in [152] Satz (19) Wielandt proved

Theorem 4.5.1. *Suppose G is a finite group and H, K are subnormal subgroups of G such that the sets of primes involved in the composition factors of prime order of H and K are disjoint. Then $[H, K] = 1$ provided $H \cap K = 1$.*

We shall not give Wielandt's proof here, but point out that this result is true even for groups with Min-*sn* as can be seen from a simple application of Lemma 4.1.8. P. Hall in Cambridge lectures (1963/64) restated Theorem 4.5.1 in a more general form.

Theorem 4.5.2. *Suppose that $\mathfrak{X}$ and $\mathfrak{Y}$ are $\{\mathrm{N}, \mathrm{s_n}\}$-closed classes of groups such that all groups in $\mathfrak{X} \cap \mathfrak{Y}$ are perfect. Then if H and K are trivially intersecting subnormal subgroups of a group G with $H \in \mathfrak{X}$ and $K \in \mathfrak{Y}$, we have $[H, K] = 1$.*

Proof. Suppose that H and K are subnormal subgroups of defects m and n, respectively, in their join J. We proceed by induction on $m+n$. If $m, n \leqq 1$, there is nothing to prove. Suppose that $n > 1$ and that the natural induction hypothesis holds. Let $K_1 = K^J$ and $L = H \cap K_1$. Then $L \lhd H$, so $L \in \mathfrak{X}$. Also L has defect at most m in K_1, K has defect $n-1$ in K_1 and $L \cap K = 1$. By the induction hypothesis $[L, K] = 1$. Hence $[L, K^h] = 1$ for all $h \in H$. Thus $[L, K_1] = 1$ and so $L (\leqq K_1)$ is abelian. But $L sn K_1 \in \mathrm{N}\mathfrak{Y} = \mathfrak{Y}$, so $L \in \mathrm{s_n}\mathfrak{Y} = \mathfrak{Y}$. Thus $L \in \mathfrak{X} \cap \mathfrak{Y}$ and L is perfect by hypothesis. Hence $L = 1$ and so we have $L = H \cap K_1 = 1$ and applying the induction hypothesis to H and K_1 we obtain $[H, K_1] = 1$ from which the result is immediate. □

Remark 3 on p. 54 is a special case of this theorem, taking for $\mathfrak{X}$ the class of semisimple groups and for $\mathfrak{Y}$ the class of all groups.

The condition on the composition factors of H and K in Wielandt's result (Theorem 4.5.1) certainly implies that no abelian homomorphic image of H is isomorphic to any subnormal subgroup of $K/\zeta_1(K)$, that is

$$\mathrm{Q}(H/H') \cap \mathrm{s_n}(K/\zeta_1(K)) = (1). \tag{1}$$

P. Hall conjectured that this condition would turn out to be central in resolving the question: when is H normalized by K in every subnormal embedding with $H \cap K = 1$? Of course if H is to be normalized by K it must permute with K and for this to be true in every subnormal embedding we must have $H \perp K$ by Theorem 4.3.8. In fact the orthogonality of H and K together with condition (1) turn out to be both necessary and sufficient as described in

Theorem 4.5.3 [127], [131]. *Suppose that H and K are groups. Then H is normalized by K in every subnormal embedding with $H \cap K = 1$ if and only if*
(1) $\mathrm{Q}(H/H') \cap \mathrm{s_n}(K/\zeta_1(K)) = (1)$, *and*
(2) $H \perp K$.

As an immediate consequence of this result we observe

Corollary 4.5.4. *A perfect subnormal subgroup is normalized by every subnormal subgroup with which it has trivial intersection. In particular, trivially intersecting perfect subnormal subgroups of a group commute elementwise.*

In order to prove Theorem 4.5.3 we need

Lemma 4.5.5 [127]. *Suppose U is generated by normal subgroups U_λ ($\lambda \in \Lambda$). If $V \lhd^r U$, then*

$$[U,{}_r V] \leqq \prod_{\lambda \in \Lambda} (V \cap U_\lambda).$$

Proof. Let $V = V_r \lhd V_{r-1} \lhd \cdots \lhd V_1 \lhd V_0 = U$ be a series connecting V to U. Suppose that for some i, $0 \leqq i < r$, we know that

$$[U,{}_i V] \leqq \prod_{\lambda \in \Lambda} (V_i \cap U_\lambda).$$

Then

$$[U,{}_{i+1} V] \leqq \left[\prod_{\lambda \in \Lambda} (V_i \cap U_\lambda),\ V_{i+1}\right] \leqq \prod_{\lambda \in \Lambda} (V_{i+1} \cap U_\lambda).$$

This is because $V \leqq V_{i+1}$ and the subgroups V_{i+1} and $V_i \cap U_\lambda$ ($\lambda \in \Lambda$) are all normal in V_i. The lemma follows since $U = [U,{}_0 V] = \prod_{\lambda \in \Lambda} U_\lambda$. □

We now deduce

Lemma 4.5.6 [127] (Lemma 2). *Let $J = HK$ with H sn J and K sn J. Suppose that $H \cap K \leqq N \lhd K$. If $\mathrm{Q}(H/H') \cap \mathrm{s_n}(K/N) = (1)$, then $K \cap H^K \leqq N$.*

Proof. We note first that H, besides having no non-trivial abelian homomorphic images in $\mathrm{s_n}(K/N)$, has no nilpotent images either. This is because, by Grün's lemma, a non-abelian nilpotent group L always has a non-trivial homomorphic image in $\zeta_1(L)$.

Let $H \lhd^n J$ and $K \lhd^s J$ and let H_i, $i = 0, \ldots, n$, be the terms of the normal closure series of H in J. Suppose that for some i, $1 \leqq i < n$, $H_i \cap K \nleqq N$ but $H_{i+1} \cap K \leqq N$. Then $\Gamma = N(K \cap H_i)/N$ is a non-trivial homomorphic image of $(K \cap H_i)/(K \cap H_{i+1})$. Now

$$H_i/H_{i+1} \cong (K \cap H_i)/(K \cap H_{i+1})$$

since

$$H_i = H(K \cap H_i) = H_{i+1}(K \cap H_i).$$

Furthermore H_i/H_{i+1} is generated by the subnormal $\mathrm{Q}(H)$-subgroups

$$H_{i+1}H^x/H_{i+1} \; (x \in H_{i-1}).$$

Hence $\mathrm{Q}(H) \cap \mathrm{s_n}(\Gamma) > (1)$. By the initial remark, Γ cannot be nilpotent. However,

$$H_i = H^{H_{i-1}} = H_{i+1}^{H_{i-1}}$$

and $H_{i-1} = H(K \cap H_{i-1})$. Thus H_i is generated by the normal subgroups

$$H_{i+1}^x \; (x \in K \cap H_{i-1}).$$

We may apply Lemma 4.5.5, since $K \cap H_i \lhd^s H_i$, to obtain

$$[H_{i,s}(K \cap H_i)] \leqq \Pi\,(H_{i+1}^x \cap K \cap H_i),$$

the product being taken over all $x \in K \cap H_{i-1}$. But $K \cap H_{i-1} \leqq N_J(K \cap H_i)$ and $K \cap H_{i+1} \leqq N \lhd K$. Therefore $[H_{i,s}(K \cap H_i)] \leqq N$ and so, in particular, the $(s+1)$th term of the lower central series of $K \cap H_i$ is in N. This shows that Γ is nilpotent, a contradiction which establishes the result. □

Roseblade points out an interesting consequence of Lemma 4.5.6 in the case where $H = H'$; then N may be taken to be the normal closure $(H \cap K)^K$ of $H \cap K$ in K.

Corollary 4.5.7. *If $H = H'\,\mathrm{sn}\,G$ and $K\,\mathrm{sn}\,G$, then $H^K = H(H \cap K)^K$.*

We may then deduce

Theorem 4.5.8. *If $H = H'\,\mathrm{sn}\,G$ and $K\,\mathrm{sn}\,G$ such that $(H \cap K)^K$ is soluble, then H is normalized by K.*

Proof. We may suppose that $G = \langle H, K \rangle$. Let $H_1 = H^K$. By induction on the defect of H in G we may assume that $K \cap H_1$ normalizes H. Since $HK = KH$, this means that $H \lhd H_1$. Also H_1, as a join of perfect subgroups, is perfect. However, H_1/H is soluble, by Corollary 4.5.7, and so $H_1 = H$, as required. □

If in Theorem 4.5.8 the group G has a finite composition series, then it is enough to assume, together with $H = H'$, that $H \cap K$ is soluble. In this form the result was proved by Wielandt [152] (Satz 24). If G has Min-*sn* and K is soluble, then the result follows from Lemma 4.1.8. It does not appear to be known whether it is sufficient to take $H \cap K$ soluble in the general case.

Proof of Theorem 4.5.3. Suppose that H and K are subnormal subgroups of a group G with $H \perp K$ and $\mathrm{Q}(H/H') \cap \mathrm{s_n}(K/\zeta_1(K)) = (1)$. It follows at once from Theorem 4.3.1 that H and K permute and hence we may apply Lemma 4.5.6 to yield $H^K \cap K \leqq Z = \zeta_1(K)$. But $K \lhd^m G$ for some $m \geqq 1$, so we have $[H, {}_m K] \leqq H^K \cap K \leqq Z$ from which we obtain $[H, {}_{m+1} K] = 1$. Finally, by Theorem 4.3.2, the fact that $H \perp K$ means that we have, for all $r, s \geqq 1$,

$$\begin{aligned}[K, H] &= [K, {}_r H]\,[H, {}_s K] \\ &= [K, {}_r H], \qquad (\text{for } s \geqq m+1) \\ &\leqq H,\end{aligned}$$

taking r so that $H \lhd^r G$. Thus $H^K = H$, as required. □

For extensions of some of these results see Schmid [132].

Returning to the situation of disjoint subnormals we now record a further result of Roseblade.

Theorem 4.5.9 [127] (Theorem B). *Suppose that H and K are subnormal subgroups of a group G. If $H \cap K = 1$, then H is normalized by K if and only if $[H, K] \cap \beta_1(G) \leqq H$.*

Here $\beta_1(G)$ denotes the Baer radical of the group G as defined on p. 86. By means of Theorem 2.5.1 (i), we obtain

Corollary 4.5.10 [131] (Korollar 4.6). *Disjoint subnormal subgroups H and K of a group G commute elementwise if and only if $[H, K]$ has no non-trivial cyclic subgroups subnormal in G.*

Note. A weaker version of this result for groups with the maximal condition for subnormal subgroups appears as (17) in [152].

In order to prove Theorem 4.5.9, Roseblade used a result which answers a question of P. Hall and is of independent interest. To describe it we define, for $r > 1$, the *rth Baer radical* $\beta_r(G)$ of a group G inductively by $\beta_r(G)/\beta_{r-1}(G) = \beta_1(G/\beta_{r-1}(G))$. The result is

Theorem 4.5.11. *Let H and K be subnormal subgroups of a group G with $H \cap K = 1$. If $H \lhd^r \langle H, K \rangle$, then $[H, K] \leqq \beta_r(G)$.*

When $r = 0$ we adopt the convention $\beta_0(G) = 1$. The case $r = 1$ is given by Corollary 2.5.5.

Question. Suppose $r > 2$. Does there exist a group $J = \langle H, K \rangle$ where H, K are disjoint subnormal subgroups with $H \lhd^r J$, but $[H, K] \nleqq \beta_{r-1}(J)$? If $r = 2$, the corresponding question has an affirmative answer, as Roseblade shows. Let $G = H \wr T$, where H is the symmetric group of degree 3 and $T \cong C_2$. Let K be the second term of the normal closure series of T in G. Then $H \lhd^2 G$ and $K \lhd^2 G$. It is easy to see that $H \cap K = 1$ and $[H, K]$ is the split extension of an elementary abelian group of order 9 by the inversion automorphism. Thus $[H, K]$ is not nilpotent and hence not in $\beta_1(G)$.

In order to prove Theorems 4.5.9 and 4.5.11 we need two preliminary lemmas.

Lemma 4.5.12. *If* $J = \langle H, K \rangle$ *and* $[H, {}_sK, H] = 1$ *for some* $s \geqq 1$, *then* $[H, K] \leqq \beta_1(J)$.

Proof. Let $Z = \zeta_1(H^K)$. Since K normalizes $[H, {}_sK]$, it follows from the hypothesis that $[H, {}_sK] \leqq Z$. Since $[H^K, {}_sK] = [H, {}_sK]$ we may apply Corollary 2.5.5 to the group ZK/Z acting on ZH^K/Z to deduce that $Z[H, K]/Z \leqq \beta_1(J/Z)$. It follows that $[H, K] \leqq \beta_1(J)$. □

Lemma 4.5.13. *If* $H_\lambda \operatorname{sn} G$ $(\lambda \in \Lambda)$ *and* $J = \langle H_\lambda | \lambda \in \Lambda \rangle$, *then* $\beta_r(J) = J \cap \beta_r(G)$, *for any* $r \geqq 0$.

Proof. Theorem 2.5.1 (i) and a straightforward induction on r show that $\beta_r(X) = X \cap \beta_r(G)$ for any subnormal subgroup X of G. Even if J is not subnormal, it is still the set-theoretic union of subnormal subgroups (see the comment on p. 32 after the statement of Theorem 1.6.14). The result follows at once. □

Proof of Theorem 4.5.11. Let J be the join of subnormal subgroups H and K of G such that $H \cap K = 1, H \lhd^r J$ and $K \lhd^n J$. We proceed by induction on r to show that $[H, K] \leqq \beta_r(G)$. If $r = 1$, $H \lhd J$ and so $[H, {}_nK] \leqq H \cap K = 1$. Hence $[H, K] \leqq \beta_1(J)$, by Lemma 4.5.12, and therefore $[H, K] \leqq \beta_1(G)$ by Lemma 4.5.13.

We now suppose $r > 1$ and assume the natural induction hypothesis. Let $H_1 = H^J$, so that $H \lhd^{r-1} H_1$ and by induction

$$[H, K \cap H_1] \leqq \beta_{r-1}(G).$$

Denoting factor groups mod $\beta_{r-1}(G)$ by bars, we have

$$[\bar{H}, {}_n\bar{K}, \bar{H}] = 1,$$

since $[H, {}_nK] \leqq K \cap H_1$. From Lemmas 4.5.12 and 4.5.13 we deduce that

$$[\bar{H}, \bar{K}] \leqq \beta_1(\bar{G}),$$

so that $[H, K] \leqq \beta_r(G)$, as required. □

Proof of Theorem 4.5.9. Let G, H, K be as in the statement of the theorem, set $X = [H, K]$ and suppose that $X \cap \beta_1(G) \leqq H$. If $K \lhd^r \langle H, K \rangle$, then

$$[H, {}_rK] \leqq (X \cap K) \cap \beta_r(G)$$

by Theorem 4.5.11. Hence

$$[H, {}_rK] \leqq \beta_r(X \cap K).$$

But

$$\beta_1(X \cap K) = X \cap K \cap \beta_1(G) \leqq H \cap K = 1$$

and so $\beta_r(X \cap K) = 1$.

Application of Lemmas 4.5.12 and 4.5.13 now gives $X \leqq \beta_1(G)$ and so $X \leqq H$, as required. □

§4.6 Universal normalizing properties: the Wielandt subgroup

For any group G we define $w(G)$, the *Wielandt subgroup* of G, to be the intersection of all the normalizers of subnormal subgroups of G. At first sight one might think that this subgroup had very little chance of being non-trivial except in very obvious and uninteresting situations. However, in 1958 Wielandt showed that this is not the case—indeed if G is a non-trivial finite group, for instance, $w(G)$ is not the identity subgroup. More precisely he proved

Theorem 4.6.1 [154]. *Suppose that G is an arbitrary group. Then $w(G)$ contains*

(a) *every simple non-abelian subnormal subgroup of G, and*

(b) *every minimal normal subgroup M of G such that M satisfies* Min-*sn*.

We note that we cannot replace 'simple' in (a) by 'perfect' as is seen by taking G itself to be a perfect group which has a subnormal subgroup which is not normal. However, see Theorem 4.6.3. We have already proved (b) for finite groups in Theorem 1.3.11. We note that the condition that M satisfies Min-*sn* cannot be dropped. Suitable examples of groups with minimal normal subgroups not contained in $w(G)$ are furnished by McLain's construction—see for example Robinson [120], p. 81, No. (iii).

Proof of Theorem 4.6.1. (a) Suppose that H is a simple non-abelian subnormal subgroup of G and K is any subnormal subgroup. If $H \cap K \neq 1$, then since $H \cap K \; sn \; H$ and H is simple we must have $H \leqq K$ and so H normalizes K trivially. If $H \cap K = 1$, then $[H, K] = 1$ by Remark 3 on p. 54.

(b) Suppose that M is a minimal normal subgroup of G which satisfies Min-*sn* and that $H \lhd^n G$. If $n \leqq 1$, then $M \leqq N_G(H)$ trivially. So suppose that $n > 1$ and proceed by induction on n. By the minimality of M we have either $H^G \cap M = 1$, in which case $[H, M] \leqq H^G \cap M = 1$ and so $M \leqq N_G(H)$, or

$M \leqq H^G$. In this latter situation suppose that N is a minimal normal subgroup of H^G contained in M. Clearly, each conjugate of N in G is a minimal normal subgroup of H^G and satisfies Min-*sn*. By the induction hypothesis each conjugate of N normalizes H. Hence $N^G \leqq N_G(H)$. But $M = N^G$ and the result follows. □

It follows at once from Theorem 4.6.1(b) that if G is a group with Min-*sn*, then the *socle* of G (that is the product of all the minimal normal subgroups of G) is contained in the Wielandt subgroup and hence, if $G \neq 1$, then $w(G) \neq 1$. However, we can say much more:

Theorem 4.6.2 [113], [123]. *Suppose that G is a group satisfying* Min-*sn*. *Then $w(G)$ has finite index in G.*

Proof. This follows at once from Theorem 1.7.10—see the remark on p. 38. □

We note that if G is the infinite dihedral group, then $w(G) = 1$; of course this group does not satisfy Min-*sn*.

In any group G it is clear that $w(G)$ is a characteristic subgroup and we can therefore define the *upper Wielandt series of* G by setting $w_1(G) = w(G)$ and, for $n > 1$, $w_n(G)/w_{n-1}(G) = w(G/w_{n-1}(G))$. It follows at once from the foregoing results that if G is a group with Min-*sn*, then $w_n(G) = G$ for some finite n and we may call the least such n the *Wielandt length* of G. In [21] Camina has investigated the relationship between the Wielandt length of certain finite groups and various other invariants of the groups. He proved, for example, that *if G is a finite soluble group with Wielandt length n and Fitting length m, then G has derived length at most $m+n-1$ if $m+n > 2$*. (The *Fitting length* of a finite soluble group G is the smallest integer m such that G has a series of length m with nilpotent factors.)

We note that if G is a group with (finite) Wielandt length n, then a trivial induction shows that subnormal subgroups of G have their defects bounded by n.

By definition $w(G)$ has all of its subnormal subgroups normal and so $w(G)$ is a $\mathfrak{T}$-group (see p. 73)—indeed $\mathfrak{T}$-groups are just those groups in which $w(G) = G$. We shall study $\mathfrak{T}$-groups in their own right in §6.4. However, we observe that $w(G)$ contains, by Theorem 4.6.1(a), every simple non-abelian subnormal subgroup of G and such subgroups are clearly perfect $\mathfrak{T}$-groups. Thus the following result of Kegel generalizes Theorem 4.6.1(a).

Theorem 4.6.3 [70]. *A perfect subnormal $\mathfrak{T}$-subgroup of a group normalizes each subnormal subgroup of the group.*

Thus, for any group G, all perfect subnormal $\mathfrak{T}$-subgroups of G are contained in $w(G)$. We shall deduce Theorem 4.6.3 from

Theorem 4.6.4 [70]. *Suppose that H is a normal $\mathfrak{T}$-subgroup of a group G. Then H normalizes every subnormal subgroup K of G for which either $K/H \cap K$ or $H/H \cap K$ is perfect.*

Proof. Assuming the notation of the statement of the theorem, we note that $H \cap K \lhd HK$ and so we may suppose that $G = HK$ and $H \cap K = 1$. If $K = K'$, the result follows at once by Corollary 4.5.4. Hence we assume that $K \neq K'$ and that $H = H' \nleqq N_G(K)$. Then by Theorem 4.5.3 there exists a subnormal subgroup D of H such that $\zeta_1(H) < D$ and $D/\zeta_1(H)$ is abelian. Therefore D is nilpotent and every subgroup of D is subnormal in the $\mathfrak{T}$-group H. Hence if $z \in D$, $\langle z \rangle \lhd H$ and so H induces an abelian group of automorphisms of $\langle z \rangle$. However, $H = H'$ and so H centralizes z. Thus $z \in \zeta_1(H)$ and $D = \zeta_1(H)$, a contradiction. □

In order to deduce Theorem 4.6.3 from Theorem 4.6.4 we need only prove

Lemma 4.6.5. *The normal closure of a perfect subnormal $\mathfrak{T}$-subgroup of a group is a perfect $\mathfrak{T}$-group.*

Proof. Suppose that H is a subnormal perfect $\mathfrak{T}$-subgroup of a group G. Then $H_1 = H^G$, as a join of perfect subgroups, is perfect. By induction on the defect of H in G we may assume that $H_2 = H^{H_1}$ is a perfect normal $\mathfrak{T}$-subgroup of H_1 and so is contained in $w(H_1)$, by Theorem 4.6.4. But $w(H_1)$ is a characteristic subgroup of H_1 and so $w(H_1) \lhd G$. Hence $H_1 = H_2^G \leqq w(H_1)$ and therefore H_1 is a $\mathfrak{T}$-group. □

We note that Kegel actually proves that it is enough to assume that the $\mathfrak{T}$-subgroup in question is ascendant, from which he deduces that *a perfect ascendant $\mathfrak{T}$-subgroup normalizes every ascendant subgroup.* In fact Hall (in his Cambridge lectures) showed that *a perfect ascendant $\mathfrak{T}$-subgroup is subnormal of defect at most 2 and normalizes every descendant subgroup.* Also *a normal $\mathfrak{T}$-subgroup normalizes every descendant perfect subgroup* (cf. the first case of Theorem 4.6.4).

Finally in this section we recall (Theorem 1.7.4) that the class $\mathfrak{Min}$-*sn* is coalescent. Therefore, by Theorem 4.6.2, *if H and K are subnormal subgroups of a group G, both of which satisfy* Min-*sn and where H has no proper subgroups of finite index, then H normalizes K* (see Robinson [113] and Roseblade [123]). Kegel [70] has extended this to

Proposition 4.6.6. *Suppose H is a normal $\mathfrak{T}$-subgroup of a group G. Then H normalizes every subnormal subgroup K of G such that* (i) *$H/H \cap K$ satisfies* Min-*sn and has no proper subgroups of finite index and* (ii) *$K/H \cap K$ is generated by subgroups which satisfy* Min-*sn*.

This result is a straightforward consequence of Theorem 1.7.10. That the condition on $K/H \cap K$ is necessary is shown by the following example

[70]. Let p be a prime and P a quasicyclic p-group and form the direct product $N = P \times Z$ of P with an infinite cyclic group $Z = \langle z \rangle$. Let σ be an isomorphism of a group Q to P and let Q operate on N as follows:

$$[Q, P] = 1,$$

$$z^q = zq^\sigma, \qquad q \in Q.$$

Then $J = \langle P, Q \rangle$ is a normal subgroup of $G = \langle Z, P, Q \rangle$, J satisfies Min-*sn* and has no proper subgroups of finite index. However, J does not normalize Z.

5
THE JOIN PROBLEM—A CRITERION

In §4.3 we have seen that if H and K are groups with

$$H/H' \otimes K/K' = 0,$$

then $HK = KH$ whenever H and K are subnormally embedded in a group G; and conversely if the above tensor product is non-zero, then there is a group G with H, K as subnormal subgroups and $HK \neq KH$. Of course $HK = KH$ is a sufficient condition for $J = \langle H, K \rangle$ to be subnormal in G, but it is not necessary. The purpose of this chapter is to find necessary and sufficient conditions for the subnormality of J whenever H and K are subnormally embedded in a group G. As in §3.3, our methods use properties of powers of augmentation ideals; and the examples exhibiting non-subnormal joins are derived from Theorem 1.5.5. The main results presented here are due to J. P. Williams [167], though the methods and examples differ in many places.

§5.1 Williams' Join Theorem

We say that

an abelian group is f.r.p.d.

if it is the direct sum of a group U of finite rank and a periodic divisible group V. Then the p-component (that is the subgroup consisting of all the elements of p-power order) of U is a direct sum of n_p cyclic and C_{p^∞}-subgroups, for each prime p, with the numbers n_p finite and bounded, and the torsion-free quotient of U embeds in a direct sum of finitely many copies of $\mathbb{Q}$. The group V is a direct sum of C_{p^∞}-subgroups, with no restriction on the number of summands or the primes p. Since a divisible subgroup of an abelian group is always a direct summand [34], the word 'direct' can be omitted from the definition of *f.r.p.d.* groups and U can always be assumed to have finite p-components. In particular the class of *f.r.p.d.* groups is Q-closed.

Williams' result is the following.

Theorem 5.1.1. *Let H, K be groups.*

(i) *If $H/H' \otimes K/K'$ is f.r.p.d., then $\langle H, K \rangle$ is subnormal in all groups in which H and K can be subnormally embedded.*

(ii) *Conversely if $H/H' \otimes K/K'$ is not f.r.p.d., then there is a group G, containing H and K as subnormal subgroups, such that $\langle H, K \rangle$ is not subnormal in G.*

We begin by finding necessary and sufficient conditions on abelian groups X and Y for $X \otimes Y$ to be *f.r.p.d.* Denote the rank of a group G by $r(G)$ and, for an abelian group G, denote the periodic subgroup by G_0 and the p-component by G_p. So $G_0 = \bigoplus_p G_p$. Still assuming G abelian, write $G_\infty = G/G_0$, the torsion-free quotient of G, and denote the rank of G_∞/pG_∞ by g_p. For each prime p, let A_p and B_p be *basic* subgroups of X_p and Y_p respectively. (See, for example, [34].) Then A_p is a direct sum of r_p (say) cyclic groups, with A_p pure in X_p and X_p/A_p divisible. (A subgroup A of an abelian group G is said to be *pure* in G if $nG \cap A = nA$ for all integers $n \geqq 1$.) Similarly B_p is a direct sum of s_p (say) cyclic groups. While A_p and B_p need not be unique, they are determined to within isomorphism [34]. (We will take care not to confuse A_p, B_p with p-components.)

Some properties of tensor products will be required.

Lemma 5.1.2. *Let X, Y be abelian groups and $T = X \otimes Y$. Then*
(i) $T_\infty \cong X_\infty \otimes Y_\infty$ *and* $r(T_\infty) = r(X_\infty).r(Y_\infty)$;
(ii) *for each prime* p, $T_p \cong (A_p \otimes B_p) \oplus (\bigoplus_{y_p \text{ copies}} X_p) \oplus (\bigoplus_{x_p \text{ copies}} Y_p)$.

Proof. (i) The natural homomorphisms $X \to X_\infty$, $Y \to Y_\infty$ define a homomorphism $T \to X_\infty \otimes Y_\infty$ with kernel T_0. (See [35], Theorem 61.5.) Since the rank of a torsion-free abelian group is the cardinality of a maximal $\mathbb{Z}$-independent set of elements, the second part follows easily.
(ii) By [35], Theorem 61.5,

$$T_0 \cong (X_0 \otimes Y_0) \oplus (X_0 \otimes Y_\infty) \oplus (X_\infty \otimes Y_0).$$

Taking p-components we have

$$T_p \cong (X_p \otimes Y_p) \oplus (X_p \otimes Y_\infty) \oplus (X_\infty \otimes Y_p).$$

By [35], Theorem 61.1, $X_p \otimes Y_p \cong A_p \otimes B_p$. Suppose that U is an abelian p-group and V is a torsion-free abelian group. Let $\{v_\lambda + pV \mid \lambda \in \Lambda\}$ be a basis of V/pV. Then the elements of $U \otimes V$ can be written uniquely as finite sums $\Sigma_\lambda u_\lambda \otimes v_\lambda$, where $u_\lambda \in U$. (See [34], Theorem 65.4.) So (ii) follows. □

An elementary result on the rank of an abelian group will also be needed. The proof is an easy exercise.

Lemma 5.1.3. *For any abelian group X,*

$$r(X) = r(X_\infty) + \max\{r(X_p) \mid \textit{all } p\}.$$

Now we can find exactly when a tensor product is *f.r.p.d.*

Lemma 5.1.4. *Let X, Y be abelian groups. Then $X \otimes Y$ is f.r.p.d. if and only if there is an integer d such that $r(X_\infty).r(Y_\infty) \leqq d$ and, for all primes p,*

$$r_p s_p,\ r_p y_p,\ x_p s_p \leqq d.$$

Proof. Suppose that $T = X \otimes Y$ is *f.r.p.d.* Then $T = U \oplus V$, where $r(U) = d$ (say) is finite and V is periodic and divisible. Since T_∞ is a homomorphic image of U, $r(T_\infty) \leqq d$. It follows from Lemma 5.1.2(i) that $r(X_\infty).r(Y_\infty) \leqq d$. For all primes p, $T_p = U_p \oplus V_p$, and $r(U_p) \leqq r(U) = d$ while V_p is divisible. Thus a basic subgroup of U_p is a basic subgroup of T_p and hence has rank at most d. However, from Lemma 5.1.2(ii), we see that a basic subgroup of T_p is isomorphic to

$$(A_p \otimes B_p) \oplus \Big(\bigoplus_{y_p \text{ copies}} A_p \Big) \oplus \Big(\bigoplus_{x_p \text{ copies}} B_p \Big)$$

of rank $r_p s_p + r_p y_p + x_p s_p$. Therefore this sum cannot exceed d.

Conversely suppose that there is an integer d such that

$$r(X_\infty).r(Y_\infty),\ r_p s_p,\ r_p y_p,\ x_p s_p \leqq d$$

for all primes p. Then, by Lemma 5.1.2(i), $r(T_\infty) \leqq d$. Also, for each p,

$$r(A_p \otimes B_p) = r_p s_p \leqq d.$$

If $y_p \geqq 1$, then $r_p \leqq d$ and so X_p splits over A_p. (If A_p has exponent n, then it is easy to see that nX_p complements A_p in X_p.) Thus

$$\overbrace{X_p \oplus \cdots \oplus X_p}^{\longleftarrow y_p \longrightarrow}$$

is the direct sum of $r_p y_p$ cyclic groups and a divisible group. Similarly

$$\overbrace{Y_p \oplus \cdots \oplus Y_p}^{\longleftarrow x_p \longrightarrow}$$

is the direct sum of $x_p s_p$ cyclic groups and a divisible group. Therefore, by Lemma 5.1.2(ii), T_p is the direct sum of at most $3d$ cyclic groups and a divisible group. It follows from Lemma 5.1.3 that T is the direct sum of a group of rank at most $4d$ and a periodic divisible group, that is T is *f.r.p.d.* □

In order to prove Theorem 5.1.1(ii), it is necessary to analyse the non-*f.r.p.d.* situation. Before we do this, consider an abelian group X and assume the notation described before Lemma 5.1.2. Then, for each prime p, $pX_0 = pX \cap X_0$ and so

$$(pX + X_0)/pX \simeq X_0/pX_0 \simeq X_p/pX_p.$$

However, $X_p = A_p + pX_p$ and $A_p \cap pX_p = pA_p$ (by the purity of A_p in X_p). Thus

$$X_p/pX_p \simeq A_p/pA_p. \tag{1}$$

Also $X/(pX+X_0) \cong X_\infty/pX_\infty$ and therefore

$$r(X/pX) = r_p + x_p. \tag{2}$$

Recall that a group G is a *subdirect sum* of groups G_λ if G embeds in the unrestricted direct (that is cartesian) sum of the G_λ in such a way that the projection of G into each G_λ is surjective. (In multiplicative terminology *sum* is replaced by *product*.)

Lemma 5.1.5. *Let H, K be groups and suppose that $H/H' \otimes K/K'$ is not f.r.p.d. Then H, K have homomorphic images $\bar{H}$, $\bar{K}$ respectively, satisfying one of the following conditions:*

(i) *For some prime p, one of $\bar{H}$, $\bar{K}$ is an infinite elementary abelian p-group and the other has order p.*
(ii) *There is an infinite set of primes $p_1, p_2, \ldots$ such that one of $\bar{H}$, $\bar{K}$ is a subdirect sum of cyclic groups, i of order p_i for each i, and the other is a subdirect sum of cyclic groups, one of order p_i for each i.*
(iii) *$\bar{H}$, $\bar{K}$ are both torsion-free abelian, one of infinite rank and one of rank* 1.

Proof. Let $X = H/H'$, $Y = K/K'$. By (2), H has an elementary abelian p-quotient with rank at least $\max\{r_p, x_p\}$ and K has a similar quotient with rank at least $\max\{s_p, y_p\}$.

Case 1. Suppose that, for some prime p, one of the products $r_p s_p, r_p y_p, x_p s_p$ is infinite. Then one factor of this product is infinite and the other is non-zero. Thus condition (i) is satisfied.

Case 2. Suppose that the set $\{r_p s_p, r_p y_p, x_p s_p \mid \text{all } p\}$ consists of finite, but unbounded, integers. Then the same is true of one of the sets $\{r_p s_p \mid \text{all } p\}$, $\{r_p y_p \mid \text{all } p\}$, $\{x_p s_p \mid \text{all } p\}$. Of the non-zero products of this particular set, the first factors or the second factors must form an unbounded set and therefore condition (ii) holds.

Case 3. Finally suppose that $r(X_\infty).r(Y_\infty)$ is infinite. So $X_\infty \neq 0$. Since X_∞ embeds in $X_\infty \otimes \mathbb{Q}$ ([34], Theorem 64.4), X_∞ has a non-zero homomorphic image in $\mathbb{Q}$. The same applies to Y_∞ and at least one of X_∞, Y_∞ has infinite rank. Therefore condition (iii) is satisfied.

According to Lemma 5.1.4, at least one of these three cases applies and so the proof is complete. □

Theorem 5.1.1(i) will be obtained from part (i) of the following result. Part (ii) emphasizes the crucial role played by the *f.r.p.d.* condition.

Theorem 5.1.6

(i) *Let J be a nilpotent group generated by subgroups H, K and suppose that $H/H' \otimes K/K'$ is f.r.p.d. Then given any integers $a, b \geq 1$, there is an integer*

$f \geqq 1$ such that

$$\mathfrak{j}^f \leqq \mathbb{Z}J\,\mathfrak{h}^a + \mathbb{Z}J\,\mathfrak{k}^b.$$

(ii) *Conversely let H, K be nilpotent groups with $H/H' \otimes K/K'$ not f.r.p.d. Then there is a nilpotent group J generated by H and K such that $\mathfrak{j}^f \nleqq \mathbb{Z}J\,\mathfrak{h}^2 + \mathbb{Z}J\,\mathfrak{k}^2$ for all $f \geqq 1$.*

The next section is devoted to the proofs of Theorems 5.1.1(ii) and 5.1.6(ii).

§5.2 Examples

Let H, K be groups with $H/H' \otimes K/K'$ *not f.r.p.d.* Then one of the three situations of Lemma 5.1.5 applies and in order to prove Theorems 5.1.1(ii) and 5.1.6(ii) we consider these cases separately.

Case (*i*). We may assume that there are epimorphisms

$$H \xrightarrow{\alpha} \bar{H}, \qquad K \xrightarrow{\beta} \bar{K}, \tag{1}$$

where $\bar{H}$ is a countably infinite, elementary abelian p-group, for some prime p, and $\bar{K}$ has order p. Let F be a field with p elements and A be a countably infinite F-space. As in Theorem 1.5.5, we take R to be the exterior algebra on A. Let $\bar{J}$ be the multiplicative group of 2×2 matrices over R generated by H_0 and K_0, where

$$H_0 = \text{all} \begin{pmatrix} 1 & a \\ 0 & 1 \end{pmatrix}, \ a \in A,$$

$$K_0 = \text{all} \begin{pmatrix} 1 & 0 \\ f & 1 \end{pmatrix}, \ f \in F.$$

Take $\bar{H} = H_0$, $\bar{K} = K_0$.

We construct a group $J = \langle H, K \rangle$, with H and K subnormal in J, such that J has a quotient isomorphic to $\bar{J}$. Then we can adapt the method of Theorem 1.5.5. First form the free product $P = H * K$. So there is an exact sequence

$$1 \to N_1 \to P \xrightarrow{\theta} \bar{J} \to 1,$$

where θ is constructed from α and β. From (3) in Section 1.5 and Fitting's Theorem 1.6.1, we see that $\bar{J}$ is nilpotent of class at most 4. Therefore

$$N = \langle [H, {}_4K], [K, {}_4H] \rangle^P \leqq N_1. \tag{2}$$

Now there is an exact sequence $1 \to [H, K] \to P \xrightarrow{\phi} H \times K \to 1$ in which ϕ maps H (as a subgroup of P) isomorphically onto H (as a subgroup of $H \times K$), that is $[H, K] \cap H = 1$ in P. Since $N \leqq [H, K]$, it follows that H embeds as HN/N in $P/N = J$, say. Similarly K embeds in J and $J = \langle H, K \rangle$. Moreover from (2) we see that

$$H \lhd^4 J, \qquad K \lhd^4 J. \tag{3}$$

If $V = R \oplus R$, then $\bar{J}$ acts in a natural way on V. Let $N_0 = N_1/N$. Thus $J/N_0 \simeq \bar{J}$ and so J also acts on V (N_0 acting trivially). Let $G = V]J$. As in Theorem 1.5.5,

$$[V, H, H] = 1 \tag{4}$$

and so $H \lhd^2 VH$. Therefore $H \lhd^6 G$, by (3), and similarly $K \lhd^6 G$. Also from Theorem 1.5.5 we know that J/N_0 is self-normalizing in G/N_0 and so

$$J \textit{ is self-normalizing in } G. \tag{5}$$

Consider the abelian group V as a $\mathbb{Z}J$-module. From (4) it follows that $V\mathfrak{h}^2 = 0$. Similarly $V\mathfrak{k}^2 = 0$. Hence if

$$\mathfrak{j}^f \leqq \mathbb{Z}J\mathfrak{h}^2 + \mathbb{Z}J\mathfrak{k}^2 \tag{6}$$

for some f, we would have $V\mathfrak{j}^f = 0$, that is $[V, {}_f J] = 1$, contradicting (5). Now suppose that H and K are nilpotent. Then $P/[H, K] \simeq H \times K$ is nilpotent and so there is an integer $l(\geqq 1)$ such that $\gamma_l(P) \leqq N_1 \cap [H, K]$. Therefore replacing N above by $\gamma_l(P)$, J becomes nilpotent. Also H and K are still embedded subnormally in G, J remains self-normalizing and (6) fails to hold for all f. This completes case (i).

Case (ii). Now we may assume that there are epimorphisms (1) with $\bar{H}$ a subdirect sum of the groups $\overset{\longleftarrow i \longrightarrow}{C_{p_i} \oplus \cdots \oplus C_{p_i}}$, all $i \geqq 1$, and $\bar{K}$ a subdirect sum of the groups C_{p_i}, all $i \geqq 1$, where $\{p_i \mid i \geqq 1\}$ is an infinite set of primes. For each $i \geqq 1$, let F_i be a field with p_i elements, A_i an i-dimensional F_i-space and R_i the exterior algebra generated by A_i over F_i. Let F, A and R be the unrestricted direct sums of the F_i, A_i and R_i, respectively. So F is a commutative ring with 1, A is an F-module and R is an A-algebra. Denote by A^*, F^* the images of $\bar{H}$, $\bar{K}$, respectively, under their natural embeddings in A, F. Then identify $\bar{H}$, $\bar{K}$ with multiplicative groups of 2×2 matrices over R:

$$\bar{H} = \text{all}\begin{pmatrix} 1 & a \\ 0 & 1 \end{pmatrix}, a \in A^*; \qquad \bar{K} = \text{all}\begin{pmatrix} 1 & 0 \\ f & 1 \end{pmatrix}, f \in F^*.$$

Let $\bar{J} = \langle \bar{H}, \bar{K} \rangle$ and denote by $\bar{J}_i$ the subgroup of $\bar{J}$ generated by the matrices

$$\begin{pmatrix} 1 & a \\ 0 & 1 \end{pmatrix}, \quad \begin{pmatrix} 1 & 0 \\ f & 1 \end{pmatrix}$$

with $a \in A_i$, $f \in F_i$, all $i \geqq 1$. Then $\bar{J}$ is a subdirect product of the subgroups $\bar{J}_i$, $i \geqq 1$. As in case (i), each $\bar{J}_i$ is nilpotent of class at most 4 and so the same is true of $\bar{J}$. Thus as before we can construct $J = \langle H, K \rangle$ with $H \lhd^4 J$, $K \lhd^4 J$, $V = R \oplus R$ and $G = V]J$. Again $H, K \lhd^6 G$.

We must show that J is not subnormal in G. (Certainly J is not self-normalizing in this case. For, if r is in the ith component of R_i, then the element

$(r, 0)$ of V centralizes J.) However, we claim that $[G, {}_iJ] \nleq J$ for any i. To see this, observe that A_{i+1}, as an $(i+1)$-dimensional F_{i+1}-space, has a basis $\{a_0, a_1, \ldots, a_i\}$, say. Let $\nu: J \to \bar{J}_{i+1}$ be the natural homomorphism and, for $1 \leqq j \leqq i$, choose $x_j \in J$ such that

$$x_j^\nu = \begin{pmatrix} 1 & a_j \\ 0 & 1 \end{pmatrix}.$$

Similarly choose $y \in J$ such that

$$y^\nu = \begin{pmatrix} 1 & 0 \\ e & 1 \end{pmatrix},$$

where e is the identity element of F_{i+1}. Let $v = (a_0, 0) \in V$. Then $[G, {}_{2i}J]$ contains $[v, x_1, y, x_2, y, \ldots, x_i, y]$. An easy calculation shows that this commutator is $(a_0 a_1 \cdots a_i, 0)$, which is not the zero of V. It follows that $[G, {}_{2i}J] \nleq J$ for any i, and so J is not subnormal in G.

The argument of case (i) also establishes Theorem 5.1.6(ii) in this case.

Case (*iii*). Finally we assume that there are epimorphisms (1) with $\bar{H}$ and $\bar{K}$ torsion-free abelian, $\bar{H}$ of infinite rank and $\bar{K}$ of rank 1. Then

$$A = \bar{H} \otimes \mathbb{Q}, \qquad F = \bar{K} \otimes \mathbb{Q}$$

are $\mathbb{Q}$-spaces of infinite dimension and dimension 1, respectively. Let R be the exterior algebra generated by A over F, and identify $\bar{H}$ and $\bar{K}$ with multiplicative groups of 2×2 matrices over R via the maps

$$\bar{h} \mapsto \begin{pmatrix} 1 & \bar{h} \otimes 1 \\ 0 & 1 \end{pmatrix}, \qquad \bar{k} \mapsto \begin{pmatrix} 1 & 0 \\ \bar{k} \otimes 1 & 1 \end{pmatrix},$$

for all $\bar{h} \in \bar{H}, \bar{k} \in \bar{K}$. Let $\bar{J} = \langle \bar{H}, \bar{K} \rangle$. As in the previous cases, $\bar{J}$ is nilpotent of class at most 4. (Of course $\bar{K}$ may not be cyclic now and the calculations of Theorem 1.5.5, showing that the normal closures of $\bar{H}$ and $\bar{K}$ have class 2, need some small modifications.) The construction of J and G proceeds in the same way as before. Since $\bar{H}$ has infinite rank, J turns out to be self-normalizing in G, as in case (i). Theorem 5.1.6(ii) follows similarly.

This completes all cases of the proofs of Theorems 5.1.1(ii) and 5.1.6(ii). □

§5.3 Persistent properties of nilpotent groups

For the proof of Theorem 5.1.6(i) we need information on how the structure of a nilpotent group is determined by its derived quotient. Thus (following Robinson [116]) a class $\mathfrak{X}$ of abelian groups is said to be

(i) *tensorial* if every homomorphic image of $A \otimes B$ belongs to $\mathfrak{X}$ whenever $A, B \in \mathfrak{X}$; and

(ii) *persistent* if, for all $i \geqq 2$, $\gamma_i(G)/\gamma_{i+1}(G) \in \mathfrak{X}$ for all groups G with $G/G' \in \mathfrak{X}$.

Let G be any group, $i \geqq 1$, $u \in \gamma_i(G)$ and $g \in G$. Then the map

$$u\gamma_{i+1}(G) \otimes gG' \mapsto [u, g]\gamma_{i+2}(G)$$

is well-defined and an epimorphism

$$\gamma_i(G)/\gamma_{i+1}(G) \otimes G/G' \to \gamma_{i+1}(G)/\gamma_{i+2}(G). \tag{1}$$

Thus

tensorial classes are persistent.

Many properties of nilpotent groups G are easy consequences of this result. For example, if G/G' is generated by p-elements, then G is a p-group, and if G/G' is finite, then so is G. Also if G has class c and G/G' is finitely generated (by d elements), then all the subgroups of G are finitely generated (that is G satisfies Max) and G has finite rank bounded by a function of c and d. Similarly if G/G' has finite rank r, then G has finite rank bounded by a function of c and r.

Whenever G is a nilpotent group, we denote the periodic subgroup of G by G_0 and the torsion-free quotient G/G_0 by G_∞ (consistent with the notation already adopted when G is abelian). We also write $r_\infty(G)$ for the *torsion-free rank* of G, that is the rank of G_∞. If G is abelian, then $r_\infty(G)$ is the dimension of the $\mathbb{Q}$-space

$$G_\infty \otimes \mathbb{Q} \cong G \otimes \mathbb{Q}.$$

(See for example [34], Theorems 64.2 and 64.4.) Suppose that $N \lhd G$. Since tensoring with $\mathbb{Q}$ preserves short exact sequences, it follows that

$$r_\infty(G) = r_\infty(N) + r_\infty(G/N). \tag{2}$$

(See [35], Theorem 60.6.) Concerning torsion-free rank, we have

Proposition 5.3.1. *If H is a nilpotent group and $r_\infty(H/H')$ is finite, then $r_\infty(H)$ is finite.*

Proof. Since $H_\infty/H'_\infty \cong H/H_0H'$, a homomorphic image of H/H', (2) shows that H_∞/H'_∞ has finite torsion-free rank. By Lemma 5.1.2(i), the class of abelian groups having finite torsion-free rank is tensorial, hence persistent, that is $\gamma_i(H)/\gamma_{i+1}(H)$ has finite torsion-free rank, for all $i \geqq 1$. For $j \geqq 0$, let $Z_j = \zeta_j(H_\infty)$. By Lemma 2.3.1, the upper and lower central series of H have isomorphic refinements and hence (by (2)) $r_\infty(Z_{j+1}/Z_j)$ is finite, all $j \geqq 0$. However, each Z_{j+1}/Z_j is easily seen to be torsion-free (see for example [119], Theorem 2.25) and so Z_{j+1}/Z_j has finite rank. Therefore H_∞ has finite rank, by Lemma 1.7.11. □

Remark. A routine check shows that $r_\infty(H)$ is bounded by some function of $r_\infty(H/H')$ and the class of H.

Now we concentrate on the p-component H_p of a nilpotent group H. A group all of whose elements have pth roots is called *p-divisible.*

Proposition 5.3.2. *Let H be a nilpotent group and suppose that, for some prime p, H/H' is p-divisible. Then H_p is abelian and divisible.*

Proof. We show that

$$H \text{ is } p\text{-divisible.} \tag{3}$$

To see this, observe that the class of p-divisible abelian groups is tensorial, hence persistent. Thus if H has class c, $\gamma_c(H)$ is p-divisible. We may assume that $c \geqq 2$ and, by induction on c, that $H/\gamma_c(H)$ is p-divisible. Since $\gamma_c(H) \leqq \zeta_1(H)$, (3) follows.

From (3) we see that H_p is divisible. To see that H_p is abelian, consider the epimorphism (1) with $G = H_p$ and $i = 1$. Then $\gamma_2(H_p) = \gamma_3(H_p) = 1$. □

A simpler proof that H_p is abelian, and even lies in the centre of H_0, follows from the next result. (Of course, knowing that H_p is abelian, the fact that it lies in the centre of H_0 is a consequence of Lemma 2.5.9.)

Lemma 5.3.3. *Let G be a nilpotent group of class at most c ($\geqq 1$) and, for any $n \geqq 1$, let $X_n = \{x \mid x \in G, x^n = 1\}$. Then the exponent of $G/C_G(X_n)$ divides n^{c-1}.*

Proof. The result is trivial if G is abelian, so we proceed by induction on c. Assume that $c \geqq 2$ and that the lemma is true for groups of class $\leqq c-1$. Then for all $g \in G$ and $x \in X_n$,

$$[x, g^{n^{c-2}}] \in \zeta_1(G).$$

Hence $[x, g^{n^{c-1}}] = [x, g^{n^{c-2}}]^n = [x^n, g^{n^{c-2}}] = 1$ as required. □

Thus we obtain

Corollary 5.3.4. *A divisible subgroup of a periodic nilpotent group G lies in the centre of G.*

The same elementary commutator argument yields

Lemma 5.3.5. *If G is a group with $G/\zeta_1(G)$ periodic, abelian and divisible, then G is abelian.*

Proof. Let $Z = \zeta_1(G)$ and $a, b \in G$. Then $a^n \in Z$, some $n \geqq 1$, and there is an element c in G such that $b \equiv c^n \bmod Z$. So $b = c^n z$, some $z \in Z$, and therefore

$$[a, b] = [a, c^n] = [a^n, c] = 1. \quad \square$$

We return to the consideration of the p-component of a nilpotent group. Observe that additive notation is adopted for abelian groups.

Proposition 5.3.6. *Let H be a nilpotent group of class c, $X = H/H'$ and p be a fixed prime. Suppose that there is a finite integer d such that $r(X/pX) \leqq d$. Then there is an integer $e = e(c, d)$ and there are normal subgroups F, D of H_p such that F is finite of rank at most e, D is abelian and divisible and $H_p = FD$.*

Proof. Notice that the commutativity of D and the normality of F and D in H_p will follow from Corollary 5.3.4.

Suppose that H is abelian and recall that A_p is a basic subgroup of X_p of rank r_p. By hypothesis and (2) in §5.1, A_p is finite of rank at most d and so X_p splits over A_p (see Lemma 5.1.4). Thus the proposition follows in this case.

Now we proceed by induction on $c\,(\geqq 2)$ and assume the truth of the proposition for groups of class $c-1$. Let D be the unique maximal divisible subgroup of H_p, which exists and lies in $\zeta_1(H_p)$, by Corollary 5.3.4. Then by Lemma 5.3.5, H_p/D has no non-trivial divisible subgroups. Since $D \lhd H$, we may factor by D and assume that

$$H_p \textit{ has no non-trivial divisible subgroups.} \tag{4}$$

We must show that H_p is finite of rank bounded by a function of c and d.

Let $\mathfrak{X}$ be the class of abelian groups Y with $r(Y/pY)$ finite. If Y_1, Y_2 are abelian groups and $T = Y_1 \otimes Y_2$, then $T/pT \cong Y_1/pY_1 \otimes Y_2/pY_2$. Therefore the class $\mathfrak{X}$ is tensorial, hence persistent and if $Z = \gamma_c(H)$, then $r(Z/pZ) \leqq d^c$. If P/Z is the p-component of H/Z, induction gives

$$P/Z = (U/Z)(V/Z),$$

where U/Z is finite of rank $\leqq e(c-1, d)$ and V/Z is abelian and divisible. By Lemma 5.3.5, V is abelian. Also $V/pV = (pV+Z)/pV \cong Z/(pV \cap Z)$, a homomorphic image of Z/pZ, and so $r(V/pV) \leqq d^c$. By the abelian case (applied to V) and (4), $H_p \cap V$ (the p-component of V) is finite of rank at most $e(1, d^c)$. Finally

$$H_p/H_p \cap V \cong H_pV/V \leqq P/V \cong U/U \cap V,$$

a homomorphic image of U/Z, and therefore $H_p/H_p \cap V$ is finite of rank $\leqq e(c-1, d)$. The proposition now follows from Lemma 1.7.11. □

For Theorem 5.1.6(i) we need

Lemma 5.3.7 [40] (§4.3, Proposition 3). *If G is a periodic divisible group, then $\mathfrak{g}^2 = \mathfrak{g}^3$.*

Proof. Let $n \geqq 1$ and $x \in G$. Then $x^{n-1} + x^{n-2} + \cdots + x + 1 \equiv n \bmod \mathfrak{g}$ and so

$$x^n - 1 \equiv n(x-1) \bmod \mathfrak{g}^2. \tag{5}$$

Now let $y \in G$. It suffices to show that $(x-1)(y-1) \in \mathfrak{g}^3$. There is an element

$z \in G$ such that $z^n = y$ and

$$(x-1)(y-1) = (x-1)(z^n-1) \equiv n(x-1)(z-1) \bmod \mathfrak{g}^3$$

by (5) (with z for x). Thus, again by (5),

$$(x-1)(y-1) \equiv (x^n-1)(z-1) \bmod \mathfrak{g}^3.$$

The result follows by taking n to be the order of x. □

Now we can establish the augmentation condition for nilpotent groups where one of the generating subgroups is periodic and divisible.

Theorem 5.3.8. *Let J be a nilpotent group of class c generated by subgroups H and K with H periodic and divisible. Then given $a, b \geqq 1$, there exists $f = f(c, b) \geqq 1$ such that* $\mathfrak{j}^f \leqq \mathbb{Z}J\mathfrak{h}^a + \mathbb{Z}J\mathfrak{k}^b$.

Proof. By Proposition 3.3.11, if Y and Z are subgroups of a nilpotent group X of class $\leqq c$ and if $X = YZ$, then there is an integer $f = f(c, b)$ such that

$$\mathfrak{x}^f \leqq \mathbb{Z}X\mathfrak{y}^2 + \mathbb{Z}X\mathfrak{z}^b. \tag{6}$$

Let $x \in \mathfrak{j}^f$. Now H is the direct product of its p-components H_p and (by Corollary 5.3.4) each H_p is the direct product of C_{p^∞}-subgroups. Therefore there are subgroups H_1 of H and K_1 of K such that H_1 is divisible, H_1 and K_1 have finite rank and $x \in \mathfrak{j}_1^f$, where $J_1 = \langle H_1, K_1 \rangle$. Then J_1/J_1' and hence J_1 have finite rank and so, by Theorem 3.3.2, there is $e \geqq 1$ such that

$$\mathfrak{j}_1^e \leqq \mathbb{Z}J_1\mathfrak{h}_1^a + \mathbb{Z}J_1\mathfrak{k}_1^b.$$

Let $N = H_1^{K_1}$. Thus $\mathfrak{n}^e \leqq \mathbb{Z}J\mathfrak{h}^a + \mathbb{Z}J\mathfrak{k}^b$. However, H_1 lies in the centre of the periodic subgroup of J_1 (by Corollary 5.3.4) and therefore N is periodic and divisible. (Lemma 1.7.8 also shows this.) Then by Lemma 5.3.7, $\mathfrak{n}^2 = \mathfrak{n}^3 = \mathfrak{n}^4 = \cdots$ and so

$$\mathfrak{n}^2 \leqq \mathbb{Z}J\mathfrak{h}^a + \mathbb{Z}J\mathfrak{k}^b.$$

Finally $J_1 = NK_1$ and from (6) we have

$$x \in \mathfrak{j}_1^f \leqq \mathbb{Z}J_1\mathfrak{n}^2 + \mathbb{Z}J_1\mathfrak{k}_1^b \leqq \mathbb{Z}J\mathfrak{h}^a + \mathbb{Z}J\mathfrak{k}^b.$$

Since x was an arbitrary element of $\mathfrak{j}^f$, the theorem is proved. □

§5.4 Proofs of the main theorems

Proof of Theorem 5.1.6(i). We consider first the case when

$$H/H' \text{ and } K/K' \text{ are not periodic.} \tag{1}$$

The method is to find periodic normal subgroups L, D of H and M, E of K with H/LD and K/ME of finite rank, L orthogonal to K, H orthogonal to M and D,

E divisible. Then we can use the augmentation condition results of Theorems 3.3.2 and 5.3.8 and Proposition 3.3.11.

Thus suppose that $J = \langle H, K \rangle$ is nilpotent of class c, $H/H' \otimes K/K'$ is *f.r.p.d.* and (1) is true. With $X = H/H'$, $Y = K/K'$, it follows that

$$X_\infty \neq 0 \neq Y_\infty.$$

From Lemma 5.1.4 (and using the notation employed there), there is an integer d such that $r(X_\infty).r(Y_\infty) \leqq d$ and for all primes p

$$r_p s_p,\ r_p y_p,\ x_p s_p \leqq d. \tag{2}$$

Therefore $1 \leqq r(X_\infty) \leqq d$ and $1 \leqq r(Y_\infty) \leqq d$ and so $x_p \leqq d$ and $y_p \leqq d$, for all p. We partition the set of all primes as follows:

$$\pi_1 = \{p | s_p > d\}, \quad \pi_2 = \{p | r_p > d\}, \quad \pi_3 = \{p | p \notin \pi_1 \cup \pi_2\}.$$

Thus (2) shows that if $p \in \pi_1$, then $r_p = x_p = 0$. Therefore X_p is divisible and X_∞ is p-divisible and so X is p-divisible. Thus H_p is divisible, by Proposition 5.3.2. Also $X \otimes K_p/K_p' = 0$ and therefore $[H, K_p] = 1$, by Corollary 4.3.7. Similarly if $p \in \pi_2$, then K_p is divisible and $[H_p, K] = 1$.

Now suppose that $p \in \pi_3$. Then $r_p \leqq d$ and so $r(X/pX) \leqq 2d$, by (2) in Section 5.1. Therefore by Proposition 5.3.6, $H_p = F_p D_p$, where F_p is a finite subgroup of rank at most $e = e(c, 2d)$ and D_p is a divisible abelian subgroup. Similarly $K_p = G_p E_p$, where G_p is finite of rank $\leqq e$ and E_p is divisible and abelian.

Let $L = \langle H_p | p \in \pi_2 \rangle$, $D = \langle H_p, D_q | p \in \pi_1, q \in \pi_3 \rangle$. Then $L, D \lhd H$, $[L, K] = 1$ and D is periodic, divisible and abelian. Also each p-component of H/LD has rank $\leqq e$ and $r_\infty(H/LD) = r_\infty(H)$ is finite and bounded by some function of c and d, by the remark after Proposition 5.3.1. Therefore if $r(H/LD) = r$, then r is finite and bounded by a function of c and d (by Lemma 1.7.11). Similarly let $M = \langle K_p | p \in \pi_1 \rangle$, $E = \langle K_p, E_q | p \in \pi_2, q \in \pi_3 \rangle$, so that $M, E \lhd K$, $[H, M] = 1$, E is periodic, divisible and abelian and $r(K/ME) = s$, say, bounded by a function of c and d.

Now let H_1/LD, K_1/ME be finitely generated subgroups of H/LD, K/ME respectively. Then there are subgroups U, V such that

$$H_1 = \langle U, LD \rangle, \qquad K_1 = \langle V, ME \rangle$$

and U can be generated by at most r elements, V can be generated by at most s elements. Write $J_1 = \langle H_1, K_1 \rangle$. Thus J is the set-theoretic union of all such subgroups J_1 and it suffices to show that there is an integer $f = f(a, b, c, d)$ such that

$$\mathfrak{j}_1^f \leqq \mathbb{Z} J_1 \mathfrak{h}_1^a + \mathbb{Z} J_1 \mathfrak{k}_1^b.$$

Let $W = \langle U, V \rangle$. Then W is finitely generated with rank bounded by a function of c and d and there exists $f_1 = f_1(a, b, c, d)$ such that

$$\mathfrak{w}^{f_1} \leqq \mathbb{Z} W \mathfrak{u}^a + \mathbb{Z} W \mathfrak{v}^b, \tag{3}$$

by Theorem 3.3.2. Write $P = \langle W, D, E \rangle$. Two applications of Theorem 5.3.8 show that

$$\mathfrak{p}^{f_2} \leqq \mathbb{Z}P\,\mathfrak{w}^{f_1} + \mathbb{Z}P\,\mathfrak{d}^a + \mathbb{Z}P\mathfrak{e}^b \tag{4}$$

for some $f_2 = f_2(a, b, c, d)$. Finally consider $J_1 = \langle P, L, M \rangle$. We have $L \lhd J$ and so $PL = LP$. Similarly $M \lhd J$ and $(PL)M = M(PL)$. Therefore two applications of Proposition 3.3.11 give an integer $f = f(a, b, c, d)$ such that

$$\mathfrak{j}_1^f \leqq \mathbb{Z}J_1\,\mathfrak{p}^{f_2} + \mathbb{Z}J_1\,\mathfrak{l}^a + \mathbb{Z}J_1\,\mathfrak{m}^b \leqq \mathbb{Z}J_1\,\mathfrak{h}_1^a + \mathbb{Z}J_1\,\mathfrak{k}_1^b,$$

by (3) and (4).

This completes the proof of Theorem 5.1.6(i) under the assumption (1). For the remaining case we may assume that $Y = K/K'$ is periodic. Since periodicity is tensorial and hence persistent, this means that

K is periodic.

This time Lemma 5.1.4 gives no information about $r_\infty(X)$ and so we have no control over $r_\infty(H)$. Instead we find normal subgroups L, D, F of K such that $K = LDF$, with L orthogonal to H, D periodic and divisible (and abelian) and F of finite rank.

Again let J have class c. We partition the primes as follows:

$$\pi_1 = \{p \mid s_p > d\}, \quad \pi_2 = \{p \mid 1 \leqq s_p \leqq d\}, \quad \pi_3 = \{p \mid s_p = 0\}.$$

If $p \in \pi_1$, then (2) shows that $r_p = x_p = 0$ and hence, as before, X is p-divisible and $[H, K_p] = 1$. If $p \in \pi_2$, then (as for $p \in \pi_3$ in the previous case) $K_p = F_p D_p$, where F_p is finite of rank at most $e = e(c, d)$ (note that $y_p = 0$) and D_p is divisible and abelian. If $p \in \pi_3$, then Y_p is divisible and so K_p is divisible and abelian, by Proposition 5.3.2.

Let $L = \langle K_p \mid p \in \pi_1 \rangle$, so that $[H, L] = 1$; let $D = \langle K_p, D_q \mid p \in \pi_3, q \in \pi_2 \rangle$, a divisible abelian group; and let $F = \langle F_p \mid p \in \pi_2 \rangle$, so that $r(F) \leqq e$. Then $L, D, F \lhd K$, the latter since, for $p \in \pi_2$, $D_p \leqq \zeta_1(K_p)$, by Corollary 5.3.4. Let F_1 be a finite subgroup of F and $W_1 = \langle H, F_1 \rangle$. Suppose that there is an integer $f_1 = f_1(a, b, c, d)$ such that

$$\mathfrak{w}_1^{f_1} \leqq \mathbb{Z}W_1\,\mathfrak{h}^a + \mathbb{Z}W_1\,\mathfrak{f}_1^b. \tag{5}$$

Since $W = \langle H, F \rangle$ is the union of all such subgroups W_1, we shall then have

$$\mathfrak{w}^{f_1} \leqq \mathbb{Z}W\mathfrak{h}^a + \mathbb{Z}W\mathfrak{f}^b.$$

Then applying Theorem 5.3.8 to $P = \langle W, D \rangle$, there exists $f_2 = f_2(a, b, c, d)$ such that

$$\mathfrak{p}^{f_2} \leqq \mathbb{Z}P\mathfrak{w}^{f_1} + \mathbb{Z}P\mathfrak{d}^b.$$

Now $P = \langle H, F, D \rangle$, $[H, L] = 1$ and $L \lhd K$. Therefore $J = \langle P, L \rangle = PL$

and, by Proposition 3.3.11, there exists $f = f(a, b, c, d)$ such that

$$\mathfrak{j}^f \leqq \mathbb{Z}J\mathfrak{p}^{f_2} + \mathbb{Z}J\mathfrak{l}^b \leqq \mathbb{Z}J\mathfrak{h}^a + \mathbb{Z}J\mathfrak{k}^b$$

as required. Thus it suffices to prove (5).

Suppose that F_1 has exponent n. By Lemma 5.3.3, $H/C_H(F_1^H) = \bar{H}$ (say) has exponent dividing n^{c-1}. However, the prime divisors p of n all belong to π_2 (by the definition of F) and therefore $r_p \leqq d$ and $x_p \leqq d$, by (2). Thus, by (2) in §5.1, $r(X/pX) \leqq 2d$. Then it is easy to see that $\bar{H}$ is finite and generated by at most $2d$ elements. Therefore there is a subgroup H_1 of H with at most $2d$ generators such that

$$F_1^H = F_1^{H_1} \leqq \langle H_1, F_1 \rangle = W_2,$$

say. So W_2 is generated by at most $e + 2d$ elements and by Theorem 3.3.2 there exists $f_3 = f_3\,(a, b, c, d)$ such that

$$\mathfrak{w}_2^{f_3} \leqq \mathbb{Z}W_2\mathfrak{h}_1^a + \mathbb{Z}W_2\mathfrak{f}_1^b.$$

Finally, $W_1 = \langle H, F_1 \rangle = HW_2$ and therefore Proposition 3.3.11 gives $f_1 = f_1\,(a, b, c, d)$ such that

$$\mathfrak{w}_1^{f_1} \leqq \mathbb{Z}W_1\mathfrak{h}^a + \mathbb{Z}W_1\mathfrak{w}_2^{f_3} \leqq \mathbb{Z}W_1\mathfrak{h}^a + \mathbb{Z}W_1\mathfrak{f}_1^b,$$

establishing (5) and the proof of Theorem 5.1.6(i). □

It is not hard to see that f in this theorem is a function of a, b, c, d.

The proof of Theorem 5.1.1(i) requires the group-theoretic analogue of the above result.

Theorem 5.4.1. *Let H, K be subnormal subgroups of their join J and suppose that $H/H' \otimes K/K'$ is f.r.p.d. Then given any integers $a, b \geqq 1$, there exists $c \geqq 1$ such that*

$$\gamma_c(J) \leqq \gamma_a(H)\gamma_b(K).$$

Proof. Since the class of *f.r.p.d.* groups is Q-closed, Theorem 5.1.6(i) shows that the nilpotent homomorphic images of J satisfy the augmentation condition and then Theorem 3.3.4 completes the proof. □

Remark. The integer c is bounded by a function of a, b, the defects of H, K in J, and d (from Lemma 5.1.4).

The strategy for proving Theorem 5.1.1(i) is very similar to that used in proving Theorem 5.1.6(i) and therefore we need a result about subnormality analogous to Theorem 5.3.8.

Proposition 5.4.2. *Let H, K be subnormal subgroups of G and let $J = \langle H, K \rangle$. Suppose that there is a subgroup $N \lhd J$ with N sn G and HN/N periodic, divisible and abelian. Then J is subnormal in G.*

Proof. Let $N = N_m \lhd N_{m-1} \lhd \cdots \lhd N_1 \lhd N_0 = G$ be the normal closure series of N in G. Then J normalizes each N_i and it suffices to show that $JN_i \, sn \, JN_{i-1}$, all i.

Now $N_i \lhd JN_{i-1}$, so if bars denote quotients modulo N_i, we must show that $\bar{J} \, sn \, \bar{J}\bar{N}_{i-1}$. We have

$$\bar{H} = HN_i/N_i \cong HN/(HN \cap N_i),$$

a homomorphic image of HN/N and so $\bar{H}$ is periodic, divisible and abelian. Therefore, by Theorem 2.2.8, $\bar{H}^{\bar{J}\bar{N}_{i-1}}$ is abelian (and periodic and divisible). Thus $\bar{H} \lhd^2 \bar{J}\bar{N}_{i-1}$ and by Corollary 1.2.2

$$\bar{J} = \langle \bar{H}, \bar{K} \rangle \, sn \, \bar{J}\bar{N}_{i-1}. \quad \square$$

Remark. The subnormal defect of J in G is bounded by a function of the defects of N and K in G.

Proof of Theorem 5.1.1(i). We have H, K subnormal in G and $T = H/H' \otimes K/K'$ *f.r.p.d.* We must show that J is subnormal in G. By Theorem 5.4.1,

$$N = \gamma_c(J) \leqq HK$$

for some c. Then $HN \leqq HK$ and hence $HN = H(HN \cap K)$. Now $HN \, sn \, J$ and so $HN \cap K \, sn \, K$. Therefore $HN \cap K \, sn \, G$ and (by Theorem 1.2.5) $HN \, sn \, G$. Similarly $KN \, sn \, G$. Let bars denote quotients modulo N. Then $\bar{H}/\bar{H}' \otimes \bar{K}/\bar{K}'$, as a homomorphic image of T, is *f.r.p.d.* and $\bar{J} = \langle \bar{H}, \bar{K} \rangle$ is nilpotent. As in Theorem 5.1.6(i), we distinguish two cases. Thus suppose first that

$$\bar{H}/\bar{H}' \textit{ and } \bar{K}/\bar{K}' \textit{ are not periodic.}$$

As in the corresponding case of Theorem 5.1.6(i), there are normal subgroups L, D of HN with $N \leqq L \cap D$, $L \lhd J$, D/N periodic, divisible and abelian and HN/LD of finite rank r. Similarly there are normal subgroups M, E of KN with $N \leqq M \cap E$, $M \lhd J$, E/N periodic, divisible and abelian and KN/ME of finite rank s.

Consider finitely generated subgroups U/N, V/N of HN/N, KN/N respectively, and let

$$J_1 = \langle U, V, LD, ME \rangle.$$

Every finitely generated subgroup of J lies in a subgroup like J_1 (for some U, V). Therefore if we prove that J_1 is subnormal in G with defect bounded independently of the choice of U, V, Lemma 1.7.5 will guarantee that J is subnormal in G. By Lemma 3.4.2, $W = \langle U, V \rangle$ is subnormal in G with defect suitably bounded. Then two applications of Proposition 5.4.2 show that $P = \langle W, D, E \rangle \, sn \, G$. Finally consider $J_1 = \langle L, M, P \rangle$. Since $L \lhd J$, it follows that $LP \, sn \, G$ and since $M \lhd J$ we obtain $J_1 = (LP)M \, sn \, G$.

For the remaining case we can assume that

$$\bar{K}/\bar{K}' \text{ and therefore } \bar{K} \text{ are periodic.}$$

Again, as in the corresponding case of Theorem 5.1.6(i), there are normal subgroups L, D and F of $KN = LDF$, with $N \leqq L \cap D \cap F$, $L \lhd J$, $\bar{D}$ periodic, divisible and abelian and $\bar{F}$ of finite rank at most e. Let $\bar{F}_1 = F_1/N$ be a finite subgroup of $\bar{F}$. As before, there is a subgroup H_1/N of $\bar{H}$, with a finite number (independent of the choice of $\bar{F}_1$) of generators, such that

$$F_1^H \leqq \langle H_1, F_1 \rangle = W_2,$$

say.

By Lemma 3.4.2, W_2 is subnormal in G with defect bounded by some function of the defects of H and K in G and the integer d (from Lemma 5.1.4). If $W_1 = \langle H, F_1 \rangle$, then $W_1 = HW_2$ is subnormal in G, by Theorem 1.2.5. Let $W = \langle H, F \rangle$. Every finitely generated subgroup of W lies in some subgroup W_1 and hence W is subnormal in G, by Lemma 1.7.5. Proposition 5.4.2 now shows that $P = \langle W, D \rangle$ is subnormal in G. Finally since $L \lhd J$, $J = PL$ is subnormal in G, as required. □

Again the subnormal defect of J in G is bounded by some function of the defects of H and K in G and the rank of $H/H' \otimes K/K'$ modulo its maximal periodic divisible subgroup.

It is easy to extend the main results of this chapter from two groups H and K to finitely many. Thus suppose that $n \geqq 2$ and $H_1, \ldots, H_n$ are groups with $H_i/H_i' \otimes H_j/H_j'$ *f.r.p.d.* for all $i \neq j$. If $H_1, \ldots, H_n$ are subnormally embedded in some group G and if $J_i = \langle H_1, H_2, \ldots, H_i \rangle$, then

$$J_i/J_i' \otimes H_{i+1}/H_{i+1}' \quad \text{is } \textit{f.r.p.d.} \tag{6}$$

for $1 \leqq i \leqq n-1$. Therefore, by Theorem 5.1.1(i) and induction on i, it follows that

$$J = J_n \text{ is subnormal in } G.$$

Similarly given positive integers $a_1, \ldots, a_n$, there exists $c \geqq 1$ such that

$$\gamma_c(J) \leqq \gamma_{a_1}(H_1) \cdots \gamma_{a_n}(H_n) \tag{7}$$

(by Theorem 5.4.1, induction on n and (6)). Finally if in addition each H_i is nilpotent, then there exists $f \geqq 1$ such that

$$\mathfrak{j}^f \leqq \mathbb{Z}J\,\mathfrak{h}_1^{a_1} + \cdots + \mathbb{Z}J\,\mathfrak{h}_n^{a_n}. \tag{8}$$

For, we may assume that $\mathfrak{j}_i^{f_i} \leqq \mathbb{Z}J_i\,\mathfrak{h}_1^{a_1} + \cdots + \mathbb{Z}J_i\,\mathfrak{h}_i^{a_i}$, for some $i \geqq 1$, and proceed by induction on i. By (7) we know that J_{i+1} is nilpotent. Therefore, by (6) and Theorem 5.1.6(i), the induction step goes through and (8) follows.

6

GROUPS WITH MANY SUBNORMAL SUBGROUPS

§6.1 Groups in which every subgroup is subnormal

The first case to consider is that of a group each of whose subgroups is normal; such groups are called *Dedekind* groups [28]. Clearly abelian groups are Dedekind groups and the structure of non-abelian Dedekind groups (*Hamiltonian* groups [171]) is given by

Theorem 6.1.1. *A non-abelian group has each of its subgroups normal if and only if it is the direct product of a quaternion group of order* 8 *and a periodic abelian group with no elements of order* 4.

Proof. Suppose first of all that G is non-abelian with all subgroups normal. Let a, b be elements of G such that $[a,b] = c \neq 1$. Let A, B, C be the subgroups of G generated by a, b, c, respectively, and set $H = \langle A, B \rangle$. Then each subgroup of H is normal. This clearly implies $C \leqq A \cap B$ and hence $C \leqq \zeta_1(H)$. Thus H is nilpotent of class 2. Furthermore, $c = a^r = b^s$, say, where r, s are integers, and hence

$$c^s = [a,b]^s = [a,b^s] = [a,c] = 1.$$

It follows that a and b have finite orders m and n, say. We choose a, b so that m and n are minimal. Let p be any prime dividing m. Then

$$1 = [a^p, b] = [a,b]^p = c^p$$

and we deduce at once that m and n are both powers of p—at least p^2, since $1 < C < A$ and $1 < C < B$.

Suppose that $a^{p^\alpha} = c^\gamma$, $b^{p^\beta} = c^\delta$, where p does not divide γ, δ. Then, by replacing a by a^δ and b by b^γ, we may assume that

$$a^{p^\alpha} = b^{p^\beta} = [a,b] = c \neq 1; \quad \text{where } \alpha \geqq \beta \geqq 0.$$

Since H is nilpotent of class 2, it is easy to see that

$$(xy)^n = x^n y^n [x,y]^{-n(n-1)/2}.$$

We exploit this relation as follows: H is clearly generated by a and $a^{-p^{\alpha-\beta}}b = b_1$ and hence the order of b_1 must be at least equal to that of b. Using the relation above, we have

$$b_1^p = a^{-p^{\alpha-\beta+1}} b^p c^{-p(p-1)p^{\alpha-\beta}/2}$$

and since a^p, b^p are in the centre of H we deduce

$$b_1^{p^\beta} = c^{-p^\alpha(p-1)/2}.$$

Therefore $\alpha = \beta = 1$ and $p = 2$ and so H is a quaternion group of order 8. We next show that $G = HD$, where $D = C_G(H)$. Suppose that $x \in G$ and $[x, a] \neq 1$. Then $a^x = a^{-1}$ and hence $[a, bx] = 1$. If $[bx, b] \neq 1$, we have $[abx, b] = 1$ and so $abx \in D$. Thus $x \in DH$, as required.

Suppose now that $y \in D$. Then $[a, by] \neq 1$ and hence by has finite order, by a previous argument. But $[b, y] = 1$ and $b^4 = 1$ so that y has finite order. If y has order 4, then $(by)^4 = 1$ and so $[a, by] = (by)^2 = b^2y^2$. However, $[a, by] = a^2 = b^2$ and therefore we obtain $y^2 = 1$, a contradiction. Thus D has no elements of order 4. Since D has all of its subgroups normal, it must be abelian—it can contain no quaternion subgroup.

The converse is an easy exercise. □

For generalizations see N.N. Vil'jams [147], D. S. Nymann [103] and I. N. Abramovskii [1] and [2]. See also §6.4 on $\mathfrak{T}$-groups.

Once we replace normality by subnormality, the situation is much more complicated, although very significant progress has been made. We have seen in Theorem 2.5.1 that Baer groups, that is groups in which every finitely generated subgroup is subnormal, are locally nilpotent. The largest obvious class of groups in which every subgroup is subnormal is the class of nilpotent groups. Of course if G is nilpotent of class c, then each subgroup of G has subnormal defect at most c. The first main development in this area came when Roseblade in [126] proved conversely that a group is nilpotent if all its subgroups are subnormal of bounded defect.

In order to describe Roseblade's result in more detail, we need to introduce some auxiliary notation. Let d be a non-negative and n a positive integer. Then $\mathfrak{U}_{d,n}$ will denote the class of groups in which every n-generator subgroup is subnormal of defect at most d. By Lemma 1.7.5 it follows that

$$\mathfrak{U}_d = \bigcap_{n=1}^{\infty} \mathfrak{U}_{d,n}$$

is the class of groups in which *every* subgroup is subnormal of defect at most d. We note that $\mathfrak{U}_d \leqq \mathrm{L}\mathfrak{N}$, by Theorem 2.2.4. Roseblade's result is then

Theorem 6.1.2 [126] (Theorem 1). *There exist functions f_1 and f_2 such that for each $d \geqq 0$,*

$$\mathfrak{U}_{d, f_2(d)} \leqq \mathfrak{N}_{f_1(d)}.$$

Although the functions f_1 and f_2 exist, they are defined in such a complicated way as to make reasonable estimates for them very difficult.

It follows at once of course that $\mathfrak{U}_d \leqq \mathfrak{N}_{f_1(d)}$ for all d, and hence we have

Corollary 6.1.3. *A group in which every subgroup is subnormal of defect at most d is nilpotent of class at most $f_1(d)$.*

In [55] Heineken proved that if a group G has all its cyclic subgroups subnormal of defect at most 2, then G is nilpotent of class $\leqq 3$ provided G is not periodic. Later Mahdavianary [92] proved that the assumption that G is not periodic is unnecessary here. (See also Stadelmann [138] and Cappit [22].)

As a run up to the proof of Theorem 6.1.2 we show that it is essentially a result about finite p-groups, which fact will help to shorten the original proof somewhat. Suppose then that $f_1(d)$ and $f_2(d)$ exist such that a finite p-group in $\mathfrak{U}_{d,f_2(d)}$ has nilpotency class at most $f_1(d)$. Let G be any group in $\mathfrak{U}_{d,f_2(d)}$ and let K be any finitely generated subgroup of G. Then $K \in \mathfrak{U}_{d,f_2(d)}$ and hence, in particular, $K \in \mathfrak{U}_{d,1}$, so that every cyclic subgroup of K is subnormal. Therefore K is certainly generated by subnormal abelian subgroups and hence it is a Baer group and, since it is also finitely generated, it is nilpotent by Theorem 2.5.1(ii). Let $\bar{K}$ be any finite homomorphic image of K. Then $\bar{K}$ is the direct product of its Sylow subgroups, each of which is in $\mathfrak{U}_{d,f_2(d)}$ and hence in $\mathfrak{N}_{f_1(d)}$ by our assumption. It follows that $\bar{K} \in \mathfrak{N}_{f_1(d)}$. But K, being a finitely generated nilpotent group, is residually finite, and therefore K is in $\mathfrak{N}_{f_1(d)}$. (See Robinson [122] 5.4.17.) It follows at once that $G \in \mathfrak{N}_{f_1(d)}$ as required.

The next step in the proof of Theorem 6.1.2 is to bound nilpotency class in terms of d and derived length.

Theorem 6.1.4. *There exists a function $f_3(d,m)$ such that for all $d > 0$ and $m \geqq 0$,*

$$\mathfrak{U}_{d,d} \cap \mathfrak{A}^m \leqq \mathfrak{N}_{f_3(d,m)}.$$

It is apparently unknown whether there exists a function $g(d)$ such that $\mathfrak{U}_{d,d} \leqq \mathfrak{N}_{g(d)}$.

Of course $\mathfrak{U}_1$ is the class of Dedekind groups and so $\mathfrak{U}_1 (= \mathfrak{U}_{1,1}) \leqq \mathfrak{N}_2$. Thus we can set $f_3(1,m) = 2$ for all $m \geqq 0$. The major part of the proof of Theorem 6.1.4 consists in establishing the metabelian case ($m = 2$) which we separate as

Theorem 6.1.5. *There exists a function $\beta(d)$ such that for all $d \geqq 0$,*

$$\mathfrak{U}_{d,d} \cap \mathfrak{A}^2 \leqq \mathfrak{N}_{\beta(d)}.$$

In order to deduce Theorem 6.1.4 from Theorem 6.1.5, we suppose inductively that $m > 1$, $f_3(d, m-1)$ is defined, and that $G \in \mathfrak{U}_{d,d} \cap \mathfrak{A}^m$. Then $G/G'' \in \mathfrak{N}_{\beta(d)}$ by Theorem 6.1.5 and by induction G' has class at most $f_3(d,m-1)$. An important result of P. Hall [46] now yields that G is nilpotent of class at most

$$f_3(d,m) = \binom{f_3(d,m-1)+1}{2}\beta(d) - \binom{f_3(d,m-1)}{2}.$$

For any group A and integer $n \geqq 1$, let $A^n = \langle a^n | a \in A \rangle$. In order to prove Theorem 6.1.5 we need, following P. Hall's account (Cambridge lectures 1963/64),

Lemma 6.1.6. *Suppose that A is an abelian normal subgroup of a group G and that $x \in G$. Then $\xi_x : a \mapsto [a, x]$ is an endomorphism of A and if $[A, G'] = 1$, then there exists $\beta_1(d)$ such that*

$$A^{\beta_1(d)} \leqq \zeta_{2^d - 1}(G),$$

provided that $\xi_x^d = 0$ for all $x \in G$.

Proof. The fact that ξ_x is an endomorphism is clear. Let $x, y \in G$ and let $\xi = \xi_x$, $\eta = \xi_y$. Then we may clearly write

$$[a, xy] = a^{\xi + \eta + \xi\eta}$$

and since $[A, G'] = 1$, we have $\xi\eta = \eta\xi$. If $d = 1$, then A lies in the centre of G and we can put $\beta_1(1) = 1$. Hence we may assume inductively that $d > 1$. Then $2d - 2 \geqq d$ and, since $\xi_{xy}^d = 0$, we have

$$(\xi + \eta + \xi\eta)^{2d-2} = 0$$

in the ring of endomorphisms of A.

Now any term of degree d or more in either ξ or η is zero and hence

$$\binom{2d-2}{d-1} \xi^{d-1} \eta^{d-1} = 0.$$

Now let

$$t = \binom{2d-2}{d-1}$$

and set $A_1 = \langle a^{\theta_x} | a \in A,\ x \in G \rangle$ where $\theta_x = t\xi_x^{d-1}$. Then A_1 is a normal subgroup of G and

$$a_1^{\xi_x^{d-1}} = 0$$

for all $a_1 \in A_1$, $x \in G$. By the induction hypothesis we have

$$A_1^b \leqq \zeta_{2^d - 2}(G) = Z,$$

say, where $b = \beta_1(d-1)$. Thus, for any $a \in A$,

$$a^{tb\xi_x^{d-1}} \in Z,$$

so that if $A_2 = A_1^{tb}$, then $a_2^{\xi_x^{d-1}} \in Z$ for all $a_2 \in A_2$. Working in G/Z and noting that $A_2 \lhd G$, a second application of the induction hypothesis shows that for all $a_2 \in A_2$,

$$Za_2 \in \zeta_{2^d - 2}(G/Z).$$

But this certainly means that $a_2^b \in \zeta_{2^d - 1}(G)$. Therefore $a^{tb^2} \in \zeta_{2^d - 1}(G)$. The result follows on defining

$$\beta_1(d) = \binom{2d-2}{d-1}\beta_1(d-1)^2. \quad \square$$

We may now proceed to the

Proof of Theorem 6.1.5. Suppose that $G \in \mathfrak{U}_{d,d} \cap \mathfrak{A}^2$ and set $A = G'$. Let $x \in G$. Then $\langle x \rangle \lhd^d G$ and so $[a, {}_d x] \leqq \langle x \rangle$, whence $[a, {}_{d+1} x] = 1$ for any $a \in A$. It follows at once from Lemma 6.1.6 that

$$A^b \leqq \zeta_{2^d}(G) = Z,$$

say, where $b = \beta_1(d+1)$. Replacing G by G/Z we may assume that $A^b = 1$. Let $x \in G$ and $y = x^k$, where $k = d!b$. Using the notation of Lemma 6.1.6, set $\xi = \xi_x, \eta = \xi_y$. Then

$$\eta = k\xi + \binom{k}{2}\xi^2 + \cdots + \binom{k}{d}\xi^d,$$

since $\xi^{d+1} = 0$. Moreover, b divides all $\binom{k}{i}$, $i = 1, \ldots, d$, and hence $\eta = 0$. In other words

$$[y^k, A] = 1, \qquad \text{for all } y \in G. \tag{1}$$

Let $x_1, \ldots, x_d$ be arbitrary fixed elements of G and set $X = \langle x, x_1, \ldots, x_d \rangle$. Then $X/C_X(A)$ is an abelian group of exponent dividing k and so has order at most k^{d+1}. Furthermore $X' \leqq A$, so X' has exponent dividing b. Also X' is an abelian group generated by the conjugates of commutators in the generators $x, x_1, \ldots, x_d$ and an easy calculation now yields that

$$|X'| \leqq b^{d(d+1)k^{d+1}/2} = \beta_2(d),$$

say.

Set $Y = \langle [a, x, x_1, \ldots, x_d] \mid a \in A \rangle$. Then it is easy to see that Y is normal in G, since G is metabelian and $A = G'$. Now $G \in \mathfrak{U}_{d,d}$ and so $[a, x, x_1, \ldots, x_{d-1}] \in \langle x, x_1, \ldots, x_{d-1} \rangle = H$, say, since H has defect at most d in G. Thus $[a, x, x_1, \ldots, x_d] \in X'$ and so $|Y| \leqq \beta_2(d)$. However, G is locally nilpotent and using the fact that a minimal normal subgroup of a locally nilpotent group is central (see [119], p. 154), an elementary argument shows that

$$Y \leqq \zeta_{\beta_2(d)}(G).$$

Now $\gamma_{d+3}(G)$ is generated by elements of the form $[u_1, u_2, x, x_1, \ldots, x_d] = [a, x, x_1, \ldots, x_d]$, where $a = [u_1, u_2] \in A$. Thus

$$\gamma_{d+3}(G) \leqq \zeta_{\beta_2(d)}(G).$$

Recalling that we have been working modulo Z, we conclude the proof by setting

$$\beta_1(d) = d+2+\beta_2(d)+2^d. \quad \square$$

In order to deduce Theorem 6.1.2 from Theorem 6.1.4, we need to show that there are functions $f(d)$ and $m(d)$ such that $\mathfrak{U}_{d,\,f(d)} \leqq \mathfrak{A}^{m(d)}$. To isolate the key steps involved in doing this, we proceed to see what is required in trying to establish Theorem 6.1.2 by induction. The essential ingredient in the induction step is

Lemma 6.1.7. *Suppose that $G \in \mathfrak{U}_{d,\,r}$ and that H is a t-generator subgroup of G, $t \leqq r$. Then*

$$H_{d-j}/H_{d-j+1} \in \mathfrak{U}_{j,\,r-t}$$

for $0 \leqq j \leqq d$. (Here H_i is the ith term of the normal closure series of H in G.)

Proof. Suppose that X/H_{d-j+1} is an $(r-t)$-generator subgroup of H_{d-j}/H_{d-j+1}. Then $X = YH_{d-j+1}$, where Y has $r-t$ generators. Set $K = \langle H, Y\rangle$. Then K has $t+(r-t) = r$ generators and hence $K \lhd^d G$ by hypothesis. Therefore $K = K_d \lhd^j K_{d-j}$ and furthermore $K \leqq H_{d-j} \leqq K_{d-j}$, so that $K \lhd^j H_{d-j}$. Finally, since $X = YH_{d-j+1} = KH_{d-j+1}$, we obtain $X/H_{d-j+1} \lhd^j H_{d-j}/H_{d-j+1}$, as required. $\square$

In order to prove Theorem 6.1.2, we have to show that $f_1(d)$ and $f_2(d)$ exist with $\mathfrak{U}_{d,\,f_2(d)} \leqq \mathfrak{N}_{f_1(d)}$. If $s = 0$, then we may clearly take $f_1(0) = f_2(0) = 0$ and if $d = 1$ we may take $f_1(1) = 2$ and $f_2(1) = 1$, using the fact that $\mathfrak{U}_{1,1} \leqq \mathfrak{N}_2$. Thus we may assume inductively that $d > 1$ and that the result holds for $d-1$.

We define $m = (d-1)([\log_2 f_1(d-1)]+1)$, [ ] denoting integral part, $f_4(d) = f_3(d, m)+1$ and $f_2(d) = f_2(d-1)+f_4(d)$. Let $G \in \mathfrak{U}_{d,f_2(d)}$ and suppose that H is a subgroup of G with $f_4(d)$ generators. If $0 \leqq j < d$, we have

$$H_{d-j}/H_{d-j+1} \in \mathfrak{U}_{j,\,f_2(d-1)}$$

by Lemma 6.1.7, since $f_2(d)-f_4(d) = f_2(d-1)$. By the induction hypothesis we now get

$$H_{d-j}/H_{d-j+1} \in \mathfrak{N}_{f_1(d-1)}$$

for $0 \leqq j < d$. It follows from this and the definition of m that $(H^G)^{(m)} = H_1^{(m)} \leqq H_d = H$. (Recall that $H_1^{(m)} \leqq \gamma_{2^m}(H_1)$; see [119], p. 51.) Moreover $f_2(d) \geqq d$, so that

$$H^G/(H^G)^{(m)} \in \mathfrak{U}_{d,\,d} \cap \mathfrak{A}^m.$$

By Theorem 6.1.4, $H^G/(H^G)^{(m)}$ is nilpotent of class at most $f_3(d, m) = f_4(d)-1$. Hence

$$\gamma_{f_4(d)}(H^G) \leqq H.$$

Thus if we define $\mathfrak{X}_r$ to be the class of all groups G such that $\gamma_r(H^G) \leqq H$ for each r-generator subgroup H, then

$$G \in \mathfrak{X}_{f_4(d)}.$$

The final element in the proof consists in establishing

Lemma 6.1.8. *There exists a function $f_5(r)$ such that*

$$\mathfrak{X}_r \leqq \mathfrak{N}_{f_5(r)}.$$

For then we take $f_1(d) = f_5(f_4(d))$ and the proof of Theorem 6.1.2 is complete.

Proof of Lemma 6.1.8. By the argument on p. 173 we may assume that G is a finite p-group. Let $G \in \mathfrak{X}_r$ and suppose that H is an r-generator subgroup of G. Then $\gamma_r(H^G) \leqq H$ and hence $H \lhd^{r-1} H^G \lhd G$ so that $H \lhd^r G$. Thus

$$\mathfrak{X}_r \leqq \mathfrak{U}_{r,r}.$$

It follows from Theorem 6.1.4 that it will be sufficient for our purposes if we can show that G is soluble of derived length at most $f_6(r)$ for some function f_6. For in that case G will be nilpotent of class at most

$$f_5(r) = f_3(r, f_6(r)).$$

In order to obtain a bound on the derived length of G, we first show that we can reduce to the case where G is a product of normal abelian subgroups. To see this we define an ascending series

$$1 = G_0 \leqq G_1 \leqq \cdots \leqq G_\alpha \leqq \cdots$$

of normal subgroups of G as follows: $G_{\alpha+1}/G_\alpha$ is the product of all the abelian normal subgroups of G/G_α. Now suppose that $x \in G$. Then $\gamma_r(\langle x \rangle^G) \leqq \langle x \rangle$ so that $[x, \gamma_r(\langle x \rangle^G)] = 1$. It follows that $\gamma_{r+1}(\langle x \rangle^G) = 1$ and hence $\langle x \rangle^G \leqq G_r$.

Thus $G_r = G$ and therefore if we can show that there exists a function $f_7 = f_7(r)$ such that G_{i+1}/G_i is soluble of derived length at most $f_7(r)$, then G is soluble of derived length at most $f_6(r) = rf_7(r)$. Thus we have our reduction to the case where G is a product of normal abelian subgroups.

Choose now a set of generators $\{x_\lambda | \lambda \in \Lambda\}$ for G with $\langle x_\lambda \rangle^G$ abelian and let X be a subgroup of G generated by some r-element subset $\{x_{\lambda_1}, \ldots, x_{\lambda_r}\}$ of these generators. Then $X^G = \langle x_1^G, \ldots, x_r^G \rangle$, a product of r abelian normal subgroups and therefore nilpotent of class at most r by Fitting's theorem. Bringing into play the fact that $G \in \mathfrak{X}_r$, we also have

$$V = \gamma_r(X^G) \leqq X.$$

The rank of X, as an r-generator group of class at most r, is at most $r + r^2 + \cdots + r^r$ (using (1) on p. 162) and so V has number of generators bounded by $r + r^2 + \cdots + r^r$. In addition V is abelian and the product of all such subgroups V is $\gamma_r(G)$.

Recalling that G is a finite p-group, we now need

Lemma 6.1.9. *Let the finite p-group G be a product of normal r-generator abelian subgroups. Then G has nilpotency class $f_8(r) + 1$ at most, where*

$$f_8(r) = r(5r - 1)/2.$$

For the group G of Lemma 6.1.8, this gives $\gamma_{k+1}(\gamma_r(G)) = 1$, where $k = f_8(r + r^2 + \cdots + r^r) + 1$. So G is soluble of derived length at most

$$f_7(r) = 3 + [\log_2 k] + [\log_2 (r - 1)],$$

as required to complete the proof of Lemma 6.1.8.

The proof of Lemma 6.1.9 depends on a property of p-groups of automorphisms of abelian p-groups due to P. Hall (in the case of finite p-groups and then extended to abelian p-groups of finite rank by Baer and Heineken [9]). We only need the finite version here and we record it as

Lemma 6.1.10. *Suppose that A is a finite abelian p-group of rank r. Then any p-subgroup of* Aut *A can be generated by $r(5r - 1)/2$ elements.*

Proof. We shall use additive notation for A throughout the proof. Let P be any p-subgroup of Aut A and let $C = C_P(A/pA)$. Then A/pA is an elementary abelian group of order p^r and P/C is isomorphic with a subgroup of a Sylow p-subgroup of the general linear group $GL(r, p)$ and so $|P/C| \leqq p^{r(r-1)}$ (Zassenhaus [171], p. 142). Hence P/C can be generated by $r(r - 1)/2$ elements. For this reason it is enough to show that C can be generated by $r(5r - 1)/2 - r(r - 1)/2 = 2r^2$ elements.

If $pA = 0$, then $C = 1$ and we are done. So we assume that the exponent p^e of A is greater than p. Suppose that $1 \neq \alpha \in C$. Then there is a largest positive integer $m = m(\alpha)$ such that $m < e$ and α centralizes $A/p^m A$. Let $a_1, \ldots, a_r$ be a basis for A, so that we can write, for $1 \leqq i \leqq r$,

$$a_i\alpha = a_i + p^m \sum_{j=1}^{r} \alpha_{ij} a_j,$$

where the integers α_{ij} are not all divisible by p. Thus we can represent α by the matrix

$$1 + p^m M,$$

where $M = M(\alpha)$ is the $r \times r$ matrix (α_{ij}). Moreover if l is the largest integer such that $lm < e$, we can represent α^{-1} by the matrix

$$1 + \sum_{i=1}^{l} (-p^m M).$$

Now let $\{\alpha_1, \ldots, \alpha_d\}$ be a minimal set of generators for C and set $M_i = M(\alpha_i)$, $m_i = m(\alpha_i)$ and

$$m(\alpha_1, \ldots, \alpha_d) = m_1 + \cdots + m_d.$$

Among all possible minimal generating sets for C, choose $\{\alpha_1, \ldots, \alpha_d\}$ to be one whose m-value $m(\alpha_1, \ldots, \alpha_d)$ is maximal. We assume that the m_i are ordered so that $m_1 \leqq m_2 \leqq \cdots \leqq m_d$. If $d \leqq 2r^2$, then there is nothing more to prove. So assume that $d > 2r^2$.

Suppose now that $m_{r^2+1} = 1$, which means that $m_j = 1$ for all $j \leqq r^2 + 1$, and let i be least such that $M_1, \ldots, M_i$ are linearly independent over $GF(p)$. Then clearly $1 < i \leqq r^2 + 1$ and there exist integers k_j such that

$$M_i \equiv \sum_{j=1}^{i-1} k_j M_j \bmod p.$$

Now set

$$\alpha = \alpha_i^{-1} \alpha_1^k \cdots \alpha_{i-1}^k.$$

Since $1 = m_1 = \cdots = m_i$, the automorphism α is represented by a matrix $1 + pM$, where

$$M = -M_i + \sum_{j=1}^{i-1} k_j M_j \equiv 0 \bmod p.$$

Therefore α centralizes A/p^2A and $m(\alpha) > 1$. But $\alpha_1, \ldots, \alpha_{i-1}, \alpha, \alpha_{i+1}, \ldots, \alpha_d$ generate C and since $m(\alpha) > 1 = m(\alpha_i)$, these elements form a minimal generating set with m-value greater than the maximum $m(\alpha_1, \ldots, \alpha_d)$. This contradiction means that $m_{r^2+1} > 1$. Now $r^2 < d - r^2$, so $M_{r^2+1}, \ldots, M_d$ are linearly dependent mod p. We let t be the least integer subject to the conditions $r^2 + 1 < t \leqq d$ and

$$M_t \equiv \sum_{j=r^2+1}^{t-1} k_j M_j \bmod p$$

for suitable integers k_j. Now set

$$\alpha = \alpha_t^{-1} \prod_{j=r^2+1}^{t-1} \alpha_j^{\kappa_j}, \qquad \text{where } \kappa_j = k_j p^{m_t - m_j},$$

remembering that $m_j \leqq m_t$ if $j \leqq t$. A binomial expansion shows that α is represented by $1 + p^{m_t} M$, where

$$M \equiv -M_t + \sum_{j=r^2+1}^{t-1} k_j M_j \equiv 0 \bmod p.$$

It follows that $m(\alpha) > m_t$ and so $m(\alpha_1, \ldots, \alpha_{t-1}, \alpha, \alpha_{t+1}, \ldots, \alpha_d) > m(\alpha_1, \ldots, \alpha_d)$, a final contradiction. □

We proceed now to deduce Lemma 6.1.9. Suppose that G is a finite p-group which is a product of normal r-generator abelian subgroups. Then there exists a subset $\{x_\lambda \mid \lambda \in \Lambda\}$ such that $X_\lambda = \langle x_\lambda \rangle^G$ is an r-generator abelian group and G is the product of the X_λ. Fix $\lambda \in \Lambda$, set $X = \langle x_\lambda \rangle^G$ and let $H = G/C_G(X)$. By Proposition 6.1.10 H can be generated by $f_8(r) = r(5r-1)/2$ elements. Therefore if $\Phi(H)$ is the Frattini subgroup of H (that is the intersection of the maximal subgroups of H), we have $H/\Phi(H)$ is an elementary abelian p-group with at most $f_8(r)$ generators. Moreover H can be generated by the $x_\mu C_G(X)$, $\mu \in \Lambda$, and hence by $f_8(r)$ of these elements by the Burnside basis theorem. Thus H is a product of $f_8(r)$ abelian normal subgroups and is therefore nilpotent of class at most $f_8(r)$. Hence $\gamma_{f_8(r)+1}(G)$ centralizes $X = X_\lambda$. But the X_λ generate G and the result is immediate. □

The proof of Roseblade's theorem (Theorem 6.1.2) is now complete and so we turn in the next section to the counterexamples constructed by Heineken, Mohamed and Smith which demonstrate the existence of groups which are not nilpotent but have every subgroup subnormal.

§6.2 Non-nilpotent groups with every subgroup subnormal

The Heineken–Mohamed group

Four years after Roseblade's theorem (Theorem 6.1.2) appeared, Heineken and Mohamed constructed a group which shows that the bound on the defects of the subgroups in Roseblade's result cannot be relaxed. Their example has each of its subgroups subnormal, yet the group is not nilpotent and so the subnormal defects are not bounded.

More precisely we have

Theorem 6.2.1 [56]. *There exists a group G with the following properties:*

(HM I) *G' is elementary abelian of exponent p and $G/G' \cong C_{p^\infty}$;*

(HM II) *every proper subgroup of G is subnormal and nilpotent;*

(HM III) *the centre of G is trivial.*

We first describe some consequences of this result. A group G is said to satisfy the *normalizer condition* if $H < N_G(H)$ for all proper subgroups H of G. If G is such a group, then we can define an ascending series $\{N_\alpha(H)\}$ from H by the rules

$$N_0(H) = H, \quad N_{\alpha+1}(H) = N_G(N_\alpha(H)) \text{ and } N_\lambda(H) = \bigcup_{\alpha<\lambda} N_\alpha(H)$$

where α is an ordinal and λ is a limit ordinal. We call this series the *ascending normalizer series* of H in G. Clearly the series reaches G after a finite or infinite number of steps so that H is ascendant in G. Conversely, a proper ascendant subgroup of a group is always distinct from its normalizer. Thus groups with the normalizer condition are precisely those groups in which every subgroup is ascendant. It is known (see [120], p. 2) that all groups satisfying the normalizer condition are locally nilpotent. The Heineken–Mohamed example demonstrates that such groups need not be hypercentral (thus answering Problem 21 of the Kuroš–Černikov Survey [73]). Furthermore, it is easy to see that the direct square of a Heineken–Mohamed group does not satisfy the normalizer condition. (See Problem 17 of [73].)

Some comments on the genesis of the example in Theorem 6.2.1 are in order. Heineken and Mohamed showed that if G is a non-nilpotent soluble group satisfying (HM II), then $G/G' \cong C_{p^\infty}$ for some prime p, G is a countable p-group and $(G')^p \neq G'$. The fact that G must be a p-group means that if a torsion-free soluble group satisfies (HM II), then it is nilpotent. It appears to be an open question whether there exist any torsion-free groups which satisfy (HM II) and yet are not nilpotent.

In [51] Hartley has constructed groups G_n, for each natural number n, which satisfy (HM II) and (HM III) and such that G' is abelian of exponent p^n. He asks whether there exist non-hypercentral groups G for which G' is an abelian p-group of infinite exponent, $G/G' \cong C_{p^\infty}$ and $HG' < G$ for every $H < G$. Such a group would have each of its proper subgroups hypercentral and ascendant.

It was subsequently shown by Heineken and Mohamed [57] and independently by Meldrum [101] that there are $2^{\aleph_0}$ pairwise non-isomorphic groups with the Heineken–Mohamed properties. In [52] Hartley obtained an example of a group G satisfying (HM I), (HM II) and (HM III), where G is a subgroup of the wreath product $C_p \wr C_{p^\infty}$, and used an approach suggested by Meldrum's paper [101] to show that G has $2^{\aleph_0}$ pairwise non-isomorphic quotients with trivial centre.

Theorem 6.2.2 (Hartley [52]). *Let p be a prime and let $W = C_p \wr C_{p^\infty}$. Then W contains a subgroup G with the following properties:*

(i) *$G/G' \cong C_{p^\infty}$ and G' is elementary abelian;*
(ii) *every proper subgroup of G is subnormal and nilpotent;*
(iii) *G has trivial centre; and*
(iv) *G has $2^{\aleph_0}$ pairwise non-isomorphic quotients, each with trivial centre.*

This result clearly supersedes Theorem 6.2.1. We content ourselves with proving (i)–(iii) in detail and sketching (iv), leaving the interested reader to consult Hartley's paper for further information.

Proof. Let $W = C \wr U$, where $C = \langle c \rangle$ is cyclic of order p and

$$U = \langle u_1, u_2, \ldots \mid u_1^p = 1, u_{i+1}^p = u_i, i = 1, 2, \ldots \rangle$$

is a quasicyclic p-group.

If we denote the base group of W by B, then $W = B \rtimes U$ and if R denotes the group ring $\mathbb{Z}_pU$ of U over $\mathbb{Z}_p$, then B can be regarded in the usual way as a (right) R-module, where U acts by conjugation. We write b^ξ for the image of an element $b \in B$ under this action of $\xi \in R$.

We now let

$$a \text{ be an element of } B^{u_2 - 1} \setminus B^{(u_1 - 1)^{p-1}}, \tag{1}$$

for example, $a = c^{u_2 - 1}$, and construct by induction a sequence of elements $a_i \in B^{u_i - 1}$, for $i = 2, 3, \ldots$, which satisfy the relations

$$(u_{i+1}a_{i+1})^p = u_i a_i a^{(u_{i+1} - 1)^{p-1}}, \qquad \text{for } i \geqq 2. \tag{2}$$

We start by taking $a_2 = 1$ and so assume that we have obtained the sequence up to and including a_i. By the induction hypothesis and the fact that $a \in B^{u_2 - 1} \leqq B^{u_i - 1}$, there exists $b \in B$ such that

$$a_i a^{(u_{i+1} - 1)^{p-1}} = b^{u_i - 1} = b^{(u_{i+1} - 1)^p}.$$

For (2) to hold we need

$$a_{i+1}^{(u_{i+1} - 1)^{p-1}} = a_i a^{(u_{i+1} - 1)^{p-1}} = b^{(u_{i+1} - 1)^p}.$$

Hence we can set $a_{i+1} = b^{u_{i+1} - 1} \in B^{u_{i+1} - 1}$. Let $v_i = u_i a_i$, for $i \geqq 2$, so that

$$v_{i+1}^p = v_i a^{(u_{i+1} - 1)^{p-1}}, \qquad i \geqq 2. \tag{3}$$

Note that

$$v_2^{p^2} = 1. \tag{4}$$

We now set $G = \langle v_i \mid i \geqq 2 \rangle$ and show that G has the properties (i)–(iv).

First of all

$$1 = [v_{i+1}, v_{i+1}^p] = [v_{i+1}, v_i a^{(u_{i+1} - 1)^{p-1}}] = [v_{i+1}, a^{(u_{i+1} - 1)^{p-1}}][v_{i+1}, v_i],$$

from which we obtain

$$[v_{i+1}, v_i] = a^{u_i - 1} \tag{5}$$

since $(u_{i+1} - 1)^p = u_i - 1$. Moreover G' is a normal subgroup of G contained in B and so G' is an R-submodule of B. Hence, since $a^{u_i - 1} \in G'$, $i \geqq 2$, we have $G' \geqq a^{\mathfrak{u}}$, where $\mathfrak{u}$ is the augmentation ideal of R and

$$a^{\mathfrak{u}} = \{a^\xi \mid \xi \in \mathfrak{u}\}.$$

On the other hand it follows from (5) that $G' \leqq a^{\mathfrak{u}}$ and so $G' = a^{\mathfrak{u}}$. From (3),

$$v_{i+1}^p \equiv v_i \bmod a^{\mathfrak{u}}, \tag{6}$$

and then from (6) we see that every subgroup of G which properly contains $a^{\mathfrak{u}}$ also contains $u_1 = v_2^p$. Hence, since $u_1 \notin G \cap B$, we have

$$G' = a^{\mathfrak{u}} = G \cap B. \tag{7}$$

Thus (i) of the theorem holds.

In order to establish (ii) we need two preliminary lemmas.

Lemma 6.2.3 [56] (Lemma 2). *Suppose that G is a non-abelian group possessing an abelian normal subgroup N of finite exponent p^k such that $G/N \simeq C_{p^\infty}$. Then*
(a) *G is not hypercentral;*
(b) *every proper subgroup of G is subnormal and nilpotent if and only if there is no proper subgroup H of G such that $HN = G$.*

Proof. We set $D = C_G(\zeta_2(G) \cap N)$. The exponent of G/D does not exceed that of N and since $N \leqq D$, G/D is a homomorphic image of G/N. Therefore $D = G$ and $\zeta_2(G) \cap N \leqq \zeta_1(G)$. If G were hypercentral we would conclude that $N \leqq \zeta_1(G)$ so that G is abelian, a contradiction. Hence (a) is true. If there is no proper subgroup of G which supplements N, then for each $V < G$, VN is an extension of an abelian group of exponent p^k by a cyclic group of order p^r, for some r. Thus VN is nilpotent (Baumslag [11]). Hence V *sn* VN and, since $VN \lhd G$, we have V *sn* G.

Conversely, suppose that every proper subgroup of G is subnormal and nilpotent. Then if $G = HN$ with $H < G$, we have that H is nilpotent and subnormal. But N is an abelian normal subgroup of G and hence G is nilpotent (by Proposition 1.6.3), in contradiction to (a). □

The other result we need is

Lemma 6.2.4 [56]. *The set of all ideals of $\mathbb{Z}_pC_{p^\infty}$ is totally ordered.*

Proof. It is clearly sufficient to show that if u and v are elements of $\mathbb{Z}_pC_{p^\infty}$ and u is not in the ideal generated by v, then v is in the ideal generated by u. Of course, u and v are both contained in the subring R_1 generated by the elements of C_{p^∞} of order p^k for some k. Now $R_1 = \mathbb{Z}_p\langle x \rangle$, for some x, and the ideals of R_1 are ordered since they are just

$$(1-x)^s R_1, \qquad \text{for } 0 \leqq s \leqq p^k,$$

as

$$0 = 1 - x^{p^k} = (1-x)^{p^k}.$$

Therefore v is contained in the ideal of R_1 generated by u and so of course the same holds in $\mathbb{Z}_pC_{p^\infty}$. □

We can now prove Theorem 6.2.2 (ii). By Lemma 6.2.3 it is enough to show that no proper subgroup of G supplements G'. Suppose then that $HG' = G$. Then H contains elements $\{v_k a^{\xi_k} | k = 2, 3, \ldots\}$, where $\xi_k \in \mathfrak{u}$, by the foregoing argument. Hence ξ_k is a sum of products of elements of the form $(1-u_j)$ and therefore is nilpotent in $\mathbb{Z}_pC_{p^\infty}$. Now by (3)

$$v_{k+1}^p = v_k a^{(u_{k+1}-1)^{p-1}},$$

so that we obtain

$$\begin{aligned}(v_{k+1}a^{\xi_{k+1}})^p &= v_{k+1}^p\, a^{\xi_{k+1}(u_{k+1}-1)^{p-1}} \\ &= v_k a^{(1+\xi_{k+1})(u_{k+1}-1)^{p-1}} \in H.\end{aligned}$$

Multiplying by $(v_k a^{\xi_k})^{-1}$ yields

$$a^{(1+\xi_{k+1})(u_{k+1}-1)^{p-1}-\xi_k} \in H.$$

We wish to show that

$$a^{u_k - 1} \in H \tag{8}$$

and we distinguish two cases for this purpose.

Case (i). Suppose that $(u_{k+1}-1)^{p-1}R \neq \xi_k R$. By Lemma 6.2.4 the ideals of $\mathbb{Z}_p C_{p^\infty}$ are nested and so if $\eta, \zeta \in R$ and $\eta R \neq \zeta R$, then the larger is $(\eta - \zeta)R$. Recall that $1 + \xi_{k+1}$ is a unit in R and choose $\eta = (1+\xi_{k+1})(u_{k+1}-1)^{p-1}$ and $\zeta = \xi_k$. Then it follows in particular that $((1+\xi_{k+1})(u_{k+1}-1)^{p-1} - \xi_k)R$ contains $(u_{k+1}-1)^p = u_k - 1$. Since $G' = a^{\mathrm{u}}$,

$$a^{(1+\xi_{k+1})(u_{k+1}-1)^{p-1}-\xi_k} \in G' \cap H \lhd G$$

and so (8) follows.

Case (ii). Now suppose that $(u_{k+1}-1)^{p-1}R = \xi_k R$. We have, writing $q = p^2$,

$$\begin{aligned} v_{k+2}^q &= (v_{k+1}\, a^{(u_{k+2}-1)^{p-1}})^p = v_{k+1}^p\, a^{(u_{k+2}-1)^{p-1}(u_{k+1}-1)^{p-1}} \\ &= v_{k+1}^p\, a^{(u_{k+2}-1)^{p-1}(u_{k+2}-1)^{q-p}} \\ &= v_{k+1}^p\, a^{(u_{k+2}-1)^{q-1}} \\ &= v_k\, a^{(u_{k+1}-1)^{p-1}+(u_{k+2}-1)^{q-1}}. \end{aligned}$$

Consequently

$$\begin{aligned} (v_{k+2}\, a^{\xi_{k+2}})^q &= v_{k+2}^q\, a^{\xi_{k+2}(u_{k+2}-1)^{q-1}} \\ &= v_k\, a^{(1+\xi_{k+2})(u_{k+2}-1)^{q-1}+(u_{k+1}-1)^{p-1}}. \end{aligned}$$

On multiplying this by $(v_k a^{\xi_k})^{-1}$ we obtain

$$a^{(1+\xi_{k+2})(u_{k+2}-1)^{q-1}+(u_{k+1}-1)^{p-1}-\xi_k} \in H.$$

As before consider the ideals generated by

$$(u_{k+2}-1)^{q-1} \qquad \text{and} \qquad (u_{k+1}-1)^{p-1} - \xi_k.$$

The former is $(u_{k+2}-1)^{q-1}R$ and the latter has the form

$$(u_{k+1}-1)^{p-1}\rho R = (u_{k+2}-1)^{q-p}\rho R,$$

for some $\rho \in R$. Since $(u_{k+2}-1)^m = 0$ if and only if $m \geqq p^{k+2}$, these two ideals are distinct. Therefore, as in case (i), the larger ideal is generated by

$$(1+\xi_{k+2})(u_{k+2}-1)^{q-1} + (u_{k+1}-1)^{p-1} - \xi_k$$

and contains

$$(u_{k+2}-1)^q = u_k - 1.$$

From (9) we obtain $a^{u_k - 1} \in H$, as required.

We see, then, that $a^{u_k - 1} \in H$, for all $k \geqq 2$, from which it follows that $a^{\mathfrak{u}} \leqq H$. Hence $G' \leqq H$ and so $G = H$, as required. This completes the proof of (ii).

Our next step is to show that U acts on $a^{\mathfrak{u}}$ faithfully and fixed-point-freely, that is no non-trivial element of $a^{\mathfrak{u}}$ is fixed by all of U. The latter part is immediate since B contains $a^{\mathfrak{u}}$ and $C_B(U) = 1$. If U does not act faithfully on $a^{\mathfrak{u}}$, then we must have

$$1 = a^{\mathfrak{u}(u_1 - 1)} = a^{(u_1 - 1)\mathfrak{u}},$$

so that $a^{u_1 - 1} \in C_B(U) = 1$. Therefore $a \in B^{(u_1 - 1)^{p-1}}$, which is not so by (1). Hence the action is faithful.

Armed with these facts about the action of U, we can proceed. First of all recall that $BG \geqq U$ and so the fixed-point-free action of U on $a^{\mathfrak{u}} = G'$ yields

$$\zeta_1(G) \cap G' = 1. \qquad (10)$$

The faithfulness of the action implies that $C_W(G') = B$, so that

$$C_G(G') = G \cap B = G', \qquad (11)$$

from (7). Putting (10) and (11) together we get $\zeta_1(G) = 1$, and (iii) is proved.

The final part of Theorem 6.2.2, showing that G has $2^{\aleph_0}$ distinct quotients, depends on a deeper analysis of $a^{\mathfrak{u}}$ which yields

Proposition 6.2.5. *There exists a set $\mathscr{J}$ of $2^{\aleph_0}$ ideals J of R such that $J \leqq \mathfrak{u}$ and*
(i) *U acts faithfully and fixed-point-freely on $a^{\mathfrak{u}}/a^J$ for all $J \in \mathscr{J}$; and*
(ii) *the annihilator in R of $a^{\mathfrak{u}}/a^J$ is J.*

We do not prove this result here (see Hartley [51] for the details), but conclude this section by showing how (iv) follows from it.

Let $J \in \mathscr{J}$ as in Proposition 6.2.5. Then a^J is a normal subgroup of G and $a^J \leqq G'$. It follows from Proposition 6.2.5 that

$$C_G(G'/a^J) = G \cap B = G',$$

so that

$$\zeta_1(G/a^J) = 1.$$

Suppose now that J_1 and J_2 are in $\mathscr{J}$ and α is an isomorphism mapping $G_1 = G/a^{J_1}$ onto $G_2 = G/a^{J_2}$. Then α induces an isomorphism of G_1/G_1' with G_2/G_2' which we may combine with the natural isomorphism between U and G_i/G_i', $i = 1, 2$, to give an automorphism θ of U which satisfies

$$(x^u)^\alpha = (x^\alpha)^{u\theta} \quad (x \in G_1', u \in U). \qquad (12)$$

If $u \in R = \mathbb{Z}_p U$, then (12) will continue to hold provided we extend θ to R linearly. It follows that θ maps the annihilator in R of G_1' onto that of G_2', so that we obtain $J_1^\theta = J_2$ from Proposition 6.2.5. Furthermore, if $R_i = \mathbb{Z}_p \langle u_i \rangle$, then θ leaves R_i invariant and so leaves all its ideals invariant since they are just the powers of the unique maximal ideal. Therefore

$$J_1^\theta = \left(\bigcup_{i=1}^{\infty} J_1 \cap R_i \right)^\theta = \bigcup_{i=1}^{\infty} (J_1 \cap R_i)^\theta = \bigcup_{i=1}^{\infty} J_1 \cap R_i = J_1.$$

Hence $J_1 = J_2$ and we are done. □

Hypercentral groups with every subgroup subnormal

Suppose that G is a non-nilpotent hypercentral group with every proper subgroup subnormal and nilpotent. Then there exists $x \in \zeta_2(G) \backslash \zeta_1(G)$ and the map $g \mapsto [x, g]$, $g \in G$, is a homomorphism with non-trivial abelian image. Hence $G' < G$ and so G' is nilpotent. Thus G is soluble and, as pointed out on p. 181, $G/G' \simeq C_{p^\infty}$ and $(G')^p \neq G'$, for some prime p. But then it follows at once from Lemma 6.2.3 (a) that G is abelian, a contradiction. In other words a non-nilpotent group in which every proper subgroup is subnormal and nilpotent cannot be hypercentral. The question naturally arises as to the existence of non-nilpotent hypercentral groups in which every subgroup is subnormal. The *hypercentral length* of a hypercentral group G is the first ordinal ρ such that $\zeta_\rho(G) = G$. Smith in [136] proves

Theorem 6.2.6. *There exists a group G with the following properties:*
(S I) *G is metabelian of rank* 2,
(S II) *every subgroup H of G is subnormal and has finite index in its second normal closure in G,*
(S III) *G is hypercentral of length $\omega + 1$.*

Of course G is a further example of a non-nilpotent group satisfying the normalizer condition. However, here we can say rather more. Suppose that H is a subgroup of an arbitrary group G. Then as on p. 180 we can form the ascending normalizer series of H in G by setting $N_0(H) = H$ and defining inductively $N_{i+1}(H) = N_G(N_i(H))$. Note that if H *sn* G, it is not in general true that this series reaches G after a finite number of steps even for a finite group G (see Camina [19] and [20]). We shall say that a group G satisfies the *strong normalizer condition* if $H \leq G$ implies $N_i(H) = G$ for some finite i. It is not hard to see that the Heineken–Mohamed examples do not satisfy the strong normalizer condition. However, Smith's group does satisfy this condition.

Before constructing Smith's group and deducing its properties, we record a number of results which show that in some sense the example is best possible. It is apparently still an open question whether groups with every subgroup subnormal can be insoluble. However, hypercentral groups with this property must be soluble.

Theorem 6.2.7 (Brookes [17]). *A hypercentral group G with all subgroups subnormal is soluble.*

Proof. We proceed by induction on the hypercentral length ρ of G. If $\rho = 0$, the group is trivial and there is nothing to prove. If $\rho > 0$ but ρ is not a limit ordinal, then $\rho = \sigma + 1$, for some σ, and by induction $\zeta_\sigma(G)$ is soluble. But $\zeta_{\sigma+1}(G)/\zeta_\sigma(G)$ is abelian and so $G = \zeta_{\sigma+1}(G)$ is soluble.

Hence we assume that ρ is a limit ordinal and that all hypercentral groups of hypercentral length less than ρ are soluble if all of their subgroups are subnormal. Let G have hypercentral length ρ and suppose that G is not soluble. We construct a subgroup H of G which is not subnormal and thus obtain a contradiction.

Let $H_1 = 1$, $n_1 = 0$, $r_1 = 0$ and $\alpha_1 = 0$ and construct a set of finitely generated subgroups H_i of G, strictly ascending sequences of integers r_i and n_i and a set of ordinals $\alpha_i < \rho, i > 1$, as follows: suppose that $H_1, \ldots, H_{i-1}$, n_{i-1}, r_{i-1} and α_{i-1} have already been defined. Let $Z_j = \zeta_j(G)$ and define

$$K_i = \bigcap_{n=1}^{\infty} (Z_{\alpha_{i-1}} G^{(n)}).$$

Then the groups $Z_{\alpha_{i-1}} G^{(n)}/Z_{\alpha_{i-1}}$ are just the terms of the derived series of $G/Z_{\alpha_{i-1}}$.

Suppose that G/K_i is soluble. Then $K_i = Z_{\alpha_{i-1}} G^{(n)}$ for some n. If $K_i > Z_{\alpha_{i-1}}$, then $K_i/Z_{\alpha_{i-1}}$ is a non-trivial hypercentral group and as such has a proper derived group $Z_{\alpha_{i-1}} G^{(n+1)}/Z_{\alpha_{i-1}}$. But this contradicts the definition of K_i and hence $K_i = Z_{\alpha_{i-1}}$. However, $Z_{\alpha_{i-1}}$ is soluble by induction and hence G is soluble, a contradiction. Thus G/K_i is not soluble. Therefore $G/Z_{\alpha_{i-1}}$ is not soluble and so neither is the n_{i-1}th derived subgroup $Z_{\alpha_{i-1}} G^{(n_{i-1})}/Z_{\alpha_{i-1}}$. Set $L_{i-1} = Z_{\alpha_{i-1}} G^{(n_{i-1})}$ and (for any s) let φ_s be the natural homomorphism

$$\varphi_s\colon L_{i-1} \to L_{i-1}/Z_{\alpha_{i-1}} L_{i-1}^{(s)}.$$

We show that there exists an s_i such that there is a finitely generated subgroup of defect greater than r_{i-1} in $L_{i-1}^{\varphi_{s_i}}$.

For suppose that this is not the case. Let f_1 and f_2 be the functions defined in Theorem 6.1.2 and put $s = f_1(r_{i-1}) + 1$. By hypothesis all finitely generated, and in particular all $f_2(r_{i-1})$-generated subgroups of $L_{i-1}^{\varphi_s}$ are subnormal of defect no more than r_{i-1}. Therefore by Theorem 6.1.2, $L_{i-1}^{\varphi_s}$ is nilpotent of class at most $f_1(r_{i-1}) = s - 1$, a contradiction, since $L_{i-1}^{\varphi_s}$ has derived length exactly s. Hence there exist s_i and a finitely generated subgroup H_i such that $H_i^{\varphi_{s_i}}$ has defect $r_i > r_{i-1}$ in $L_{i-1}^{\varphi_{s_i}}$. We may of course take H_i to be a subgroup of $G^{(n_{i-1})}$. Let $n_i = n_{i-1} + s_i > n_{i-1}$. Then $\langle H_1, \ldots, H_i \rangle$ is a finitely generated subgroup of G and so there is an $\alpha_i < \rho$ such that $\langle H_1, \ldots, H_i \rangle \leqq Z_{\alpha_i}$. We have now defined H_i, r_i, n_i and α_i. The final step is to show that the subgroup

$$H = \langle H_i | i \geqq 1 \rangle$$

cannot be subnormal in G, and thus obtain a contradiction.

Consider, then, the image $\bar{H}$ of H in $\bar{G} = G/Z_{\alpha_{t-1}} G^{(n_t)}$, for some t. Now

$H_i \leqq Z_{\alpha_{t-1}}$ if $i \leqq t-1$, $H_i \leqq G^{(n_i-1)}$ for all i, and $n_t < n_i$ if $t < i$. Consequently, $H_i \leqq G^{(n_t)}$ if $i \geqq t+1$. So $\bar{H}_i$ is trivial except for $i = t$. Thus $\bar{H} = \bar{H}_t$. Now $\bar{H}_t$ was defined to have defect at least r_t in $\bar{L}_{t-1}$ which means that H has defect no less than r_t in G. But $r_1 < r_2 < \cdots$ is an unbounded sequence and so H cannot be subnormal. □

Brookes deduces Theorem 6.2.7 from a stronger result which Casolo [24] has improved to: *A group, in which every non-trivial section* (*that is quotient of a subgroup*) *is not perfect and in which every subgroup is subnormal, is soluble.*

It is a well-known result of Hall (see [46]) that if $N \lhd G$ and N and G/N' are nilpotent, then G is nilpotent. An easy application of this fact, together with the foregoing theorem, shows that if a hypercentral non-nilpotent group with all subgroups subnormal exists, then there is such a group which is metabelian.

We note from Theorem 6.2.6 that Smith's group has rank 2. In general a locally nilpotent group G has finite rank if and only if the torsion subgroup T of G has finite rank and G/T is nilpotent of finite rank. Such a group is hypercentral (see [120], p. 38). The further fact that a group of finite rank, in which every finitely generated subgroup is subnormal, is nilpotent if it possesses elements of only finitely many distinct prime orders (Gruenberg [39]) means that a non-nilpotent hypercentral group of finite rank with all subgroups subnormal must have elements of infinitely many distinct prime orders. On the other hand such a group cannot be periodic, again in contradistinction to the Heineken–Mohamed group, for we have

Proposition 6.2.8 (Smith [136]). *A periodic group with finite rank and all subgroups subnormal is nilpotent.*

Proof. Suppose that G is such a group and is not nilpotent. By the preceding remarks G is a direct product of p-groups G_p and each G_p is nilpotent. Since G is not nilpotent, the G_p must have unbounded class. Thus, given n, there is a prime $p(n) = p$ such that G_p has class greater than n. It follows from Theorem 6.1.2 that, given m, there is a prime $p(m) = p$ such that G_p has a subgroup H_p of defect at least m. The subgroup H generated by the H_p cannot be subnormal, a contradiction. □

Again we note that Smith's group has hypercentral length $\omega + 1$. That this is best possible (for non-nilpotent groups of finite rank) follows from

Proposition 6.2.9 (Smith [136]). *A hypercentral group of hypercentral length at most ω, which has finite rank and all subgroups subnormal, is nilpotent.*

Proof. Suppose that G satisfies the hypotheses of the proposition. If T is the torsion subgroup of G, then T is nilpotent, by Proposition 6.2.8, and G/T is nilpotent, by Gruenberg's result referred to above. Therefore G is soluble. By Lemma 2.3.5, the upper central series and the derived series of G have isomorphic refinements. Then it is easy to see that all but a finite number of the factors of the upper central series are periodic, that is there is a finite integer i such that $G/\zeta_i(G)$ is periodic and hence nilpotent, again by Proposition 6.2.8. Thus G is nilpotent. □

Construction of Smith's group (Theorem 6.2.6)

For each positive integer n, let p_n denote the nth prime and define $X_n = \langle x_n \rangle$, a cyclic group of order p_n^n. Set $X = \operatorname{Dr}_{n=1}^{\infty} X_n$. Let $Z = \langle z \rangle$ be an infinite cyclic group and put $A = X \times Z$. We now set $G = \langle A, y_1, y_2, \ldots \rangle$ subject to

(R1) $[x_m, y_n] = 1$, for $m \neq n$,
(R2) $[x_n, y_n] = x_n^{p_n}$,
(R3) $y_n^{p_n^{n-1}} = zx_n$,
(R4) $[y_1, y_n] = x_n^{-p_n}$, $\quad n \neq 1$,
(R5) $[y_m, y_n] = x_m^{p_m u_{mn}} x_n^{-p_n u_{nm}}$, $\quad$ for $m \neq n$ and $m, n \neq 1$, where, for $i \neq j$, $u_{ij} p_j^{j-1} \equiv 1 \bmod p_i^i$.

Then G is an extension of A and it is clear that $X \lhd G$, G/X is isomorphic to a subgroup of $\mathbb{Q}$ via $Xy_n \mapsto 1/p_n^{n-1}$, $Xz \mapsto 1$, and G/A is a direct product of cyclic groups of order p_n^{n-1} for different primes p_n. Also rank $G = 2$. Fixing $n > 1$ and adjoining the relations

$$x_m = 1, \qquad \text{for } m \neq n,$$

and

$$zx_n = 1,$$

it is not hard to see by routine calculations that G has a nilpotent image of class exactly n for all n. Thus G is not nilpotent.

We now show that every subgroup H of G is subnormal in G. If $H \cap A \leqq X$, we have $H/H \cap X = H/H \cap A \cong HA/A$ which is periodic. Then, since X is periodic, so too is H and since G/X is torsion-free, this forces $H \leqq X$. Hence $H \lhd X \lhd G$, that is $H \lhd^2 G$. So we may assume that $H \cap A > H \cap X$. Then $xz^r \in H$ for some $x \in X$, $r > 0$, and since x has finite order and A is abelian, we have

$$z^s \in H,$$

for some $s > 0$. We set

$$X^* = \langle x_n | p_n \nmid s \rangle$$

and show that $H \lhd HX^*$.

Let $x_n \in X^*$, $g \in H$. Since G/A is abelian, we may write

$$g = a y_{\alpha_1}^{\beta_1} \cdots y_{\alpha_t}^{\beta_t},$$

where the α_j are distinct and the $\beta_j \neq 0$. If none of the $\alpha_j = n$, then $[x_n, g] = 1$

by (R1). If $n = \alpha_t$, say, then writing $\beta_t = \beta$,

$$[x_n, g] = [x_n, y_n^\beta],$$

which element we now show is contained in H.

From (R3) it follows that, for some p_n'-number u,

$$g^u = a' y_n^\beta,$$

for some $a' \in A$. Then z^s, g^u are in H and therefore so is

$$[z^s, g^u] = [z, g^u]^s.$$

Now

$$\begin{aligned} [z, g^u] &= [z, y_n^\beta] \\ &= [y_n^{\beta p_n^{n-1}} x_n^{-1}, y_n^\beta] \qquad \text{(by (R3))}, \\ &= [x_n^{-1}, y_n^\beta] \in X_n \qquad \text{(by (R2))}, \end{aligned}$$

and, since s and p_n are coprime, we see that

$$[x_n^{-1}, y_n^\beta] \in H.$$

But then

$$[x_n, y_n^\beta] = [x_n^{-1}, y_n^\beta]^{-1} \in H,$$

as required.

Thus $H \lhd X^*H = H^*$, say. Now let X^0 be the subgroup of X generated by all X_m such that p_m divides s. Then $X = X^* \times X^0$. Further let p_d be the largest prime divisor of s. Then it follows from (R1) and (R2), by an easy calculation, that

$$X_e \leqq \zeta_d(G),$$

for all $e \leqq d$. Hence $X^0 \leqq \zeta_d(G)$ and so

$$H^* \lhd^d H^* X^0 = HX \lhd G,$$

since G/X is abelian. We therefore have

$$H \lhd^{d+2} G.$$

We have seen that $X_e \leqq \zeta_d(G)$, for all $e \leqq d$, and it is clear from this that

$$X \leqq \zeta_\omega(G),$$

so that G, being non-nilpotent, has hypercentral length precisely $\omega + 1$, by Proposition 6.2.9.

We next show that G satisfies the strong normalizer condition. If $H \cap A \leqq X$, then exactly as on p. 189, $H \leqq X$, whence $N_1(H) = N_G(H) \geqq A$ and so $N_2(H) = G$. So we may assume that $H \cap A > H \cap X$ and adopt the notation used above. It is clear, by induction on r, that $\zeta_r(G) \leqq N_r(H)$, for $r = 1, 2, \ldots$. Moreover $X^0 \leqq \zeta_d(G)$ and $H \lhd HX^*$. Hence

$$HX = HX^* X^0 \leqq N_1(H)\zeta_d(G) \leqq N_d(H).$$

But G/X is abelian and hence

$$N_{d+1}(H) = G.$$

Our final task is to show that if $H \leqq G$, then $|H_2 : H|$ is finite, where H_2 is the second normal closure of H in G. This is trivial if $H \leqq X$. If not, we have, again with the familiar notation,

$$HX^0 \lhd HX \lhd G.$$

Hence $H_2 \leqq HX^0$. But $|HX^0 : H|$ is finite and the result follows. □

§6.3 Groups in which every subgroup is almost subnormal

We say that a subgroup H of a group G is *almost subnormal* in G if $|H_n : H|$ is finite for some n, where H_n is the nth normal closure of H in G. Under these circumstances $H_r = H_{r+1}$ for some finite r, and we call this subgroup the *subnormal closure* of H in G. It is clearly the smallest subnormal subgroup of G containing H. If r is chosen least such that $H_r = H_{r+1}$ and $|H_r : H| = s$, then we call the ordered pair (r, s) the *near defect* of H in G.

In light of §§6.1 and 6.2, the question arises: what do groups look like if all their subgroups are almost subnormal of bounded near defect—in particular, how far can they be from nilpotent? In [76] Lennox proved that they are finite-by-nilpotent. More precisely we prove

Theorem 6.3.1. *Suppose that G is a group in which each finitely generated subgroup is almost subnormal of bounded near defect. Then G is finite-by-nilpotent. Furthermore there is a function μ such that if $|H_n : H| \leqq m$ for all finitely generated subgroups H of G, then $|\gamma_{\mu(m+n)}(G)| \leqq m!$.*

The case $m = 1$ is essentially Roseblade's theorem (Theorem 6.1.2) and indeed $\mu(m+n) = f_1(m+n)+1$. The case $n = 1$ had been anticipated by I. D. Macdonald in [87]. He proves the stronger result that *if the cyclic subgroups of a group G are of finite index, bounded by m, in their normal closures, then $|G'| \leqq m^k$, where $k = 900 (\log_2 m)^3$.*

The Heineken–Mohamed examples (§6.2) show that even if $m = 1$, the bound n cannot be relaxed. If $n = 1$, the bound m cannot be relaxed, since FC-groups (that is groups in which each element has only finitely many conjugates) are precisely those groups in which each finitely generated subgroup is of finite index in its normal closure, and not all FC-groups are finite-by-nilpotent. We also note that Neumann [102] has shown that if G is a group with $|H_1 : H|$ finite for all (that is not only finitely generated) subgroups H of G, then G' is finite and so G is finite-by-abelian. However, Smith's example (Theorem 6.2.6) shows that there exist groups G which are not finite-by-nilpotent, but have $|H_2 : H|$ finite for all $H \leqq G$. Smith's group is hypercentral and this leads to the question of how nearly hypercentral are groups G such that, for some fixed n, $|H_n : H|$ is finite for all $H \leqq G$.

It is easy to see that a finite-by-nilpotent group G has each of its subgroups almost subnormal of bounded near defect. For, $\gamma_{n+1}(G)$ is finite, for some n, and if $H \leqq G$ and we set $K = H\gamma_{n+1}(G)$, then $K \lhd^n G$, so H_n is contained in K and hence $|H_n : H| \leqq |\gamma_{n+1}(G)|$. Thus Theorem 6.3.1 gives a characterization of finite-by-nilpotent groups. It is fruitful to compare this with a very well-known earlier characterization due to P. Hall. In [45] he showed that *finite-by-nilpotent groups are precisely those groups which have some term of their upper central series of finite index*. Thus if G is such a group with $|G : \zeta_r(G)| = s$, say, and $H \leqq G$, then clearly H is subnormal of defect at most r in a subgroup, namely $H\zeta_r(G)$, of finite index at most s in G. This property also characterizes finite-by-nilpotent groups.

Theorem 6.3.2. *Suppose that G is a group and that there exist integers r, s such that if $H \leqq G$ and H is finitely generated, then H is subnormal of defect at most r in some subgroup of finite index at most s in G. Then G is finite-by-nilpotent and* $|G : \zeta_{\mu(r+s)}(G)| \leqq (s!)^{\lambda(s)}$.

Here $\lambda(s)$ is the number of primes less than s and μ is as in Theorem 6.3.1. It is not known whether $\lambda(s)$ can be replaced by 1.

In the case $r = 1$, the hypothesis amounts to saying that every finitely generated subgroup of G has at most s conjugates. In [102] Neumann proved that *a group in which every subgroup has only a finite number of conjugates has its centre of finite index*. Macdonald then showed in [87] that *if in G no subgroup has more than s conjugates, then $|G : \zeta_1(G)| \leqq s^\alpha$, where* $\alpha = 81(\log_2 s)^2$. Lennox has shown in [76], by giving an alternative proof of Neumann's theorem, that in order to obtain Macdonald's conclusion, it is enough to assume that no finitely generated subgroup of G has more than s conjugates.

We noted in §6.1 that Roseblade's theorem reduces quite rapidly to the finite case. In order to reduce Theorems 6.3.1 and 6.3.2 to the finite case, we need the following weaker version of both of these results for finitely generated groups.

Theorem 6.3.3. *Suppose that G is a finitely generated group and that to each cyclic subgroup X of G there corresponds a series of the form*

$$X = X_n \leqq X_{n-1} \leqq \cdots \leqq X_1 \leqq X_0 = G,$$

where, for each $i = 0, 1, \ldots, n-1$, either X_{i+1} is ascendant in X_i or X_{i+1} is of finite index in X_i. Then G is finite-by-nilpotent.

In the language of Phillips [108], the cyclic subgroups X are *f-ascendant* in G and so Theorem 6.3.3 has the corollary that *a group with every cyclic subgroup f-ascendant is locally finite-by-nilpotent*. We also note two further easy corollaries of Theorem 6.3.3, omitting the simple deductions.

Corollary 6.3.4. *A finitely generated group is finite-by-nilpotent if and only if it has only a finite number of self-normalizing subgroups.*

Corollary 6.3.5. *A finitely generated group is finite-by-nilpotent if and only if each self-normalizing subgroup is of finite index.*

Comparing the forms of Theorems 6.1.2, 6.3.2 and 6.3.3, it is natural to ask whether the first two can be deduced from a result analogous to the last. In this connection we have

Theorem 6.3.6. *Suppose that G is a group such that if H is a finitely generated subgroup of G, then there is a finite series*

$$H = H(\alpha) \leqq \cdots \leqq H(0) = G$$

of length $\alpha = \alpha(G)$ *and where, for some subset* $J \subseteq \{0, 1, \ldots, \alpha-1\}$, *we have* $H(i+1) \lhd H(i)$, *if* $i \notin J$, *and*

$$\sum_{i \in J} |H(i): H(i+1)| \leqq f = f(G).$$

Then G is finite-by-nilpotent and $|G: \zeta_{\mu(1+f)}(G)| \leqq (f!)^{\lambda(f)}$.

The proof is an immediate consequence of the arguments which we shall adduce for Theorem 6.3.2, and we omit it.

Proof of Theorem 6.3.3. Suppose first of all that G is any group (not necessarily finitely generated) in which to each cyclic subgroup X there corresponds a series of the type stated in the hypothesis. Write H *asc* K whenever H is an ascendant subgroup of K. Then

$$X^r \text{ asc } G, \qquad \text{for some } r. \tag{1}$$

(Here X^r is the subgroup generated by the rth powers of the elements of X.) For, X^s *asc* X_1 for some s, by induction on the length of the series. If X_1 *asc* G, we are home, so we may assume that $|G: X_1|$ is finite. Then the core Y of X_1 in G also has finite index in G. But then $Y \cap X^s$ *asc* G and since $Y \cap X^s = X^r$, for some r, we have (1).

It follows from (1) that if R is the *Gruenberg* radical of G (that is the join of the ascendant abelian subgroups of G; see [119], p. 61), then G/R is periodic. The next step is to show that

$$G/R \text{ is locally finite.} \tag{2}$$

In order to do this we may assume that $R = 1$, so that G is periodic. By Corollary 2.2.3 (i), it suffices to show that, with X as above, $\bar{X} = X^G$ is locally finite. We may suppose inductively that $U = X^{X_1}$ is locally finite. If X_1 *asc* G, then U *asc* G and so $U^G = \bar{X}$ is locally finite (see [119], p. 20). Hence we may assume that $|G:X_1|$ is finite and so $|G:Y|$ is finite. Thus $|U:U \cap Y|$ is finite and since $U \cap Y \, asc \, G$, we have $U \cap Y \leqq L$, the locally finite radical of G. Since the class of locally finite groups is P-closed ([133]; see also [119], p. 35), we may suppose that $L = 1$. Then U is finite and $|G:N_G(U)|$ is finite. Hence U^G is finite, by Dietzmann's lemma ([30]; see also [119], p. 45). Therefore $\bar{X} = U^G \leqq L = 1$ and (2) is established.

Suppose now that G satisfies the hypotheses of Theorem 6.3.3. Then G is finitely generated and it follows from (2) that G/R is finite. But then R is finitely generated and since R is locally nilpotent (see [119], p. 61), we have that R is in fact nilpotent. Hence G is nilpotent-by-finite and so satisfies the maximal condition for subgroups, since it is finitely generated. Since the hypotheses are inherited by all homomorphic images of G, we can make a noetherian induction reduction to the case where $1 \neq N \lhd G$ implies that G/N is finite-by-nilpotent, but G is not. Also, since G satisfies the maximal condition, the hypothesis now becomes: if X is a cyclic subgroup of G, then for each $i = 0, 1, \ldots, n-1$, either $X_{i+1} \lhd X_i$ or $|X_i : X_{i+1}|$ is finite.

Let $A = \zeta_1(R)$. Since the periodic part P of A is finite, we have G/P, and hence G, finite-by-nilpotent if $P \neq 1$. Therefore we may assume that A is torsion-free. Let $a \in A$, $x \in G$. We show that

$$[a, {}_r x] = 1, \qquad \text{for some } r, \tag{3}$$

that is a is a *right Engel* element of G. Set $X = \langle x \rangle$. Suppose that, for some i, $j(\neq 0)$, k exist such that $[a^j, {}_k x] \in X_i$. This is clear for $i = 0$. Then if $|X_i : X_{i+1}|$ is finite, we have $[a^j, {}_k x]^t \in X_{i+1}$ for some t. But A is an abelian normal subgroup of G and hence $[a^{jt}, {}_k x] \in X_{i+1}$. If, on the other hand, $X_{i+1} \lhd X_i$, then

$$[a^j, {}_k x, x] = [a^j, {}_{k+1} x] \in X_{i+1}.$$

Thus by induction there exist integers $u(\neq 0)$, v such that $[a^u, {}_v x] \in X_n = X$. Therefore $[a^u, {}_{v+1} x] = 1$. Set $r = v + 1$. Again since A is abelian and normal, we obtain $[a, {}_r x]^u = 1$ and so $[a, {}_r x] = 1$ as A is torsion-free. Thus (3) is established.

Hence a is a right Engel element of G and as such lies in the hypercentre of G by a result of Baer ([7], see also [120], p. 52). (The *hypercentre* of a group is the limit of the upper central series.) Then $A \leqq \zeta_r(G)$, for some r, since G satisfies the maximal condition. Moreover $A \neq 1$, since G is not finite, and so G/A is finite-by-nilpotent. By Hall's theorem, some term Z/A of the upper central series of G/A is of finite index in G/A, so that $[Z, {}_s G] \leqq A$, for some s. But then $[Z, {}_{r+s} G] = 1$ and hence $|G:\zeta_{r+s}(G)|$ is finite. Thus G is a finite-by-nilpotent group, a contradiction. □

Proof of Theorem 6.3.1. We shall use Theorem 6.3.3 to reduce to the finite case. For notational convenience let $\mathfrak{X}(m, n)$ denote the class of groups G all of whose finitely generated subgroups H have $|H_n : H| \leqq m$. Suppose that we have proved the finite version of the theorem. Let $G \in \mathfrak{X}(m, n)$ and suppose that K is an arbitrary finitely generated subgroup of G. Then $K/N \in \mathfrak{X}(m, n)$, for all $N \lhd K$, and in particular if K/N is finite. By the finite version, $|\gamma_d(K/N)| \leqq m!$, where $d = \mu(m+n)$. Hence

$$|\gamma_d(K)N/N| \leqq m!. \tag{4}$$

By Theorem 6.3.3, K is finite-by-nilpotent and since it is also finitely generated, it is nilpotent-by-finite and hence residually finite. By (4), $|\gamma_d(K)| \leqq m!$.

We now select K to be a finitely generated subgroup of G with $|\gamma_d(K)|$ maximal. Let $x \in \gamma_d(G)$. Then $x \in \gamma_d(H)$ for some finitely generated subgroup H of G. If we put $J = \langle H, K \rangle$, then $x \in \gamma_d(J)$. However, $\gamma_d(J) \geqq \gamma_d(K)$ and the maximality condition forces equality. Hence $\gamma_d(G) = \gamma_d(K)$ and the result follows.

It remains to prove Theorem 6.3.1 for finite groups. In order to do this, we need

Lemma 6.3.7. *Let G be a finite group and p be a prime. Then there exists a soluble p'-subgroup H of G such that the subnormal closure of H in G is normal and is the smallest normal subgroup of G modulo which G is a p-group.*

The relevance of this to the investigation in hand is seen in

Corollary 6.3.8. *Suppose that G is a finite group and that all subgroups of G have index at most m in their subnormal closures. Then the smallest term U of the lower central series of G has order at most $m!$.*

For, if p is a prime divisor of $|U|$, Lemma 6.3.7 gives a p'-subgroup H with $U \leqq H_1$ and $|H_1 : H| \leqq m$. Thus if p^α is the highest power of p dividing $|U|$, $p^\alpha \leqq m$, since H is a p'-group. This is the case for all p dividing $|U|$ and so $|U| \leqq m!$, as required. □

Suppose now that G is a finite group satisfying the hypotheses of Theorem 6.3.1. By Corollary 6.3.8, the smallest term of the lower central series of G, $\gamma_d(G)$ say, has order at most $m!$. It remains to bound d and for that purpose we may assume that $\gamma_d(G) = 1$. It follows from the hypotheses and Theorem 6.1.2 that the class of G is at most $f_1(m+n)$ and we may take $d = \mu(m+n) = f_1(m+n) + 1$. Hence the proof of Theorem 6.3.1 is complete subject to Lemma 6.3.7.

Proof of Lemma 6.3.7. Let G be a finite group and p be a prime. We proceed by induction on $|G|$, the result being trivial for $|G| = 1$, and assume the natural induction hypothesis. If G is a p-group, set $H = 1$ and we are done. So we may assume that G has a non-trivial Sylow q-subgroup Q, where q is a prime distinct from p. Let $N = N_G(Q)$. Then $G = NQ_1$ (where $Q_1 = Q^G$), by the Frattini argument, and so the subnormal closure of Q is normal in G, that is it is Q_1.

Now $|N/Q| < |G|$, so there exists a soluble p'-subgroup H/Q of N/Q such that the subnormal closure of H/Q in N/Q is normal in N/Q and is the *p-residual* of N/Q, that is the smallest normal subgroup of N/Q modulo which N/Q is a p-group. Clearly H is a soluble p'-group. We show that H is the subgroup we want. First of all it is evident that the subnormal closure of H in N is normal in N and the subnormal closure of H in $G = NQ_1$ contains Q_1, since $H \geqq Q$. Hence the subnormal closure of H in G is normal in G. Finally suppose that $L \lhd G$ and G/L is a p-group. Then $N/N \cap L$ is a p-group and so $H \leqq N \cap L$. Therefore $H_1 \leqq L$. But $H_1 \geqq Q_1$ and hence $G/H_1 = NQ_1/H^N Q_1$, which is a p-group. Thus H_1 is the p-residual of G. The proof is complete. □

Proof of Theorem 6.3.2. It is convenient to prove first the result for finite groups. This ultimately depends on

Lemma 6.3.9. *Suppose that G is a finite group in which each self-normalizing subgroup has index at most s. Then the index of the hypercentre of G is at most* $(s!)^{\lambda(s)}$.

Proof. Let p be any prime divisor of $|G|$ and S be any Sylow p-subgroup of G. Set $N = N_G(S)$. Then N is self-normalizing and so $|G:N| \leqq s$. By Sylow's theorem, $|G:N| \equiv 1 \bmod p$, so that if $p \geqq s$, we must have $G = N$, that is $S \lhd G$. Hence the intersection of the normalizers of *all* the Sylow p-subgroups of G is the intersection of those with $p < s$.

Now for fixed p, the Sylow p-subgroups of G are all conjugate. It follows that the intersection of the normalizers of these subgroups is just the core of any one of them and hence is of index at most $s!$ in G. Therefore the intersection of the normalizers of all Sylow subgroups of G is of index at most $(s!)^{\lambda(s)}$. But this, by a result of Baer ([4], Corollary 3), is the hypercentre of G. Indeed it is $\zeta_c(G)$, where c is the largest nilpotency class of a Sylow subgroup of G [76]. □

Proof of the finite version of Theorem 6.3.2. Suppose that G is a finite group with every subgroup subnormal of defect at most r in some subgroup of index at most s. Then all self-normalizing subgroups have index at most s and therefore the hypercentre Z of G has index at most $(s!)^{\lambda(s)}$, by Lemma 6.3.9. Let S be a Sylow p-subgroup of G. Then S is nilpotent and has each of its subgroups subnormal of defect at most $r+s$, since S inherits the hypothesis on G. By Theorem 6.1.2, S has nilpotency class less than $\mu(r+s)$. By the remark at the end of the proof of Lemma 6.3.9, $Z = \zeta_{\mu(r+s)}(G)$ and the proof is complete.

We may now finish the

Proof of Theorem 6.3.2. Suppose that G is a group in which every finitely generated subgroup is subnormal of defect at most r in some subgroup of index at most s. Set $\mu = \mu(r+s)$ and $t = (s!)^{\lambda(s)}$. Let L be an arbitrary finitely generated subgroup of G. Then, since the hypotheses are subgroup-inherited, we can apply Theorem 6.3.3 to L and conclude that L is a finite-by-nilpotent group. Since it is also finitely generated, it is residually finite (see p. 195). Let x be any element of the hypercentre Z of L and suppose that $\{M_\lambda\}$ is a set of normal subgroups of finite index in L intersecting in the unit subgroup. Setting $\bar{L} = L/M_\lambda$, we have $\bar{x} \in \zeta_\mu(\bar{L})$, by the last step in the proof of the finite version of the theorem. Therefore

$$[x, {}_\mu L] \leqq \bigcap_\lambda M_\lambda = 1.$$

It follows that $x \in \zeta_\mu(L)$, and so $Z = \zeta_\mu(L)$.

Suppose now that $z \in \zeta_\omega(G)$ and let $g_1, \ldots, g_\mu$ be arbitrary elements of G. By the above argument, if $L = \langle z, g_1, \ldots, g_\mu \rangle$, we have $\zeta_\omega(L) = \zeta_\mu(L)$, and since $z \in \zeta_\omega(G)$ implies $z \in \zeta_\omega(L)$, we have

$$[z, g_1, \ldots, g_\mu] = 1.$$

It follows that $\zeta_\omega(G) = \zeta_\mu(G)$. Furthermore, if L is again any finitely generated subgroup of G, then $L/\zeta_\mu(L)$ is a finite group with trivial centre and so, by the finite version of the theorem, $|L : \zeta_\mu(L)| \leqq t$.

Choose a finitely generated subgroup L of G with $|L : \zeta_\omega(L)|$ maximal. If H is any finitely generated subgroup of G with $H \geqq L$, then

$$|H : \zeta_\omega(H)| \geqq |L : \zeta_\omega(H) \cap L| \geqq |L : \zeta_\omega(L)|,$$

and so, by definition of L, we must have equality throughout. It follows that $H = L\zeta_\omega(H)$ and $\zeta_\omega(L) \leqq \zeta_\omega(H)$, and it is now easy to see that $\zeta_\omega(L) \leqq \zeta_\omega(G)$. But H has the same properties as L, and so, by the same token, $\zeta_\omega(H) \leqq \zeta_\omega(G)$. It is now apparent that $G = L\zeta_\omega(G)$. Then $G/\zeta_\omega(G) \simeq L/\zeta_\omega(L)$ has order at most t as required. □

We conclude the section by pointing out that Mann has proved analogous results for the class of finite-by-locally nilpotent groups, of which the following is an example: *a group G is finite-by-locally nilpotent if and only if there exists a natural number m such that, to each subgroup H of G, there corresponds a subgroup $K \geq H$ with $|K : H| \leq m$ and K is locally ascendant in G.* Here K *locally ascendant* means that, given any finitely generated subgroup L of G, $K \cap L \operatorname{asc} L$.

§6.4 Groups with all subnormal subgroups of bounded defect

In §6.1 we were interested in groups in which every subgroup is subnormal and has bounded defect. Here we turn our attention to the related class $\mathfrak{B}$ of groups in which every subnormal subgroup has bounded defect. Thus if $\mathfrak{B}_n$ denotes the class of all groups in which no subnormal subgroup has defect exceeding n, then we have

$$\mathfrak{B} = \bigcup_{n=0}^{\infty} \mathfrak{B}_n.$$

Clearly nilpotent groups of class c belong to $\mathfrak{B}_c$ and Roseblade's theorem tells us that nilpotent $\mathfrak{B}_n$-groups have class bounded in terms of n. All simple groups are $\mathfrak{B}$-groups, indeed $\mathfrak{B}_1$-groups, and hence it is usual in studying $\mathfrak{B}$-groups to impose some extra condition (solubility for example) in order to ensure that there is some texture to the subnormal structure of the group.

The class $\mathfrak{B}_1$: groups in which normality is a transitive relation

We consider first the class $\mathfrak{B}_1$. This is obviously the class $\mathfrak{T}$ (defined on p. 73) of all groups in which the relation of normality is transitive (or equivalently of all groups G for which the Wielandt subgroup $w(G) = G$). Such groups have been subjected to fairly intensive investigation since Best and Taussky [13] showed that a finite group, all of whose Sylow subgroups are cyclic, is a $\mathfrak{T}$-group. We have already seen that (i) direct products of non-abelian simple groups and (ii) groups satisfying Min-*sn* with no proper subgroups of finite index are $\mathfrak{T}$-groups (Proposition 2.2.9 and (11) on p. 73, respectively). It is clear from the definition that normal subgroups and homomorphic images of $\mathfrak{T}$-groups are $\mathfrak{T}$-groups. However, it is not true in general that subgroups of $\mathfrak{T}$-groups are $\mathfrak{T}$-groups. Indeed, as remarked above, every simple group is a $\mathfrak{T}$-group, whereas any group can be embedded in a simple group. Finite groups with all their subgroups $\mathfrak{T}$-groups have been studied in [1]. We can, however, characterize $\mathfrak{T}$-groups as follows.

Proposition 6.4.1. *A group G is a $\mathfrak{T}$-group if and only if every finite subset of G is contained in a $\mathfrak{T}$-subgroup of G.*

This result, which was proved by Robinson [110], means that the class $\mathfrak{T}$ is L-closed. It follows from the fact that $\mathfrak{T}$-groups can be characterized in terms of the normal closures of their cyclic subgroups:

Lemma 6.4.2. *A group G is a $\mathfrak{T}$-group if and only if for each cyclic subgroup X of G we have $Y = X^Y$, where $Y = X^G$.*

Proof. If G is a $\mathfrak{T}$-group and X is a cyclic subgroup of G, then $X^Y \lhd Y \lhd G$ and so $X^Y \lhd G$ by transitivity. Hence $Y = X^Y$. Conversely, suppose that the condition is satisfied in a group G. It is sufficient to show that $H \lhd G$ follows from $H \lhd K \lhd G$. Let $x \in H$ and $X = \langle x \rangle$. Then $Y = X^G \leq K$. So $Y = X^Y \leq H$. Hence $H \lhd G$. □

Proof of Proposition 6.4.1. The necessity is clear. Suppose that every finite subset of a group G is contained in a $\mathfrak{T}$-subgroup of G. Let $x \in G$, $X = \langle x \rangle$ and

$Y = X^G$. If $g \in G$, then by hypothesis there is a $\mathfrak{T}$-subgroup H of G which contains both x and g. Let $Z = X^H$. Then $x^g \in Z = X^Z$, by Lemma 6.4.2, and so $x^g \in X^Y$. Since g is arbitrary, we have $Y = X^Y$ and so, again by Lemma 6.4.2, G is a $\mathfrak{T}$-group. □

Since every subgroup of a nilpotent group is subnormal, nilpotent $\mathfrak{T}$-groups have every subgroup normal and hence are precisely the Dedekind groups of Theorem 6.1.1. In particular they are of class at most 2. We use this fact to show that $\mathfrak{T}$-groups possess a unique maximal normal nilpotent subgroup. In fact we have

Lemma 6.4.3 [110]. *Let G be a $\mathfrak{T}$-group and let $L = \gamma_3(G)$. Then $C = C_G(G')$ is the unique maximal nilpotent normal subgroup of G. Also $C = C_G(L)$ and C is a Dedekind group.*

Proof. First of all $[C', C] \leqq [G', C] = 1$, so C is nilpotent and since it is normal in the $\mathfrak{T}$-group G, it is a Dedekind group. Let N be any nilpotent normal subgroup of G and let $x \in N$. If $X = \langle x \rangle$, then we have $X \text{ sn } G$ so that $X \lhd G$. Now $G/C_G(X) \hookrightarrow \text{Aut } X$, so that $G' \leqq C_G(X)$ since $\text{Aut } X$ is abelian. Hence $x \in C$ and so $N \leqq C$. Therefore C is the unique maximal normal nilpotent subgroup of G. Moreover $C_G(L)$ is nilpotent, since G/L is, and thus $C_G(L) \leqq C$. But $C \leqq C_G(L)$ and hence $C = C_G(L)$. □

Armed with this fact, we may now prove that the class of soluble $\mathfrak{T}$-groups is restricted.

Theorem 6.4.4 [110]. *Soluble $\mathfrak{T}$-groups are metabelian.*

Proof. If the result is false, then it is clear, since $\mathfrak{T}$ is Q-closed, that there is a counterexample G with $1 \neq G''$ abelian. It follows from Lemma 6.4.3 that $[G'', G'] = 1$ and hence G' is nilpotent and of course normal in G. Therefore $G' \leqq C_G(G')$, again by Lemma 6.4.3, and so $G'' = 1$, a contradiction. □

We note in passing that the above theorem yields

Corollary 6.4.5. *Soluble $\mathfrak{T}$-groups are locally supersoluble.*

The fact that finite soluble $\mathfrak{T}$-groups are supersoluble was pointed out by Gaschutz in [36].

Proof. Suppose that G is a soluble $\mathfrak{T}$-group and let H be any finitely generated subgroup of G, with generators $x_1, x_2, \ldots, x_n$. Let $c_{ij} = [x_i, x_j]$. Then since G' is abelian (by Theorem 6.4.4), we have $\langle c_{ij} \rangle \lhd G$. Now H' is generated by the c_{ij} and it follows easily that H is supersoluble. □

As a further useful consequence of Theorem 6.4.4 we derive the fact that any $\mathfrak{T}$-group has a maximal normal soluble subgroup, which we identify in

Theorem 6.4.6. *Suppose that G is a $\mathfrak{T}$-group. Then the derived series of G terminates with G'' and $S = C_G(G'')$ is the unique maximal soluble normal subgroup of G. Moreover S is metabelian and $[S, G', G'] = 1$.*

Proof. $G/G^{(n)}$ is a soluble $\mathfrak{T}$-group for $n \geqq 2$. Hence it is metabelian by Theorem 6.4.4, and so G'' is the terminal of the derived series. Suppose that H is any soluble normal subgroup of G. Then H' is an abelian normal subgroup of G and hence, by Lemma 6.4.3, is contained in $C_G(G')$. Thus $[H', G'] = 1$. The same argument applied to G/H' yields $[H, G'] \leqq H'$ and therefore we have $[H, G', G'] = 1$. By Lemma 4.3.4, $[G'', H] = 1$, so that $H \leqq S = C_G(G'')$. On the other hand S is soluble and normal in G and the result follows. □

Suppose now that G is any $\mathfrak{T}$-group and that S is the maximal soluble normal subgroup of G as given by Theorem 6.4.6. Then G has the series $G \geqq G''S \geqq S \geqq 1$, with $G''S/S$ perfect and S metabelian. Note further that $G''S/S$ has no non-trivial abelian normal subgroups. Suppose that G is finite and, for the moment, that we know about finite soluble $\mathfrak{T}$-groups. We focus attention therefore on the perfect $\mathfrak{T}$-group $G''S/S$ whose description is given by the following result.

Theorem 6.4.7. *A finite perfect $\mathfrak{T}$-group G which has no non-trivial abelian normal subgroups is a direct product of non-abelian simple groups and thus is semisimple.*

Proof. We may clearly assume that G is directly indecomposable. Let $R = S_1 \times \cdots \times S_n$ be the socle of G, that is the product of the minimal normal subgroups of G. Here the S_i are non-abelian simple subgroups (see Theorem 2.2.10). Moreover $G/C_G(S_i)$ embeds in $\operatorname{Aut} S_i$ and writing $K_i = S_i C_G(S_i)$, since $K_i/C_G(S_i)$ corresponds to the inner automorphisms of S_i under this map, it is clear that G/K_i embeds in $\operatorname{Out} S_i$, the group of outer automorphisms of S_i. By a famous conjecture of Schreier, which is seen to be true as a result of the classification of all finite simple groups, $\operatorname{Out} S_i$ is soluble. Hence $G/K_i = 1$, since it is both soluble and perfect. Therefore $G = K_i = S_i \times C_G(S_i)$. The result follows by the direct indecomposability of G. □

Finite soluble $\mathfrak{T}$-groups were first classified by Gaschütz [36]. (For another approach see Zacher [170].) This classification depends on the special way in which a $\mathfrak{T}$-group acts by conjugation on a normal abelian subgroup and we begin by considering this situation. Suppose that G is any $\mathfrak{T}$-group and that A is a normal abelian subgroup of G. Then every subgroup of A is normal in G and so the action of G on A by conjugation gives rise to automorphisms of A which leave each subgroup of A fixed. Such an automorphism clearly maps each element of A to one of its powers and for this reason is called a *power automorphism of A*. We denote the set of all such by Paut A. Then

$$\operatorname{Paut} A = \{\alpha \mid \alpha \in \operatorname{Aut} A \quad \text{and} \quad H^\alpha = H \text{ for all } H \leqq A\}.$$

The information on power automorphisms which we need in order to classify all finite soluble $\mathfrak{T}$-groups is contained in

Proposition 6.4.8. *Suppose that A is an abelian group and $\alpha \in \operatorname{Paut} A$.*

(i) *If A is not periodic, then α is either the identity or inverts all elements of A. Thus $|\operatorname{Paut} A| = 2$.*

(ii) *If A is a p-group of finite exponent, then there is a positive integer l such that $a^\alpha = a^l$, for all $a \in A$. If α is non-trivial and has order prime to p, then α acts fixed-point-freely on A.*

Proof. (i) Suppose that a is an element of infinite order in A and that b is an arbitrary element of A. Then $a^\alpha = a^l$, $b^\alpha = b^m$ and $(ab)^\alpha = (ab)^n$ for some integers l, m, n, where $l = \pm 1$. Since α is an automorphism, we have $(ab)^\alpha = a^\alpha b^\alpha$, which entails $(ab)^n = a^l b^m$. Hence $a^n b^n = a^l b^m$. If $\langle a \rangle \cap \langle b \rangle = 1$, then this means that $a^n = a^l$ and $b^n = b^m$. But a has infinite order and so $n = l$. Therefore $b^\alpha = b^l$. On the other hand, if $\langle a \rangle \cap \langle b \rangle \neq 1$, then $1 \neq a^r = b^s$, for some r, s. Applying α to this equation, we get

$$a^{rl} = b^{sm} = a^{rm},$$

so that $l = m$ and again $b^\alpha = b^l$. Since $l = \pm 1$, the result follows.

(ii) An abelian p-group of finite exponent is a direct product of cyclic groups, since a basic subgroup must be the whole group (see p. 156). Thus we may choose $\langle a \rangle$ to be a cyclic direct factor of maximal order and $\langle b \rangle$ to be any cyclic direct factor of the complement of $\langle a \rangle$ in A. There are integers l, m and n such that $a^\alpha = a^l$, $b^\alpha = b^m$ and $(ab)^\alpha = (ab)^n$. Therefore $a^n b^n = a^l b^m$ and if a and b have orders p^r and p^s, respectively, then we must have

$$l \equiv n \bmod p^r, \qquad m \equiv n \bmod p^s.$$

By choice of a, $s \leqq r$ and so $l \equiv m \bmod p^s$. Hence $b^\alpha = b^l$, as required.

Finally suppose that $\alpha \neq 1$ and that α has a fixed point c. Then $c^l = c \neq 1$ and so $l \equiv 1 \bmod p$. Hence

$$l^{p^{e-1}} \equiv 1 \bmod p^e,$$

where p^e is the exponent of A. But this means that $\alpha^{p^{e-1}} = 1$. □

We now state and prove the theorem of Gaschütz.

Theorem 6.4.9. *Suppose that G is a finite soluble $\mathfrak{T}$-group. Then $L = \gamma_3(G)$ is the smallest term of the lower central series of G, it is abelian and G/L is a Dedekind group. Furthermore $|L|$ is odd and is coprime to $|G:L|$, so that L has a complement in G.*

Proof. $G/\gamma_4(G)$ is a nilpotent $\mathfrak{T}$-group and so has class at most 2. Hence $L = \gamma_4(G)$. Since G is metabelian by Theorem 6.4.4, L is abelian. We now show that $|L|$ is odd. Clearly L^2 (the subgroup generated by the squares of the elements of L) is normal in G and we consider the $\mathfrak{T}$-group G/L^2. The elements of G act as power automorphisms on L/L^2. However, this quotient is an elementary abelian 2-group and hence the only power automorphism is the identity. Thus $[L, G] \leqq L^2$. But $L = [L, G]$ so $L = L^2$. Therefore $|L|$ is odd.

For the last part we show that G/L has no elements of order a prime p dividing $|L|$. The existence of the complement will then follow from the Schur–Zassenhaus theorem [72]. Note that p must be odd. Let S/L be the Sylow p-subgroup of G/L and $L_{p'}$ be the p-complement of L, that is the join of the q-components of L for all $q \neq p$. Then $S/L_{p'}$ is a p-group. It is therefore a Dedekind group and since it has odd order, it must be abelian, by Theorem 6.1.1. Let $C = C_G(L_p)$, where L_p is the p-component of L. Since $L_p \leqq S$ and $S/L_{p'}$ is abelian, we have $[S, L_p] \leqq L_p \cap L_{p'} = 1$. Hence $S \leqq C$. This means that G/C is a p'-group and since L_p is abelian, G/C is isomorphic with a subgroup of $\operatorname{Paut} L_p$. If $G = C$, then $L = [L, G] = [L_{p'}, G] \leqq L_{p'} < L$, a contradiction. Hence we can choose $x \in G \backslash C$ and consider the action of x on L_p by conjugation. By Proposition 6.4.8 this action is $a \mapsto a^m$, where $m \not\equiv 1 \bmod p$. Moreover L_p and $L/L_{p'}$ are isomorphic as $\langle x \rangle$-modules and so x induces in $S/L_{p'}$ a power automorphism $a \mapsto a^n$, where $n \not\equiv 1 \bmod p$. Hence $S = [S, \langle x \rangle] L_{p'}$. But S/L is in the centre of the Dedekind group G/L since p is odd. Therefore $[S, \langle x \rangle] \leqq L$ and it now follows that $S = L$. □

Thus far we have that if G is a finite soluble $\mathfrak{T}$-group, then G has an abelian normal subgroup L, of odd order, with a nilpotent complement (in G) which is a Dedekind group of order coprime to that of L, acting on L as a subgroup of $\operatorname{Paut} L$. We now use these facts to construct all finite soluble $\mathfrak{T}$-groups.

Suppose that A is a finite abelian group of odd order and B is a finite Dedekind group of order coprime to that of A. Suppose further that there is given a homomorphism $\theta: B \to \operatorname{Paut} A$, such that for each prime p dividing $|A|$, there is an element $b_p \in B$ with b_p^θ acting non-trivially on the p-component of A. Form the split extension

$$G(A, B, \theta) = A]B.$$

This group is clearly soluble. That it is a $\mathfrak{T}$-group follows from

Lemma 6.4.10. *Let N be a normal subgroup of a finite group G such that*
(i) *G/N is a $\mathfrak{T}$-group,*
(ii) *H sn N implies that $H \lhd G$ and*
(iii) *$|N|$ and $|G:N|$ are coprime.*
Then G is a $\mathfrak{T}$-group.

Proof. Suppose that $H \lhd K \lhd G$. We show that $H \lhd G$. Now $H \cap N$ sn N and so $H \cap N \lhd G$, by (ii). We may pass to factor groups and assume that $H \cap N = 1$. Thus $(|H|, |N|) = 1$. Let $M = K \cap (HN) = H(K \cap N)$. Then $M \lhd G$, by (i), and $H \lhd M$ since $H \lhd K$. Therefore if π is the set of all prime divisors of $|G:N|$, then H is the unique maximal normal π-subgroup of M and so is normal in G. □

Thus $G(A, B, \theta)$ is a $\mathfrak{T}$-group and conversely, by Theorem 6.4.9, every finite soluble $\mathfrak{T}$-group is a $G(A, B, \theta)$. Note that if $G = G(A, B, \theta)$, then $A = [A, G] = \gamma_3(G)$, by the construction.

As previously noted, not every subgroup of a $\mathfrak{T}$-group is a $\mathfrak{T}$-group. However, the structure theorem of Gaschütz shows that finite soluble $\mathfrak{T}$-groups are well-behaved in this respect.

Theorem 6.4.11. *Subgroups of finite soluble $\mathfrak{T}$-groups are $\mathfrak{T}$-groups.*

Proof. Let $L = \gamma_3(G)$ and H be a subgroup of G. By Theorem 6.4.9, $|L|$ and $|G:L|$ are coprime and therefore so are $|H \cap L|$ and $|H:H \cap L|$. Furthermore $H/H \cap L \cong HL/L \leq G/L$, this last quotient being a Dedekind group. Hence $H/H \cap L$ is a $\mathfrak{T}$-group. Finally, subgroups of $H \cap L$ are normal in G and so, *a fortiori*, in H. Therefore H is a $\mathfrak{T}$-group by Lemma 6.4.10. □

A different characterization of finite soluble $\mathfrak{T}$-groups, in terms of Sylow subgroups, is given by Robinson in [115]. He proves that *a finite group G, in which every subgroup of a Sylow p-subgroup P is normal in $N_G(P)$, is a finite soluble $\mathfrak{T}$-group, and conversely*. This leads to the further fact that *a finite group G is a soluble $\mathfrak{T}$-group if and only if each of its p-subgroups is pronormal in G* [105]. For this we recall that a subgroup H of a group G is *pronormal* in G if, for any $g \in G$, H and H^g are conjugate in their join. Rose proved that a finite group has each of its p-subgroups pronormal if and only if every subgroup of a Sylow p-subgroup P is normal in $N_G(P)$, and this result provides the link. We note in passing that pronormality and subnormality imply normality.

The classification of infinite soluble $\mathfrak{T}$-groups is much more complicated than the finite case, and indeed is not complete, although considerable progress has been made. Robinson shows in [110] that countable periodic soluble $\mathfrak{T}$-groups, like finite soluble $\mathfrak{T}$-groups, split over the third term of the lower central series, and this fact, together with a more detailed description of power automorphisms of periodic abelian groups, leads to a complete classification. If a soluble $\mathfrak{T}$-group has elements of infinite order, this tends to have a great effect on the structure of the group. Robinson distinguishes groups G of types I and II according to whether $C_G(G')$ is non-periodic or periodic. He proves

Theorem 6.4.12. *Suppose that the group G has an abelian subgroup C and an element $z \notin C$ such that*

(i) $|G:C| = 2$,

(ii) *for all* $c \in C$, $c^z = c^{-1}$ *and*

(iii) $\langle z^2, C^2 \rangle = \langle z^2, C^4 \rangle$.

Then G is a soluble $\mathfrak{T}$-group; if C is not periodic, then G is of type I. *Conversely every non-abelian soluble $\mathfrak{T}$-group of type* I *has this structure.*

Proof. We show first that a group with the hypotheses of the theorem is a $\mathfrak{T}$-group. Let $H \lhd K \lhd G$. Every subgroup of C is normal in G, so we may assume that H contains an element zc, $c \in C$. Then by (ii), $[C, zc] = [C, z] = C^2 \leqq K$ and so $[C^2, zc] = C^4 \leqq H$. Hence H contains $\langle (zc)^2, C^4 \rangle$. However, $(zc)^2 = zczc = z^2$, by (ii), and so $H \geqq \langle z^2, C^2 \rangle$, by (iii). But $G' = [C, \langle z \rangle] = C^2$ and so $H \lhd G$. Thus G is a $\mathfrak{T}$-group, G is clearly soluble and, provided C is not periodic, $C = C_G(G')$, so that G is of type I.

Conversely, suppose that G is a non-abelian soluble $\mathfrak{T}$-group of type I and set $C = C_G(G')$. Then C is a Dedekind group, by Lemma 6.4.3, and so C is abelian since it is non-periodic. Let c be an element of infinite order in C, $d \in C$ and $z \in G$. Then $\langle c \rangle \lhd G$ and $\langle d \rangle \lhd G$, so that $c^z = c^r$ and $d^z = d^s$ for some integers r, s. If $\langle c \rangle \cap \langle d \rangle \neq 1$, then $c^l = d^m$, where l and m are non-zero. Hence $c^{lr} = d^{ms} = c^{ls}$, from which we get $r = s$ and $d^z = d^r$, since c has infinite order. Suppose that $\langle c \rangle \cap \langle d \rangle = 1$. Now $\langle cd \rangle \lhd G$ and therefore, for some integer t, $(cd)^z = (cd)^t$. Hence $c^r d^s = c^t d^t$ and $c^r = c^t$, $d^s = d^t$. This implies that $r = t$ and $d^z = d^r$. Therefore $d^z = d^r$ for all $d \in C$.

Since G is not abelian, we can choose $z \in G \backslash C$. Then $[z, C] \neq 1$, otherwise $[z, G'] = 1$ and $z \in C$, a contradiction. Hence $r \neq 1$ and since c has infinite order, we must have $r = -1$. Thus $d^z = d^{-1}$ for all $d \in C$. From $G' \leqq C$ we have $|G : C| = 2$ and $G = \langle z, C \rangle$. Furthermore $z^2 \in C$ and $(z^2)^z = z^{-2}$. Therefore $z^4 = 1$. Also $[C, z] = C^2$ and $[C^2, z] = C^4$, so that $\langle z \rangle^G = \langle z, C^2 \rangle$ and $\langle z \rangle^{\langle z, C^2 \rangle} = \langle z, C^4 \rangle$. But G is a $\mathfrak{T}$-group and so $\langle z, C^4 \rangle = \langle z, C^2 \rangle$. □

As a consequence of this result, we show that the class of finitely generated soluble $\mathfrak{T}$-groups is surprisingly small; indeed non-abelian groups of this type are finite.

Theorem 6.4.13 [110]. *Finitely generated soluble $\mathfrak{T}$-groups are finite or abelian.*

Proof. Suppose that G is a non-abelian finitely generated soluble $\mathfrak{T}$-group. Then G is supersoluble, by Corollary 6.4.5, and so every subgroup of G is finitely generated. Let $C = C_G(G')$ and suppose that C is not periodic. Then G is a soluble $\mathfrak{T}$-group of type I. Let $G = \langle z, C \rangle$ as in Theorem 6.4.12, and let T be the periodic subgroup of C. Then $T \neq C$ and C/T is free abelian of finite rank. However, $z^2 \in T$ since $z^4 = 1$ and hence, by Theorem 6.4.12, $(C/T)^2 = (C/T)^4$, which contradicts the freeness of C/T. Therefore C is periodic and even finite, being finitely generated. But $G' \leqq C = C_G(G')$ and so $|G : C|$ is finite. It follows that G is finite, as required. □

Soluble $\mathfrak{T}$-groups of type II are much more difficult to handle and in fact their classification is incomplete (see [110]).

Robinson also studies soluble $\overline{\mathfrak{T}}$-groups, that is soluble groups all of whose subgroups are $\mathfrak{T}$-groups. The non-abelian ones are periodic and are precisely the periodic soluble $\mathfrak{T}$-groups with the property that the 2-component $L_2 = 1$

and $\pi(L) \cap \pi(G/L)$ is empty, where $L = \gamma_3(G)$. Note that Theorem 6.4.11 shows that the classes $\mathfrak{T}$ and $\bar{\mathfrak{T}}$ coincide for finite soluble groups. Robinson also proves that a locally soluble $\bar{\mathfrak{T}}$-group is soluble and therefore metabelian.

To round off our discussion of $\mathfrak{T}$-groups, we mention that Robinson in [118] has classified the finite minimal non-$\mathfrak{T}$-groups, that is the finite groups which are not $\mathfrak{T}$-groups, but all of whose proper subgroups are $\mathfrak{T}$-groups. Such groups are all soluble and fall into 7 natural types. In [121] Robinson has also studied *just non-$\mathfrak{T}$-groups*, that is groups which are not $\mathfrak{T}$-groups, but all of whose proper homomorphic images are $\mathfrak{T}$-groups. The soluble just non-$\mathfrak{T}$-groups are classified (9 types) and in addition all soluble just non-$\bar{\mathfrak{T}}$-groups and all finite just non-$\bar{\mathfrak{T}}$-groups are given. Finally we note that *a finitely generated soluble group with all its finite homomorphic images $\mathfrak{T}$-groups is itself a $\mathfrak{T}$-group* [121].

Soluble $\mathfrak{B}_2$-groups

In Theorem 6.4.4 we saw that soluble groups in the class $\mathfrak{B}_1$ have derived length at most 2, and it might be hoped, perhaps in view of Theorem 6.1.2 bounding the class of nilpotent $\mathfrak{B}_n$-groups in terms of n, that there is some bound on the derived lengths of soluble $\mathfrak{B}_n$-groups. However, this is not the case and Robinson constructs examples of iterated complete wreath products which are torsion-free $\mathfrak{B}_2$-groups of derived length n, for any prescribed n (see p. 221). Also Hawkes [54] shows that any finite soluble group can be embedded in a finite soluble $\mathfrak{B}_3$-group.

However, in 1975 Camina and Renouf (unpublished) proved that a finite soluble $\mathfrak{B}_2$-group G has Fitting length at most 7. It follows from Theorem 6.1.2 that the derived length is also bounded. Indeed (as already observed on p. 173) Mahdavianary [92] has shown that nilpotent $\mathfrak{B}_2$-groups have class at most 3. Using this result and the argument of McCaughan and Stonehewer [86], we can prove

Theorem 6.4.14. *A finite soluble $\mathfrak{B}_2$-group has derived length at most* 8 *and Fitting length at most* 5.

The following example [86] shows that derived length 5 and Fitting length 4 are attainable. Let V be a vector space of dimension 2 over $GF(3)$, $H \cong GL(2, 3)$ and form $G = V]H$ in the natural way. Since all the non-trivial subgroups of H contain the central involution z and $H/\langle z \rangle$ is isomorphic to the symmetric group of degree 4, it follows that $H \in \mathfrak{B}_2$. Similarly if K *sn* G and $K \nleq V$, then $z \in VK$ and therefore $V = [V, z] \leq K$. Hence $G \in \mathfrak{B}_2$. Theorem 6.4.16 below shows that 5 and 4 are indeed the upper bounds of derived and Fitting lengths respectively of finite soluble $\mathfrak{B}_2$-groups.

We point out in passing that, in contrast to the class of finite soluble $\mathfrak{B}_1$-groups, the group $GL(2, 3)$ shows that the class of finite soluble $\mathfrak{B}_2$-groups is not subgroup-closed. For, the Sylow 2-subgroup of $GL(2, 3)$ is the semi-dihedral group of order 16, having a subgroup of order 2 and defect 3.

Theorem 6.4.14 is deduced from a property of chief factors of finite soluble $\mathfrak{B}_2$-groups.

Theorem 6.4.15. *Let F be the Fitting subgroup of a finite soluble $\mathfrak{B}_2$-group G. Then every chief factor of G avoided by F has rank at most* 2.

(The *Fitting* subgroup is the nilpotent radical and a factor H/K is *avoided* by a subgroup F if $H \cap F \leqq K$.)

Note first of all that, to within isomorphism, the chief factors referred to in the theorem are precisely those which lie above F in any chief series through F. For, if H/K is a chief factor of G with $F \leqq K < H$, then F avoids H/K. Conversely, if H/K is a chief factor of G avoided by F, then

$$FH/FK \overset{G}{\cong} H/(FK \cap H) = H/(F \cap H)K = H/K.$$

Thus FH/FK is a chief factor of G.

Deduction of Theorem 6.4.14 *from Theorem* 6.4.15. Suppose that G is a finite soluble $\mathfrak{B}_2$-group, F is the Fitting subgroup of G and $X = G/F$. The Fitting subgroup of a finite group is the intersection of the centralizers of the chief factors of the group (see [63]). Therefore the Fitting subgroup F_1 of X is the intersection of the centralizers of the chief factors H/K of X and each H/K has rank at most 2, by Theorem 6.4.15. By a theorem of Zassenhaus, it follows that the derived length of $X/C_X(H/K)$ is bounded. Indeed the derived length is at most 4 and the Fitting length at most 3 (see [31]). Hence the same bounds apply to X/F_1. It follows that G has Fitting length at most 5. By Mahdavianary's result, F and F_1 are metabelian, so that G has derived length at most 8.

Proof of Theorem 6.4.15. We suppose that the result is false and take G to be a counterexample of least order. Let M be a minimal normal subgroup of G. Then M is abelian, $M \leqq F$ and G/M is not a counterexample. Also if F_1/M is the Fitting subgroup of G/M, then $F_1 \geqq F$. Now it is easy to see, using the Jordan–Hölder theorem and the observation following the statement of the present theorem, that there is an "offending" chief factor of G (that is one with rank $\geqq 3$ and avoided by F) lying between F and F_1. Now M is an elementary abelian q-group, for some prime q, and if we denote the Sylow q-subgroup of F_1/M by Q/M, then Q is a normal q-subgroup of G and therefore, being nilpotent, is contained in F. Hence F_1/F is a q'-group.

Thus, for some prime $p \neq q$, there is an offending p-chief factor between F and F_1. Let P/M be the Sylow p-subgroup of F_1/M, so that $P \lhd G$ and there is an offending chief factor between F and FP. Hence there is an offending chief factor between $F \cap P$ and P. Let $C = C_P(M)$. Then $C \lhd G$ and we claim that

$$M = F \cap P = C. \tag{1}$$

For, since F is nilpotent and M is abelian, we certainly have $M \leqq F \cap P \leqq C$. On the other hand, suppose that $M < C$. If P_0 is the Sylow p-subgroup of C, then $1 \neq P_0 \lhd G$. Denote the Fitting subgroup of G/P_0 by F_2/P_0. Then $F_2 \cap P/P_0$ is nilpotent and so, since P/M is a p-group, $(F_2 \cap P)/(M \cap P_0) = F_2 \cap P$ is also nilpotent. Therefore $F_2 \cap P \leqq C \leqq F \cap P$, since C is nilpotent. Hence F_2 avoids $P/F \cap P$. But G/P_0 is not a counterexample to the theorem and there is an offending chief factor between $F \cap P$ and P. This contradiction shows that $M = C$ and (1) holds.

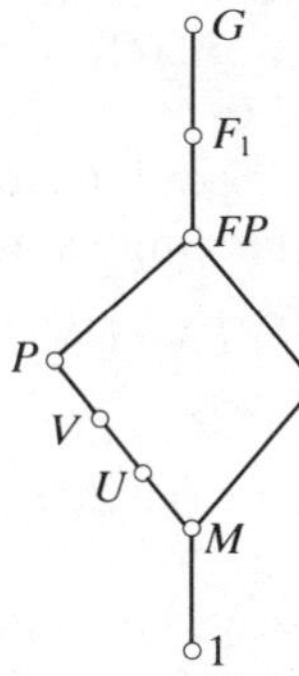

It is now evident that all chief factors of G of rank $\geqq 3$ which lie between M and P are offending and we choose one such chief factor V/U with $|V|$ as small as possible. Then

$$V = \langle v^G \rangle, \qquad \text{for all } v \in V \backslash U. \tag{2}$$

For, let $X = \langle v^G \rangle$. By (1), $1 \neq [X, M] \leqq X \cap M$ and therefore $X \geqq M$. Then

$$V/U = XU/U \overset{G}{\cong} X/X \cap U,$$

and so this last quotient is an offending chief factor between M and P. Hence $X = V$, by choice of V.

We now show that

$$V/M \text{ is an elementary abelian } p\text{-group.} \tag{3}$$

In order to see this, we choose $W \geqq M$ with W minimal such that V/W is an elementary abelian p-group. Then $W \leqq U$. Suppose that $W > M$. Let W/Y be a chief factor of G with $Y \geqq M$. We prove that

$$\text{there is an element } v \in V \backslash U \text{ with } v^p \in Y. \tag{4}$$

Accepting this for the moment, we have $\langle v^G \rangle = V$, by (2). Let bars denote subgroups and elements modulo Y. Then $\langle \bar{v} \rangle$ *sn* $\bar{G}$. But $\bar{G} \in \mathfrak{B}_2$ and hence $\langle \bar{v} \rangle \lhd \bar{V}$. It follows from (4) that $\bar{V}$ is an elementary abelian p-group. However, this contradicts our choice of W and so (3) holds.

We must now prove (4). There are two cases, according to whether p is odd or even.

Case 1. *p is odd.* Define a map $\theta: \bar{V} \to \bar{W}$ by $x \mapsto x^p$. Since $\bar{V}$ has class at most 2, $(xy)^p = x^p y^p$, for all $x, y \in \bar{V}$, and so θ is a homomorphism. By choice of V/U, $\bar{W}$ has rank at most 2. But V/U has rank at least 3 and so the kernel of θ cannot be contained in $\bar{U}$. Hence (4) holds.

Case 2. $p = 2$. Let V_0/W be a complement to U/W in V/W. It follows from (2) that each subgroup of $\bar{V}_0$ which does not lie in $\bar{W}$, being subnormal of defect at most 2 in $\bar{G}$, is normal in $\bar{V}$ and therefore in $\bar{V}_0$. On the other hand, $\bar{W}$ is clearly central in $\bar{V}_0$ (even in $\bar{P}$) and so $\bar{V}_0$ is a Dedekind group. Suppose that (4) is false, so that all elements of order 2 in $\bar{V}$ lie in $\bar{U}$. Then all elements of order 2 in $\bar{V}_0$ lie in $\bar{W}$. If $\bar{V}_0$ is abelian, this implies that

$$r(V_0/W) \leqq r(\bar{W}) \leqq 2,$$

whereas $r(V_0/W) = r(V/U) \geqq 3$. If $\bar{V}_0$ is not abelian, then by Theorem 6.1.1 it is either the quaternion group Q_8 or the direct product of Q_8 and a group of order 2. In each case $r(V_0/W) \leqq 2$, giving the same contradiction as before.

We have now reached the final stage of the proof. Let V_1 be a complement to M in V and set $U_1 = U \cap V_1$. Choose $v \in V_1 \setminus U_1$ with $C_M(v) = M_0$, say, as large as possible. Since an irreducible abelian group of linear maps is cyclic (see for example [63]) and $r(V_1/U_1) \geqq 3$, we have

$$M_0 \neq 1. \tag{5}$$

Setting $M_1 = [M, v]$, we obtain the V-decomposition $M = M_0 \times M_1$. Let D be any subgroup of M_0. Then

$$D \lhd V. \tag{6}$$

For, if $H = \langle v \rangle$, then $DM_1H \lhd MH \lhd^2 G$, since $[M, H] = M_1$, and so $DM_1H \lhd^2 G$. Therefore $DM_1H \lhd (DM_1H)^G = V$, by (2). Thus

$$DM_1H \cap M_0 = D \lhd V$$

as required.

Hence V acts on M_0 as a group of power automorphisms. By Proposition 6.4.8, the power automorphisms of M_0 are precisely the maps $a \mapsto a^r$, $0 < r < q$, and so form a cyclic group. Therefore $V/C_V(M_0)$ is cyclic. Set $C_1 = C_V(M_0)$. By (1), $M_0 \neq M$. Hence by (5) and the fact that M is a minimal normal subgroup of G, there is an element $g \in G$ such that $M_0^g \neq M_0$. Now $C_1^g = C_V(M_0^g)$ and V/C_1^g is cyclic. Therefore

$$r(V/(C_1 \cap C_1^g)) \leqq 2$$

and $C_1 \cap C_1^g \nleqq U$. Thus there is an element belonging to $V_1 \setminus U_1$ which centralizes $\langle M_1, M_0^g \rangle$, contradicting the choice of v. □

In [23] Casolo has improved Theorem 6.4.14 to

Theorem 6.4.16. *A finite soluble $\mathfrak{B}_2$-group has derived length at most* 5 *and Fitting length at most* 4.

Casolo's method depends on a more delicate arithmetic analysis of the orders of chief factors of such a group, based in part on the argument of Theorem 6.4.15. Thus let G be a finite soluble $\mathfrak{B}_2$-group, put $F_0 = 1$ and define F_i/F_{i-1} ($i \geqq 1$) inductively to be the Fitting subgroup of G/F_{i-1}. Suppose that F_1 is a p-group for some prime p. Then Casolo shows that, for all primes q not dividing $p-1$, all q-chief factors of G, lying between F_1 and F_2, are cyclic. This result leads in turn to the fact that if G is any finite soluble $\mathfrak{B}_2$-group, then F_3/F_2 is abelian and G/F_2 is supersoluble and metabelian. Thus G has Fitting length at most 4. From the example given after the statement of Theorem 6.4.14, it is clear that Casolo's result is best possible.

It is convenient to set a wider discussion of the class $\mathfrak{B}$ in the context of the somewhat larger class of groups with the subnormal intersection property. To this we turn in the next section.

§6.5 Groups with the subnormal intersection property

It is immediate from Proposition 1.1.2(ii) that in any group the intersection of finitely many subnormal subgroups is again subnormal. However, the conclusion is in general false for intersections of infinitely many subnormal subgroups. For example, if $G = \langle a, x \mid a^x = a^{-1}, x^2 = 1 \rangle$, the infinite dihedral group, and if $A_n = \langle a^{2^n}, x \rangle$, then A_n *sn* G since $G/\langle a^{2^n} \rangle$ is a finite 2-group. But $\langle x \rangle = \bigcap_{n=1}^{\infty} A_n$ is not subnormal in G.

We use the symbol $\mathfrak{S}_\infty$ to denote the class of groups with the *subnormal intersection property*, that is groups in which the intersection of an arbitrary collection of subnormal subgroups is again subnormal. The fact that $\mathfrak{B} \leqq \mathfrak{S}_\infty$ follows at once from Proposition 1.1.2(ii) and the situation is made particularly transparent by the next result.

Lemma 6.5.1 [119]

(i) *A group G belongs to the class $\mathfrak{S}_\infty$ if and only if the normal closure series of every subgroup of G becomes stationary after a finite number of steps.*

(ii) *A group G belongs to the class $\mathfrak{B}$ if and only if the normal closure series of every subgroup of G becomes stationary after a bounded finite number of steps.*

Notation. In order to avoid confusion in this section, we shall denote the nth term of the normal closure series of a subgroup H of a group G by $H^{G,n}$ and not by H_n as elsewhere in the book.

Proof of Lemma 6.5.1. (i) Suppose that the condition is satisfied and let $(K_\lambda)_{\lambda\in\Lambda}$ be a set of subnormal subgroups of G. Let $K = \bigcap_{\lambda\in\Lambda} K_\lambda$. By hypothesis, $K^{G,n} = K^{G,n+1} = \cdots$, for some $n = n(K)$. Now each K_λ is subnormal in G, so $K_\lambda = K_\lambda^{G,n(K_\lambda)}$ and therefore we have

$$K^{G,n} \leqq \bigcap_\lambda K_\lambda^{G,n(K_\lambda)} = \bigcap_\lambda K_\lambda = K.$$

Hence $K = K^{G,n}$ and K *sn* G.

Conversely suppose that G is an $\mathfrak{S}_\infty$-group and $H \leqq G$. Then $H^{G,n}$ *sn* G, for each n, and hence $H^{G,\omega} = \bigcap_{n=1}^{\infty} H^{G,n}$ is subnormal in G. But this means that, for some n,

$$H^{G,\omega} = (H^{G,\omega})^{G,n} \geqq H^{G,n} \geqq H^{G,\omega}.$$

Therefore $H^{G,n} = H^{G,\omega}$, so that $H^{G,n} = H^{G,n+1} = \cdots$.

(ii) Suppose that $H \leqq G$ implies that $H^{G,n} = H^{G,n+1} = \cdots$, where n is independent of H. Then if H *sn* G, we have $H = H^{G,n}$ and hence $H \lhd^n G$. Thus $G \in \mathfrak{B}_n$. The converse is clear. □

However, $\mathfrak{B} \neq \mathfrak{S}_\infty$. This is the content of

Proposition 6.5.2. *Suppose that p is a prime, $X \cong C_p$ and $Y \cong C_{p^\infty}$. Then $G = X \wr Y$ is an $\mathfrak{S}_\infty$-group which does not belong to the class $\mathfrak{B}$.*

Proof. Let B be the base group of G. Then $G = BY$, $B \lhd G$ and $B \cap Y = 1$. Now B is an elementary abelian p-group and hence if $y \in Y$, then regarding $y - 1$ as an endomorphism of B we get

$$(y-1)^p = y^p - 1. \tag{1}$$

Suppose that y has order p^m. Then, by (1), $(y-1)^{p^m} = 0$ and $(y-1)^{p^{m-1}} \neq 0$. But since $\langle y \rangle^G = \langle y \rangle^B$, we see that $\langle y \rangle$ *sn* G with defect exceeding p^{m-1}. The fact that Y contains elements of arbitrarily high order p^m shows that $G \notin \mathfrak{B}$.

In order to show that $G \in \mathfrak{S}_\infty$, we use Lemma 6.5.1(i). Suppose then that $H \leqq G$. If $BH < G$, then BH/B is cyclic and so there is an element $y \in Y$ such that $H \leqq \langle B, y \rangle \lhd G$. For some m, $(y-1)^{p^m} = 0$ as before, and so $\langle B, y \rangle$ is nilpotent, from which it follows that H is subnormal in $\langle B, y \rangle$ and hence in G. Thus the normal closure series of H becomes stationary after a finite number of steps. Hence we may assume that $G = BH$. Our objective will be to show that $[B, H] = [B, {}_pH]$, so that $[B, H] = [B, H, H]$ and therefore

$$H^G = H^B = H^{B,2} = H^{G,2}$$

as required. To achieve this, it is sufficient to show that any element $z = [b,h]$, where $b \in B$, $h \in H$, belongs to $[B,_pH]$.

Let $h = b'y$, $b' \in B$, $y \in Y$. Then $z = [b,y]$. Since Y is divisible, there is an element $y' \in Y$ such that $(y')^p = y$. By (1) we have

$$z = [b,(y')^p] = b^{(y')^p - 1} = b^{(y'-1)^p} = [b,_py'].$$

But $G = BH$, and so $y' = b''h'$, for some $b'' \in B$, $h' \in H$, and then

$$z = [b,_ph'] \in [B,_pH]. \quad \square$$

We note that this example has appeared on p. 72, where it is shown not to belong to the class $\mathfrak{S}^\infty$. Thus $\mathfrak{S}_\infty \nleq \mathfrak{S}^\infty$.

In connection with a natural question arising from Proposition 6.5.2, Robinson has given a complete solution to the problem of when the wreath product of two nilpotent groups is contained in either of the classes $\mathfrak{B}$ and $\mathfrak{S}_\infty$.

Theorem 6.5.3. *Suppose that H and K are nilpotent groups. Let $W = H \wr K$ and $\bar{W} = H \bar{\wr} K$.*

(A) *If W or $\bar{W}$ belongs to the class $\mathfrak{S}_\infty$, then*

(A1) *for each prime $p \in \pi(K)$, there is an integer $i = i(p) \geqq 0$ such that $H^{p^i} = H^{p^{i+1}} = \cdots$, and*

(A2) *for all but a finite number of primes p in $\pi(K)$, $H = H^p$; and if p is one of these exceptional primes, then the p-component of K satisfies the minimal condition for subgroups.*

In the case where $W \in \mathfrak{S}_\infty$, we have in addition

(A3) *either $H = 1$ or K is periodic.*

(B) *If W or $\bar{W}$ belongs to $\mathfrak{B}_n$, then for any prime $p \in \pi(K)$ for which $H \neq H^p$, we have*

(B1) *the p-component of K is finite and*

(B2) *$p \leqq n$.*

Conversely, if H and K satisfy (A1) *and* (A2), *then $\bar{W}$ belongs to $\mathfrak{S}_\infty$; if* (B1) *is also satisfied, then $\bar{W}$ belongs to $\mathfrak{B}$. If H and K satisfy* (A1), (A2) *and* (A3), *then W is in $\mathfrak{S}_\infty$; if further* (B1) *holds, then W belongs to $\mathfrak{B}$.*

The proofs of these results are to be found in [114].

It has already been mentioned (p. 205) that the subgroups of finite soluble $\mathfrak{B}_3$-groups account for all finite soluble groups. A consequence of Theorem 6.5.3 confirms the complexity of the class of soluble $\mathfrak{B}$-groups. Indeed *any soluble group of derived length n can be embedded in a soluble $\mathfrak{B}_n$-group of the same derived length*; also *for each integer $n \geqq 0$ there exists a torsion-free soluble $\mathfrak{B}_2$-group of derived length n* [114].

The metabelian group $C_p \wr C_{p^\infty}$ of Proposition 6.5.2 is not finitely generated. That this is no accident is shown by the fact that, for finitely generated soluble groups, the classes $\mathfrak{B}$ and $\mathfrak{S}_\infty$ coincide with the class of finite-by-nilpotent groups. This is a consequence of the fact that finite-by-nilpotent groups are $\mathfrak{B}$-groups (p. 192) and of

Theorem 6.5.4. *A finitely generated soluble group belongs to the class $\mathfrak{S}_\infty$ if and only if it is finite-by-nilpotent.*

The proof of this result of Robinson [112] is given in his book [122] and so we do not reproduce it here.

There have been several attempts to study the structure of rather wider classes of soluble groups with the subnormal intersection property. For example, McCaughan [83] proves that *the classes $\mathfrak{B}$ and $\mathfrak{S}_\infty$ coincide for soluble minimax groups.* (The class of *minimax* group is $\mathbf{P}(\mathfrak{Max} \cup \mathfrak{Min})$.) For further results in this area, see McDougall [88], McCaughan [82] and McCaughan and McDougall [84].

7
CRITERIA FOR SUBNORMALITY

To know that a group possesses a non-trivial proper subnormal subgroup is of course to know that the group is not simple and therefore criteria for the subnormality of a subgroup are of considerable importance in the study of the normal structure of a group.

§7.1 Permutability

Historically the first criteria for the subnormality of a subgroup of a group arose in the attempt to see what implication there was for the normal structure of a group G when G possessed a *permutable* (or *quasinormal*) subgroup, that is a subgroup H such that $HK = KH$ for all subgroups K of G. We denote this relation by H *per* G. Of course all normal subgroups are permutable. Not all permutable subgroups are, however, normal. The smallest example is the group $G = A]H$ where A is a cyclic group of order 8 with generator a say and H is a cyclic group of order 2 with generator h and the action is given by $h: a \mapsto a^5$. It is not hard to see that H is a non-normal permutable subgroup of G. (See Maier [93] for a family of such examples.) We record the fact that Huppert [62] has shown that if $G = AB$ where A, B are cyclic p-groups (p an odd prime), then every subgroup of G is permutable.

However, certain permutable subgroups are normal.

Theorem 7.1.1 (Ore [104]). *A maximal permutable subgroup of a group is normal.*

Proof. Suppose H is a maximal permutable subgroup of a group G which is not normal. Then $K \neq H$ for some conjugate K of H. By the maximality, $G = HK$, since HK *per* G, and so $K = H^{hk}$ for some $h \in H, k \in K$. But this gives the contradiction $H = K$. □

This theorem leads at once to a subnormality result for finite groups.

Theorem 7.1.2 (Ore [104]). *A permutable subgroup of a finite group is subnormal.*

Proof. Suppose that H *per* G. Then we may construct a chain $H = H_0 < H_1 < \cdots < H_n = G$ where H_i is a maximal permutable subgroup of H_{i+1}. Hence $H_i \lhd H_{i+1}$ by Theorem 7.1.1 and thus H *sn* G. □

Corollary 7.1.3. *A finite simple group cannot have a proper non-trivial permutable subgroup.*

Our main objective in this section is to prove (Theorem 7.1.12) that this criterion for simplicity is valid for all groups. We point out at once that subnormal subgroups are not, in general, permutable. To see this suppose that G is a dihedral group of order 8 generated by subgroups H and K of order 2. Then $HK \neq KH$ since $|HK| = 4$ and $G \neq HK$. However, G is nilpotent and so both H and K are subnormal.

In fact permutability is a very much stronger property than subnormality in finite groups. We recall that if H is a subgroup of a group G, then H_G, the intersection of the conjugates of H in G, is called the *core* of H in G. Also H is said to be *core-free* if $H_G = 1$. In [64] Itô and Szép proved

Theorem 7.1.4. *A core-free permutable subgroup H of a finite group G is nilpotent.*

In order to prove this, Itô and Szép first showed that H is soluble. However, the solubility of H is not dependent on the finiteness of G as was shown much later by Gross [37].

Theorem 7.1.5. *Let H be a permutable subgroup of a group G. Then if H is core-free and subnormal in G of defect n ($\geqq 1$), H is soluble of derived length at most $n-1$.*

That all derived lengths can occur is shown by Stonehewer in [143].

Proof. If $n = 1$ there is nothing to prove as $H = 1$. So suppose that $n > 1$ and that the natural induction hypothesis holds. Assume that $H^{(n-1)} \neq 1$. Then, since H is core-free, there is an $x \in G$ such that $H^{(n-1)} \nleqq H^x$. If M is the core of H in $H\langle x \rangle$, then $H\langle x \rangle / M$ is a counterexample to the theorem. We may therefore assume that $G = H\langle x \rangle$. Let L be the $(n-1)$th term of the normal closure series of H in G. Since $H \leqq L$ we have $L = H\langle y \rangle$ for some $y \in \langle x \rangle$ so that

$$[G, \langle y \rangle] = [\langle x \rangle H, \langle y \rangle] = [H, \langle y \rangle] \leqq [H, L] \leqq H.$$

But $[G, \langle y \rangle] \lhd G$ and so y is in the centre of G. Hence L *per* G and if K is the core of L in G, we have by the induction hypothesis that $L^{(n-2)} \leqq K$. Clearly $y \in K$ so that $K = \langle y \rangle (H \cap K)$ and $H \cap K \lhd K$. Therefore $K' \leqq H \cap K$. However, $K' \lhd G$ so $K' = 1$. Hence $H^{(n-1)} \leqq L^{(n-1)} \leqq K' = 1$, a contradiction. □

Proof of Theorem 7.1.4. Suppose G is not a p-group and let $p_i^{\alpha_i}$, $p_i^{\beta_i}$ be the orders of the Sylow p_i-subgroups of G and H respectively. Then $\alpha_i > \beta_i$, all i. For if not, suppose $\alpha_i = \beta_i = \alpha$ ($\geqq 1$), say, for some i and set $p_i = p$. Let G_p be a

Sylow p-subgroup of G. Then $HG_p = G_pH$ and a consideration of the order of a Sylow subgroup shows that $G_p \leqq H$. Hence all the Sylow p-subgroups of G are contained in H and so $\text{core}_G(H) \neq 1$, a contradiction.

Suppose now that $\beta_i > 0$ and $p = p_i$. Suppose $k \neq i$ and set $q = p_k$. We consider the group $K = HG_q$. Every Sylow p-subgroup of H is a Sylow p-subgroup of K, in fact H contains every Sylow p-subgroup of K and so the subgroup $S = H_p^H$ of H generated by all Sylow p-subgroups of H is normal in K. In particular S is normalized by G_q and therefore by every Sylow q-subgroup of $G, q \neq p$. It follows at once that S is normalized by the subgroup Q of G generated by all p'-elements of G. Since H is core-free, $Q < G$. Clearly $Q \lhd G$.

We now prove that S is a Sylow p-subgroup of H. This will complete the proof since p is any prime dividing the order of H and a finite group with its Sylow p-subgroups normal is nilpotent. Suppose S is not a Sylow p-subgroup of H. Then, since S is soluble (H is soluble by Theorem 7.1.5), there exist Sylow p-complements and so we can write $S = H_pT$, where $(|T|, |H_p|) = 1$. Now Q normalizes S and so H contains every conjugate of T by an element of Q. If R is the subgroup generated by all p'-elements of H, then R is normalized by all Sylow p-subgroups of G by the above argument. Since $T^Q \leqq R$, it follows that $H \geqq R \geqq T^{QG_p} = T^G \neq 1$, contradicting $H_G = 1$. □

We point out here that Maier and Schmid in [96] have improved Theorem 7.1.4 in showing that a *core-free permutable subgroup of a finite group G is contained in the hypercentre of G.*

When we leave the class of finite groups the situation becomes more complicated. Iwasawa [66] demonstrated that permutable subgroups need not be subnormal in general. His examples are constructed as follows: let p be a prime, A an abelian group of type C_{p^∞} and α a p-adic integer such that $\alpha \equiv 1 \bmod p$ ($\alpha \equiv 1 \bmod 4$ if $p = 2$). Then $a \mapsto a^\alpha$, $a \in A$, defines an automorphism of A. We form $G = A]H$, where $H = \langle \alpha \rangle$, and it turns out that every subgroup of G is permutable in G. However, if $\alpha \neq 1$, then H is not subnormal in G. For examples where a core-free permutable subgroup need not even be locally soluble (cf. Theorem 7.1.5) see Gross [38]. Still, all is not lost, since we have

Theorem 7.1.6 (Stonehewer [140]). *A permutable subgroup of a group is ascendant.*

In order to prove this result we need the following lemma which shows the interesting fact that a permutable subgroup is always normalized by an infinite cyclic subgroup disjoint from it.

Lemma 7.1.7 [140]. *Suppose that H is a permutable subgroup of a group $G = HK$, where K is an infinite cyclic group with $H \cap K = 1$. Then $H \lhd G$.*

Proof. Let p be any prime. Then HK^p is a subgroup of index p in G. We write $X = X_p$ for its core in G. Then $|G:X|$ divides $p!$ and so HX/X is a p'-subgroup of G/X. Working modulo X we have that G is finite and so H *sn* G by Theorem 7.1.2. Hence, since H is a p'-group, H^G is also a p'-group, from which it follows that $H^G \leq HK^p$. Bringing X and p into play once more we have $(HX)^G \leqq X$ so that $H^G \leqq X_p$ for all p. If we now put $N = \bigcap_p X_p$, it is clear that $|G:N|$ is infinite and $N \geqq H$. Hence $N = H$. But $N \geqq H^G$ and so $H = H^G \lhd G$. □

An alternative proof can be found in [14].

We may now prove Theorem 7.1.6. Let H *per* G. We assert that there exists an ordinal ρ and an ascending series

$$H = H_0 \lhd H_1 \lhd \cdots \lhd H_\alpha \lhd H_{\alpha+1} \lhd \cdots \lhd H_\rho = H^G$$

such that H_α *per* G and $H_{\alpha+1}/H_\alpha$ is a finite cyclic group for all ordinals $\alpha < \rho$. For, suppose that no such series exists. Then by Zorn's lemma we can construct a non-extendable series of type

$$H = H_0 \lhd H_1 \lhd \cdots \lhd H_\alpha \lhd H_{\alpha+1} \lhd \cdots \lhd H_\rho < H^G$$

where H_α *per* G and $H_{\alpha+1}/H_\alpha$ is finite cyclic. Thus there is no subgroup $H_{\rho+1}$ *per* G where $H_\rho \neq H_{\rho+1}$ and $H_{\rho+1}/H_\rho$ is finite cyclic.

Set $U = H_\rho$. Since $U < H^G$, there exists $g \in G$ such that $U < UU^g$. Since U *per* G, we must have $V = UU^g = U\langle g^n \rangle$ for some positive integer n. If $|V:U|$ is infinite, then $|U\langle g \rangle : U|$ is infinite so that $\langle g \rangle \cap U = 1$ and then $U^g = U$ by Lemma 7.1.7, a contradiction. Hence $|V:U|$ is finite and U *sn* V by Theorem 7.1.2. Let W be the penultimate term (before reaching U) in the normal closure series of U in V. Then $U \lhd W$, $U \neq W$ and W *per* G since W is generated by conjugates of U. Furthermore $W \leqq H^G$ and W/U is clearly finite cyclic, which contradicts the non-extendability of the chain. □

It is not difficult to show that there is always an ascending series from H to G of length at most $\omega + 1$. Whether there is always such a series of length ω appears to be an open question. For finitely generated groups the situation is rather better. In fact Stonehewer [141] has generalized Ore's theorem (Theorem 7.1.2) in showing that a permutable subgroup of a finitely generated group is always subnormal. This follows at once from

Theorem 7.1.8. *A permutable subgroup of a group is locally subnormal.*

By H *locally subnormal* in G we mean H is subnormal in every subgroup of form $\langle H, g_1, \ldots, g_n \rangle$. We note that this result is independent of Theorem 7.1.7 since ascendant subgroups need not be locally subnormal, and conversely. The natural semidirect product $(\ldots \wr C_p \wr C_p \wr \ldots)]C_\infty$ illustrates the former statement; the latter is a consequence of the existence of locally nilpotent groups with proper self-normalizing subgroups (Kargapolov [67]; see also [120], pp. 27–9).

Theorem 7.1.8 follows at once from Theorem 7.1.2 together with

Lemma 7.1.9. *Suppose that* $G = \langle H, g_1, \ldots, g_n \rangle$ *and* H *per* G. *Then* $|H^G : H|$ *is finite.*

Proof. Suppose that K is an infinite cyclic subgroup of G such that $H \cap K = 1$. Then we claim that $H \cap K^g = 1$ for all $g \in G$ ([140] Lemma 2.2). For, if $H^g \cap K \neq 1$ and $K = \langle k \rangle$, then $k^n \in H^g \leqq H\langle g \rangle$ for some positive integer n. By hypothesis $|H\langle k^n \rangle : H|$ is infinite so $|H\langle g \rangle : H|$ is infinite. But then g normalizes H by Lemma 7.1.7, and so $k^n \in H$, a contradiction. It follows that if L is the subgroup generated by all infinite cyclic subgroups of G which intersect H trivially, then $L \lhd G$.

Thus $G = \langle L, g_1, \ldots, g_n, H \rangle$ where we may assume $g_i \notin L, i = 1, 2, \ldots, n$. This means that each g_i has a positive power in H. We now set $U = \langle g_1, \ldots, g_n, H \rangle$ and prove by induction on $\sum_{i=1}^{n} |H\langle g_i \rangle : H|$ that $|H^U : H|$ is finite. The result then follows since $H^U = H^G$. Suppose the natural induction hypothesis. We may assume that $H^U \neq H$ so that $H^{g_i} \neq H$ for some i. But then $H < HH^{g_i}$ *per* G and by the induction it follows that $|(HH^{g_i})^U : HH^{g_i}|$ is finite. But $(HH^{g_i})^U = H^U$ and $|HH^{g_i} : H| \leqq |H\langle g_i \rangle : H| < \infty$. Hence $|H^U : H| < \infty$ as required. □

We mention here the fact that Lennox has shown [77] that if H is a core-free permutable subgroup of a finitely generated group G, then H^G is nilpotent of finite exponent and each Sylow p-subgroup of H is permutable in G. It would be of interest to know whether H is actually contained in the hypercentre of G, generalizing the Maier–Schmid result [96]. For progress in this direction, see Busetto [18] and Lennox [78].

It is of course an immediate corollary of Theorem 7.1.8 that Corollary 7.1.3 extends to finitely generated groups. However, it is in fact true that no simple group can have a proper non-trivial permutable subgroup (Theorem 7.1.12) although this is not a mere deduction from Theorem 7.1.7. In order to establish it we need to know more about the structure of core-free permutable subgroups of an arbitrary group.

Theorem 7.1.10 [140]. *Suppose that* H *is a core-free permutable subgroup of a group* G. *Then*

(i) H *is a subdirect product of finite nilpotent groups; and*
(ii) *every finitely generated subgroup of* H^G *is a subdirect product of finite nilpotent groups.*

From the first part of this theorem we have at once

Corollary 7.1.11. *A perfect permutable subgroup of a group is always normal.*

From the second part we may deduce the promised criterion for simplicity.

Theorem 7.1.12. *A simple group cannot have a proper non-trivial permutable subgroup.*

Proof. Suppose that G is simple and H *per* G with $1 \neq H \neq G$. Then $H^G = G$ and H is core-free, so by Theorem 7.1.10 G is locally a residually nilpotent group. However, by a theorem of Mal'cev [97] the class of all groups with an abelian (generalized) series of normal subgroups (and residually nilpotent groups are such) is L-closed. Thus G is abelian, a contradiction. □

Proof of Theorem 7.1.10. (i) Suppose that H is a core-free permutable subgroup of G. Let $g \in G$ and set $X = H\langle g \rangle$, $C = C(g) = H_X$. If $|X:H|$ is finite, then H/C is a permutable subgroup of the finite group X/C and hence H/C is nilpotent by Theorem 7.1.4. On the other hand if $|X:H|$ is infinite, we have $C = H \lhd X$ by Lemma 7.1.7. Thus in both cases H/C is a finite nilpotent group. But H is core-free in G and so

$$\bigcap_{g \in G} C(g) = 1.$$

It follows at once that H is residually a finite nilpotent group.

For the proof of (ii) we need

Lemma 7.1.13 [140]. *Let H per G, $g_1, \ldots, g_n \in G$ and $J = H^{g_1} \cdots H^{g_n}$. If $C_i = (H^{g_i})_J$, then J/C_i is finite and nilpotent, $1 \leqq i \leqq n$.*

Proof. We may suppose $g_1 = 1$ and that some positive power of each g_i belongs to H (otherwise $g_i \in N_G(H)$ by Lemma 7.1.7). We prove that J/C_1 is nilpotent—it is finite by Lemma 7.1.9. Suppose that $n \geqq 2$ and let

$$K = H^{g_3} \cdots H^{g_n}, \qquad L = HK\langle g_2 \rangle \geqq J.$$

Clearly $|L:J|$ is finite. Let $M = H_L$. Since H *per* L, it follows from Theorem 7.1.4 that H/M is nilpotent. Therefore H^{g_2}/M is nilpotent and we may apply Fitting's theorem (see Proposition 1.6.3) to obtain

$$HH^{g_2}/M \text{ is nilpotent.}$$

Now $M \leqq C_1$ and so HH^{g_2}/C_1 is nilpotent. Similarly HH^{g_i}/C_1 is nilpotent for $3 \leqq i \leqq n$. Applying Fitting's theorem once more yields that

$$(HH^{g_2}/C_1) \cdots (HH^{g_n}/C_1) = J/C_1$$

is nilpotent. □

Proof of Theorem 7.1.10(ii). Let H be a core-free permutable subgroup of G and suppose that F is a finitely generated subgroup of H^G. We have to show that F is residually a finite nilpotent group.

Clearly there are elements $g_1, \ldots, g_n$ in G such that $F \leqq H^{g_1} \cdots H^{g_n} = K$. Let $g \in G$. Then $F \leqq H^g K = J$, say. By Lemma 7.1.13, there is a subgroup $C \lhd J$ such that $C \leqq H^g$ and J/C is finite and nilpotent. Hence $F/F \cap C$ is finite and nilpotent. But $\bigcap_{g \in G} H^g = 1$ and so $\bigcap_g C^g = 1$. The result follows. □

We conclude this section with a result due to Kegel [69] who strengthened Ore's theorem (Theorem 7.1.2) to

Theorem 7.1.14. *Suppose that G is a finite group and H is a subgroup of G which permutes with all Sylow subgroups of G. Then H is subnormal in G.*

Proof. Suppose the result to be false. Then there exists a group G of least order and a subgroup H demonstrating the fact. We may assume that H is contained in a proper subgroup V of G which is maximal with respect to permuting with all Sylow subgroups of G. Let V_p be a Sylow p-subgroup of V. Then $V_p \leqq G_p$, a Sylow p-subgroup of G. Now $HG_p = G_pH$ and $(HG_p) \cap V = H(G_p \cap V) = HV_p$. Hence H permutes with all Sylow subgroups of V. By the minimality of G, H *sn* V. Hence we may assume $H = V$ and the result is completed by the following analogue of Theorem 7.1.1.

Lemma 7.1.15. *Suppose that H is a proper subgroup of a finite group G and that H is maximal with respect to permuting with all Sylow subgroups of G. Then $H \lhd G$.*

Proof. Suppose that H is not normal in G. Then there exists an element g of prime power order p^α which does not normalize H. We claim that $|G:H| = p^\beta$, for some β. Suppose then that $|G:H| = p^\beta m$, $(p, m) = 1$ and $m > 1$. If S is a Sylow p-subgroup of G containing g, then $HS = SH < G$, since $m > 1$. Furthermore if $K = \langle H, H^g \rangle$, then K clearly permutes with all Sylow subgroups of G and $H < K \leqq HS < G$. The maximality of H gives a contradiction. Hence $m = 1$ and $|G:H| = p^\beta$.

It follows that H is normalized by the normal subgroup N generated by all p'-elements of G. Thus $H \lhd HN$, so $HN < G$. However, HN permutes with all Sylow subgroups of G. It follows that $HN = H$, so $N \leqq H$. We may therefore assume $N = 1$ and G is a p-group. But then $H^G < G$ and H^G trivially permutes with all Sylow subgroups of G. By the maximality of H we obtain $H = H^G \lhd G$, as required. □

Kegel [69] and Deskins [29] proved that if H is a core-free subgroup of a finite group G and if H permutes with every Sylow subgroup of G, then H (and therefore H^G) is nilpotent (compare Theorem 7.1.4). However, H is not necessarily contained in the hypercentre of G and so the Maier–Schmid theorem (referred to on p. 215) does not generalize here.

Before leaving subnormality criteria connected with Sylow sugbroups we mention a problem due to Kegel [69]. If G is a finite group and H *sn* G, it is not hard to see that if S is a Sylow p-subgroup of G, then $H \cap S$ is a Sylow p-subgroup of H. Is the converse true? Kegel has given an affirmative answer in the case when H is soluble.

§7.2 Permutability of conjugates of subgroups

Suppose that G is a finite group and that A, B are subgroups of G with the property that $AB^x = B^xA$ for all $x \in G$. The implication of this condition for the subnormal structure of G has been the subject of several papers, the first of which concerns the case $A = B$.

Theorem 7.2.1 (Szép [145]). *Suppose that G is a finite group and that A is a subgroup of G with $AA^x = A^xA$ for all $x \in G$. Then A sn G.*

We remark that the converse implication does not hold. Let

$$G = \langle a, t \mid a^t = a^{-1}, a^8 = t^2 = 1 \rangle$$

be a dihedral group of order 16. Then the subgroup $\langle t \rangle$ is subnormal but does not permute with its conjugate under a. However, see Theorem B of Hartley and Peng [53].

In 1961 Kegel [68] proved a very useful criterion for the non-simplicity of a finite group which he exploited to demonstrate that any product $G = HK$ of finite nilpotent groups H, K is always soluble.

Theorem 7.2.2. *Suppose that G is a finite group and that A, B are subgroups of G with $A^xB = BA^x$ for all $x \in G$. If $G > AB$, then either A or B is contained in a proper normal subgroup of G.*

Szép's result follows at once on putting $A = B$ and using a simple induction on $|G|$.

In [156] Wielandt proved

Theorem 7.2.3. *Suppose that G is a finite group and that $A \leqq B \leqq G$ with $AB^x = B^xA$ for all $x \in G$. Then A sn G if A sn B.*

This result leads at once to the following subnormality criterion.

Corollary 7.2.4 (Wielandt [158], p. 71). *A subgroup A of a finite group G is subnormal if and only if there exists a subgroup B of G containing A such that A sn B and $AB^x = B^xA$ for all x in G.*

The sufficiency follows from Theorem 7.2.3 and in order to see necessity, set $B = G$.

In [161] Wielandt improved on the foregoing results by simultaneously generalizing them in

Theorem 7.2.5. *Suppose that G is a finite group, A and B are subgroups of G and* $AB^x = B^xA$ *for all x in G. Then the following hold:*
(a) *If* $G = AB^G = BA^G$, *then* $G = AB$.
(b) $A^B \cap B^A$ *sn G.*
(c) *If* $AB \leqq H \leqq G$, *then* $A^H \cap B^H$ *sn G.*
(d) *If X, Y are arbitrary subsets of G, then* $[A^X, B^Y]$ *sn G.*

Proof. (a) Suppose that $G = AB^G = BA^G$. We proceed by induction on $|G:A|$. If $A \lhd G$, then $G = A^GB = AB$. If this is not the case then, since $G = AB^G$, there are elements $b \in B$, $g \in G$ such that $b^g = h \notin N_G(A)$. The subgroups $A^* = \langle A, A^h \rangle$ and $B^* = B^g$ satisfy the same hypotheses as A and B. Since $|G:A^*| < |G:A|$, we have

$$G = A^*B^* = \langle A, A^h, B^g \rangle = \langle A, B^g \rangle = AB^g.$$

Hence $g \in Ag^{-1}Bg$ so that $g \in AB$. It follows that $G = AB$.

(b) We induct on $|G|$. If $G = AB$, then $A^B \cap B^A = A^G \cap B^G \lhd G$. Otherwise by (a) one of the subgroups AB^G or BA^G is smaller than G. Suppose $AB^G < G$. Then by induction $A^B \cap B^A$ *sn* AB^G. But $A^B \cap B^A \leqq B^G$ so that $A^B \cap B^A$ *sn* $B^G \lhd G$. Hence $A^B \cap B^A$ *sn G*, as required.
(An alternative proof is given by Wielandt in [158], §15.)

(c) The subgroups $K = A^H$, $L = B^H$ satisfy the same hypotheses as A, B. By (b) we get $K \cap L = K^L \cap L^K$ *sn G.*

(d) The subgroups $K = A^X$, $L = B^Y$ again satisfy the same hypotheses. We have $[K, L] \lhd K^L \cap L^K$ and so by (b) we obtain $K^L \cap L^K$ *sn G.* □

As a contribution to the case where more than two subgroups are involved, Wielandt [161] proved the following generalization of Theorem 7.2.5.

Theorem 7.2.6. *Let G be a finite group, suppose that* $A_1, \ldots, A_n \leqq H \leqq G$ *and set* $B_j = \langle A_i | i \neq j \rangle$. *If* $A_jB_j^x = B_j^xA_j$ *for* $1 \leqq j \leqq n$, *all* $x \in G$, *then* $B_1^H \cap B_2^H \cap \cdots \cap B_n^H$ *sn G.*

Proof. Set $C_j = A_j^H \cap B_j^H$. By Theorem 7.2.5(c) we have C_j *sn G*. It follows by Theorem 1.2.1 that $C_1C_2 \cdots C_n$ *sn G*. This product is calculated step by step by means of Dedekind's distribution formula as follows. Using the fact that $A_j^H \leqq B_i^H$ for $i \neq j$, we get

$$C_1C_2 = C_1(A_2^H \cap B_2^H) = C_1A_2^H \cap B_2^H = A_1^HA_2^H \cap B_1^H \cap B_2^H,$$

$$C_1C_2 \cdots C_k = A_1^HA_2^H \cdots A_k^H \cap B_1^H \cap \cdots \cap B_k^H \qquad (k = 1, \ldots, n).$$

Since $B_j^H \leqq A_1^HA_2^H \cdots A_n^H$, we obtain for $k = n$

$$B_1^H \cap B_2^H \cap \cdots \cap B_n^H = C_1C_2 \cdots C_n \text{ sn } G. \quad \square$$

For the case $A = B$ Koppe [71] has given the following generalization of Theorem 7.2.1 to infinite groups.

Theorem 7.2.7. *Suppose that G is any group and that $A \leqq G$ with $AA^x = A^xA$, all x in G. Then A is a serial subgroup of G.*

In order to prove this we need the criterion for seriality due to Hickin and Phillips, Theorem 2.4.1, that A is serial in G if and only if, for every finitely generated subgroup H of G with $H \leqq A^H$, we have $H \leqq A$.

Proof of Theorem 7.2.7. If A is not serial in G, there is, by the Hickin–Phillips criterion, a finitely generated subgroup H of G with $H \leqq A^H$ and $H \nleqq A$. So there is a minimal finite number n ($\geqq 1$) of elements $h_1, \ldots, h_n$ of H such that $C = AA^{h_1} \cdots A^{h_n}$, say, contains H. Let $B = AA^{h_1} \cdots A^{h_{n-1}}$. Then there exist elements $b \in B$, $a \in A$ such that $h_n = ba^{h_n} = bh_n^{-1}ah_n$. Hence $h_n = ab \in B$ and $B = C$, a contradiction. □

§7.3 The Wielandt maximizer lemmas

Suppose that G is a group and A is a subgroup of G. Then it is clear that each of the following conditions is both necessary and sufficient for A to be normal in G:

(1) $A \lhd \langle A, g \rangle$, all $g \in G$;

(2) $[g, a] \in A$, all $g \in G$, $a \in A$.

It is also clear that the analogous conditions

(3) A *sn* $\langle A, g \rangle$, all $g \in G$;

(4) whenever $g \in G$, $a \in A$, then $[g, {}_na] \in A$ for some positive integer n (here $[g, {}_1a] = [g, a]$ and $[g, {}_{i+1}a] = [[g, {}_ia], a]$ for $i \geqq 1$)

are necessary for the subgroup A to be subnormal in G. In this section we wish to describe the pioneering work of Wielandt [162] in developing techniques to show that conversely, for finite groups at least, (3) and (4) (and other related conditions) are also each sufficient to ensure that A is subnormal in G.

If we want to prove that (3) is a sufficient condition for A *sn* G, then in the case where G is a finite group it is natural to proceed by assuming that this is false and looking at a counterexample of least order. Thus A is not subnormal in G but A *sn* M for every maximal subgroup M of G which contains A. Wielandt's results depend on a careful analysis of such extreme situations, an analysis which yields two lemmas which have been styled the 'maximizer lemmas' by Hartley and Peng [53]. We shall first of all state these and show how to deduce a whole series of subnormality criteria from them, leaving their proofs until the end of the section.

Lemma 7.3.1 (First maximizer lemma). *Let G be a finite group and let A be a subgroup of G. Suppose that A is not subnormal in G, but A sn H for all proper subgroups H of G containing A. Then*

(i) *A is contained in a unique maximal subgroup M, and*

(ii) *if $g \in G$, then $A^g \leqq M$ if and only if $g \in M$.*

M is called the *Wielandt maximizer* of A. This lemma has a close connection with the concept of subnormalizer (see §7.7).

Lemma 7.3.2 (Second maximizer lemma). *With the hypotheses of the first maximizer lemma assume further that $A \cap H$ sn H for all $H < G$. Let M be the Wielandt maximizer of A. Then A contains an element a of prime power order such that, if $g \in G$, then $a^g \in M$ if and only if $g \in M$.*

Corollaries of the first maximizer lemma

Theorem 7.3.3 (Wielandt). *Suppose that A is a subgroup of a finite group G. Then each of the following conditions is equivalent to A sn G:*

(i) *A sn $\langle A, g \rangle$, for all g in G;*
(ii) *A sn $\langle A, A^g \rangle$, for all g in G;*
(iii) *A sn $\langle A, A^{a^g} \rangle$, for all a in A, g in G.*

Proof. Clearly A *sn* $G \Rightarrow$ (i) $\Rightarrow$ (ii) $\Rightarrow$ (iii), so it is enough to show that (iii) $\Rightarrow A$ *sn* G. Assume that this is not the case and take G to be a counterexample of least order. By the first maximizer lemma there exists a Wielandt maximizer M of A. Since A *sn* $\langle A, A^{a^g} \rangle$, we must have $\langle A, A^{a^g} \rangle \leqq M$ and by (ii) of the same lemma this means that $a^g \in M$ for all $a \in A$, $g \in G$, which in turn entails that $g \in M$ for all $g \in G$, a contradiction. □

It is not difficult to deduce from (ii) that A^G is nilpotent whenever $\langle A, A^g \rangle$ is nilpotent for all $g \in G$. This is a special case of a result of Baer ([7], Satz L). Another special case is that A^G is a p-group whenever $\langle A, A^g \rangle$ is a p-group for all $g \in G$. (See Alperin and Lyons [3].) Since the conditions (i), (ii) and (iii) are all inherited under homomorphisms of G, we shall see that Theorem 7.3.3 extends to the class of all polycyclic-by-finite groups (via Theorem 7.5.1). For a proof of (ii) in the case of groups with Max, see §7.4.

Closely related to the foregoing we have

Theorem 7.3.4. *Suppose that A is a subgroup of a finite group G. Then each of the following conditions is equivalent to A sn G:*

(i) *$g \in G$ and $g \in \langle A, A^g \rangle \Rightarrow g \in A$;*
(ii) *$a \in A$, $g \in G$ and $g \in \langle A, a^g \rangle \Rightarrow g \in A$;*
(iii) *$a \in A$, $g \in G$ and $g \in \langle A, A^{a^g} \rangle \Rightarrow g \in A$.*

Proof. Observe first of all that A *sn* G implies (i). For, if $g \in \langle A, A^g \rangle$ and A *sn* G, then g belongs to the normal closure of A in $\langle A, g \rangle$. Hence $\langle A, g \rangle = A$. Clearly (i) $\Rightarrow$ (ii) $\Rightarrow$ (iii), so it remains only to show that (iii) $\Rightarrow A$ *sn* G. We proceed by induction on $|G|$. Let $H = \langle A, A^{a^g} \rangle$. If $H < G$, then A *sn* H by the induction hypothesis. *If* $H = G$, then $g \in A$ by (iii) so that $G = A$. Hence in either case A *sn* $\langle A, A^{a^g} \rangle$ and the result follows by Theorem 7.3.3(iii). □

By contrast with Theorem 7.3.3, if $G = \langle x, a | x^a = x^{-1}, a^2 = 1 \rangle$ is the infinite dihedral group and if $A = \langle a \rangle$, then $g \in \langle A, A^{a^g} \rangle$ implies that $g \in A$. However, A is not subnormal in G and so 7.3.4(i) fails here as a subnormality criterion in the class $\mathfrak{Max}$. Of course A is descendant in G and in view of Theorem 2.4.1 it would be of interest to know for which classes of groups (i) implies seriality.

From (i) we have at once

Theorem 7.3.5. *Suppose that A is a subgroup of a finite group G. If for every $g \in G$ either $A^g A = AA^g$ or A sn $\langle A, A^g \rangle$, then A sn G.*

This yields another proof of Theorem 7.2.1. Also it is not difficult to see that if A is a subgroup of a finite group G and $AA^{a^g} = A^{a^g} A$ for all $a \in A$ and $g \in G$, then A *sn* G.

In [162] Wielandt shows that the hypothesis in Theorem 7.3.4(i) is unnecessarily sharp when we are considering it as a sufficient condition for subnormality. He demonstrates the existence of 'test sets' $T \subseteq G$ with the property that if $g \in T$ and $g \in \langle A, A^g \rangle$ implies that $g \in A$, then A *sn* G. As examples of such test sets we may take, for any n, the set T_n of all commutators $[g, e_1, \ldots, e_n]$, where the e_i belong to a set of generators of A and $g \in G$. Let $T_\infty = \bigcap_{n \geqq 1} T_n$. Then we obtain

Theorem 7.3.6. *Let A be a subgroup of a finite group G. Then each of the following conditions is equivalent to A sn G:*

(i) *A sn $\langle A, A^g \rangle$ for all $g \in T_\infty$;*

(ii) *$T_n \subseteq A$ for some n.*

We leave the proof as an exercise. In fact it is not difficult to deduce directly from Lemma 7.3.1 that if G is a finite group with a subgroup A and if for each $g \in G$ there exist subsets $A_1, \ldots, A_n$ of A such that $A = \langle A_i \rangle$ and $[g, A_1, \ldots, A_n] \leqq A$, then A *sn* G. Hartley and Peng [53] have generalized this to groups with Min.

Theorem 7.3.7. *Let G be a group satisfying* Min *and let A be a subgroup of G. Then*

(i) *if, for each $g \in G$, there exists $n = n(g) \geqq 0$ and a sequence $A_1, A_2, \ldots, A_n$ of generating sets of A such that $[g, A_1, \ldots, A_n] \leqq A$, then A is an ascendant subgroup of G; and*

(ii) *A sn G if and only if there exists an integer $n \geqq 0$ and for each $g \in G$ a sequence $A_1, \ldots, A_n$ of generating sets of A such that $[g, A_1, \ldots, A_n] \leqq A$.*

We omit the proof. The case where A is the cyclic subgroup generated by a single left Engel element a is of independent interest. (This means that, for each $g \in G$, there exists $n \geqq 0$ such that $[g, {}_n a] = 1$. If n can be chosen independently of g, then a is called a *bounded* left Engel element.) Wielandt points out in [162] that in the finite case (ii) leads to the well-known fact that every left Engel element of a finite group lies in the Fitting subgroup (Baer [7]). Hartley and Peng deduce from (ii) a generalization of Baer's result, originally proved under weaker hypotheses by Martin and Pamphilon [100].

Corollary 7.3.8. *Let the group G satisfy* Min. *Then*

(i) *every left Engel element of G lies in the Hirsch–Plotkin (that is the locally nilpotent) radical of G, and*

(ii) *every bounded left Engel element of G lies in a finite nilpotent normal subgroup of G.*

Again we omit the proof. Hartley and Peng use this result as the basis for a unified approach to the work of Martin and Pamphilon and others on Engel elements.

Wielandt has proved the following (unpublished) analogue of Theorem 7.3.7(ii) for groups with Max:

Theorem 7.3.9. *Suppose that G is a group with* Max *and $A \leqq G$. Let A_0 be a set of generators of A with $A_0 = A_0^{-1}$. Then A sn G if and only if for all $g \in G$ there exists n such that $[g, {}_n A_0] \leqq A$.*

The proof depends on results in §7.4 and may be left as an exercise.

Corollaries of the second maximizer lemma

We shall state the main result of this section in its fullest generality, as described in [53]. However, we shall prove it only for the finite case, following Wielandt. A group element of prime power order is said to be *primary*, and we interpret $[g, {}_n a]$ as g if $n = 0$.

Theorem 7.3.10. *Let the group G satisfy* Min *and let $A \leqq G$. Then*

(i) *A is ascendant in G if and only if, for each $g \in G$ and each primary $a \in A$, there exists $n = n(g, a) \geqq 0$ such that A permutes with $A^{[g, {}_n a]}$, and*

(ii) *A is subnormal in G if and only if there exists an integer $n \geqq 0$ such that A permutes with $A^{[g, {}_n a]}$ for all $g \in G$ and primary $a \in A$.*

It is immediately clear that this result is a very satisfying generalization of the Ore–Szep theorems in that it yields both necessary and sufficient conditions for subnormality (and ascendancy). As a direct corollary we have the following further subnormality criteria.

Theorem 7.3.11. *Let A be a subgroup of a group satisfying* Min. *Then A sn G if and only if there exists an integer $n \geqq 0$ with at least one of the following properties:*

(i) *For all $g \in G$ and primary $a \in A$, $[g, {}_n a] \in A$.*
(ii) *For all $g \in G$ and primary $a \in A$, $A \cap \langle a, g \rangle \lhd^n \langle a, g \rangle$.*
(iii) *For all $g \in G$, $A \lhd^n \langle A, A^g \rangle$.*
(iv) *For all $g \in G$ and primary $a \in A$, $A \lhd^n \langle A, A^{a^g} \rangle$.*

We note that in the finite case $\lhd^n$ may be replaced by *sn* and that (ii), (iv) are weaker hypotheses than Theorem 7.3.3(i) and (iii), respectively. Theorem 7.5.1 will give an immediate extension to the class of all polycyclic-by-finite groups.

Deduction of Theorem 7.3.11 *from Theorem* 7.3.10. Clearly $A \lhd^n G$ implies (i)–(iv). Also (i) implies A *sn* G by Theorem 7.3.10(ii). Furthermore (i) evidently follows from (ii) and since $[g, a, a] = a^{-[g,a]} a \in \langle A, A^{a^{-g}} \rangle$, (iv) leads to $[g, {}_{n+2}a] \in A$ for all $g \in G$ and primary $a \in A$ and hence implies A *sn* G via (i). Finally (iv) follows from (iii) and hence so also does (i). □

Deduction of Theorem 7.3.10 *from the second maximizer lemma* (*in the finite case*). It is in fact most convenient to prove first of all that Theorem 7.3.11(i) is equivalent to A *sn* G. Necessity is clear, so we suppose that given $g \in G$ and primary $a \in A$ there exists n such that $[g, {}_n a] \in A$. (Note that we may rephrase the condition this way since G is finite.) Suppose $H \leqq G$ and $g \in H$. Then if a is a primary element of $A \cap H$, we have $[g, {}_n a] \in A \cap H$ for some n by hypothesis. Thus in the minimal counterexample situation we may assume that A is not subnormal in G but that $A \cap H$ *sn* H for any $H < G$. Then, by the second maximizer lemma, A has a Wielandt maximizer M and there exists a primary element $a \in A$ such that $a^g \in M$ if and only if $g \in M$ $(g \in G)$.

Choose $x \in G \setminus M$. By hypothesis $[x, {}_n a] \in A \leqq M$ for some $n \geqq 0$. If we now choose m minimal subject to the condition $[x, {}_m a] \in M$, we then have $m > 0$ and if we set $g = [x, {}_{m-1} a]$, then $g \notin M$, but $[g, a] \in M$. Hence $a^{-g} \in M$ and so $a^g \in M$. But then $g \in M$, a contradiction. We have therefore proved that

$$\textit{if } [g, {}_n a] \in A \textit{ for all } g \in G,\ a \in A, \textit{ then } A \textit{ sn } G. \qquad (*)$$

We proceed now to establish the finite version of Theorem 7.3.10. Clearly we need only prove sufficiency. If this is not the case we may assume that G is a minimal counterexample and hence we have that A is not subnormal in G but $A \leqq H < G$ implies A *sn* H and also if $g \in G$ and a is a primary element of A, then A permutes with $A^{[g, {}_n a]}$, Let g, a be two such elements and consider $B = AA^{[g, {}_n a]}$. Then B is a subgroup and since a group cannot be the product of two conjugate proper subgroups, we must have $B < G$. By minimality, A *sn* B. Now $[g, {}_{n+1} a] \in B$ and since A *sn* B we must have $[g, {}_{n+m} a] \in A$ for some $m \geqq 1$. Since this holds for all $g \in G$ and primary $a \in A$, we can now apply $(*)$ to yield A *sn* G, as required. □

It is also clear from the proofs that Theorems 7.3.10 and 7.3.11 remain true if $[g, {}_n a]$ is replaced by $u_a^n(g)$, where u_a is any map of G into itself sending g to a word $w(g, a)$ such that $\langle a, a^g \rangle = \langle a, w(g, a) \rangle$—for example, $u_a(g) = a^g$—and where u_a^n is the result of n applications of u_a.

Wielandt has also pointed out the following additional subnormality criteria which may also be deduced from the foregoing results.

Theorem 7.3.12. *Suppose that G is a finite group and $A \leqq G$. Then each of the following conditions is equivalent to A sn G.*
(i) *Whenever a is a primary element of A and $g \in G$, then $g = [g, {}_n a]$, where $n = n(g, a)$, implies $g \in A$.*
(ii) *If a is primary in A and a is conjugate to b in $\langle a, b \rangle$, then $b \in A$.*

We omit the proof, but note that it is easy to see that (ii) implies (i) and that the infinite dihedral group demonstrates that (ii) is not a sufficient condition for subnormality in groups with Max.

Bartels in [10] has generalized Theorem 7.3.12(ii). He uses $X \sigma Y$ to denote that the subsets X, Y of G are conjugate in $\langle X, Y \rangle$ and proves the following conjecture of Wielandt.

Theorem 7.3.13. *For any subset X of a finite group G, the smallest subnormal subgroup of G containing X is $\langle Y \mid Y \subseteq G, X \sigma Y \rangle$.*

This leads to the following criterion for subnormality.

Theorem 7.3.14 [10]. *Suppose that A is a subgroup of a finite group G and X is a subset of A with $X^A = A$. Then each of the following conditions is equivalent to A sn G:*
(i) *If $X \sigma Y$ for $Y \subseteq G$, then $Y \subseteq A$.*
(ii) *If $y \in G$ and $\{x\} \sigma \{y\}$ for some $x \in X$, then $y \in A$.*

We shall not prove these two theorems.

Further extensions of the theory

Wielandt has asked the question: Suppose that A is a subgroup of a finite group G and S is a set of generators of A with the property that for each $g \in G$ and $s \in S$, $[g, {}_n s] \in A$ for some n. Is A sn G? Peng in [107] has given a positive answer in the case of finite soluble groups. For an extension of Peng's result to finitely generated subgroups of soluble minimax groups, see Whitehead [151].

For further extensions of variants of Theorem 7.3.11(i) to various classes of infinite soluble groups see Peng [106] and McCaughan and McDougall [85]. Wehrfritz [148] studies linear groups and proves the following: *Let G be a subgroup of $GL(n, \mathbb{Q})$ and A be a soluble-by-finite subgroup of G. If for each $a \in A$ there exists a positive integer m such that $[g, {}_m a] \in A$ for all $g \in G$, then A sn G.*

Proofs of the maximizer lemmas

First maximizer lemma. Suppose that G is a finite group and $A \leqq G$ such that A is not subnormal in G but $A \leqq H < G \Rightarrow A \text{ sn } H$. We require to prove (i) A is contained in a unique maximal subgroup M of G and (ii) if $A^g \leqq M, g \in G$, then $g \in M$. We note first of all that (i) $\Rightarrow$ (ii). For, $A^g \leqq M \Rightarrow A \leqq M^{g^{-1}}$. Hence $M = M^{g^{-1}}$ by (i). Thus if $g \notin M$, then $M \lhd \langle M, g \rangle = G$ and hence $A \text{ sn } G$, a contradiction. Therefore we need only establish (i). In order to do this we need

Lemma 7.3.15. *Suppose that H is an arbitrary group and that $A \leqq H$. Let $\mathscr{S}$ be the set of all subgroups of H which are generated by conjugates of A and in which A is contained as a subnormal subgroup. Suppose that $\mathscr{S}$ satisfies the maximal condition with respect to inclusion and that if $K_1 \in \mathscr{S}$, $K_2 \in \mathscr{S}$ and $K_1 \leqq K_2$, then $K_1 \text{ sn } K_2$. Then among all subgroups of H which contain A as a subnormal subgroup, there is a unique maximal one. More precisely, if $L = \langle K \mid K \in \mathscr{S} \rangle$ and $M = N_H(L)$, then* (a) $L \in \mathscr{S}$, (b) $A \text{ sn } M$, (c) *if $A \text{ sn } X$, then $X \leqq M$.*

Proof. (a) Clearly $A \in \mathscr{S}$, so $\mathscr{S} \neq \varnothing$. Hence $\mathscr{T}$, the set of maximal elements of $\mathscr{S}$, is also non-empty. The assertion $L \in \mathscr{S}$ is equivalent to the assertion $|\mathscr{T}| = 1$. Suppose then that $|\mathscr{T}| > 1$. Among all pairs (L_1, L_2) with $L_i \in \mathscr{T}$, $L_1 \neq L_2$, we choose one for which the join B of the conjugates $A^h (h \in H)$ which lie in $L_1 \cap L_2$ is maximal. Thus $A \leqq B \leqq L_1 \in \mathscr{S}$, whence $A \text{ sn } B \in \mathscr{S}$.

We cannot have both $B \lhd L_1$ and $B \lhd L_2$ since that would mean $A \text{ sn } B \lhd \langle L_1, L_2 \rangle \in \mathscr{S}$ and so $L_1 = L_2$ by choice of the L_i. We may assume that $B \ntriangleleft L_1$. Then, since $B \text{ sn } L_1$, there exists a subgroup B_1 with $B \lneqq B_1 \leqq L_1$ which is generated by suitable conjugates of B and hence belongs to $\mathscr{S}$ (for example B_1 could be the penultimate term of the normal closure series of B in L_1). It now turns out that $B \ntrianglelefteq L_2$, since otherwise $B \lhd \langle B_1, L_2 \rangle \in \mathscr{S}$ so that $B_1 \leqq L_2$ and therefore $B_1 \leqq L_1 \cap L_2$, which contradicts the maximality property of B. Hence there exists a subgroup $B_2 \in \mathscr{S}$ such that $B \lneqq B_2 \leqq L_2$. Set $C = \langle B_1, B_2 \rangle$. Then $A \text{ sn } B \lhd C$ so that $C \in \mathscr{S}$. Since $\mathscr{S}$ satisfies the maximal condition there is an $L_3 \in \mathscr{T}$ with $C \leqq L_3$. Now we have $B < B_1 \leqq L_1 \cap L_3$ and so, by the maximality property of B, we obtain $L_1 = L_3$ and similarly $L_2 = L_3$ so that $L_1 = L_2$, which contradicts our assumption. Hence $|\mathscr{T}| = 1$. Thus the unique maximal subgroup in $\mathscr{S}$ contains all $K \in \mathscr{S}$ and so is their join.

(b) A is subnormal in M since $A \text{ sn } L \lhd M$.

(c) Let $A \text{ sn } X \leqq H$. Then $N = A^X \in \mathscr{S}$. For every $x \in X$ we have $A \text{ sn } N = N^x \text{ sn } L^x \in \mathscr{S}$, so that $L^x \leqq L$. The same holds for x^{-1} and hence $x \in M$. $\square$

We may now prove Lemma 7.3.1(i). The hypotheses of Lemma 7.3.15 are fulfilled for $H = G$: for if $K \in \mathcal{S}$, then $K < H$ since A is not subnormal in G, and hence $A^h \, sn \, K$ for every $A^h \leqq K$; by Wielandt's join theorem (1.3.1), $K_1 \leqq K_2$ implies $K_1 \, sn \, K_2$. By Lemma 7.3.15 A lies in a unique maximal subgroup M of G and $A \, sn \, M$. □

Second maximizer lemma

We first of all need

Lemma 7.3.16. *Suppose that G is a finite group and that $S \, sn \, G$, $S \leqq M$, a maximal subgroup of G. Then $S \leqq M_G$.*

Proof. Since S^M is a join of subnormal subgroups of the finite group G, it is subnormal in G and hence we may suppose $S \lhd M$. Let r be minimal with $S_{r+1} \leqq M$, $S_r \nleqq M$. Here S_r is the rth term of the normal closure series of S in G. Then $G = \langle M, t \rangle$ where $t \in S_r \backslash M$. However, both M and t normalize S_{r+1} and so $S_{r+1} \lhd G$. Thus $S \leqq S_{r+1} \leqq M_G$, as required. □

Suppose now that G is finite and that A is not subnormal in G but $A \cap H \, sn \, H$ for all $H < G$. By the first maximizer lemma, A has a Wielandt maximizer M, say. We proceed by induction on $|G| + |A|$. If $M_G \neq 1$, we have the result holding in G/M_G and we can recover it for G. So we may assume that $M_G = 1$. By Lemma 7.3.16, if $B \, sn \, G$ and $B \leqq M$, then $B = 1$.

Now $|A| > 1$. If A is not simple, take $1 \neq B \lneqq\!\!\lhd A$. Then B is not subnormal in G but $B \cap H \, sn \, H$ for all $H < G$ by the hypothesis on A. Hence B has a Wielandt maximizer in G and this must be M. By the induction hypothesis we have the result.

Thus A is simple. If $A' = 1$, the result is immediate from the first maximizer lemma since A is cyclic of order a prime p. Hence $A = A'$. Let $g \in G$ and suppose $A^g \cap A \neq 1$. Then $A^g \cap A \, sn \, A$ by hypothesis and so $A^g \cap A = A$ since A is simple. Hence $A^g = A$ and so $g \in M$ by definition of M. Suppose now that $A \cap M^g \neq 1$. Now $A \cap M^g \, sn \, M^g$ and $A^g \, sn \, M^g$. So A^g normalizes $A \cap M^g$ (see Theorem 4.5.1). Then $b \in A^g$ implies $A \cap A^b \neq 1$ (since $A^b \geqq A \cap M^g \neq 1$). Hence $b \in M$ by the foregoing. It follows that $A^g \leqq M$ and hence $g \in M$ by the first maximizer lemma. Thus $A \cap M^g \neq 1$ implies $g \in M$. Therefore if $g \in G \backslash M$, we obtain $A^g \cap M = 1$. So any $a \in A$, $a \neq 1$, satisfies $a^g \notin M$ if $g \notin M$. □

§7.4 Groups satisfying the maximal condition

The theory of subnormality criteria for groups with Max is not nearly so far developed as that for groups with Min. However, in unpublished work Wielandt has obtained the following extension of Theorem 7.3.3(ii).

Theorem 7.4.1. *Suppose that G is a group with* Max *and* $A \leqq G$. *Then* A *sn* G *if and only if* A *sn* $\langle A, A^g \rangle$ *for all* $g \in G$.

In order to develop his ideas in greatest generality and pave the way for future applications, Wielandt starts with the notion of a *Sylow property*: $\mathscr{P}$ is a Sylow property of groups if for any group G there is a set $\mathscr{P}G$ of subgroups of G such that

(1) if $A \in \mathscr{P}G$ and ϕ is a homomorphism of G, then $A^\phi \in \mathscr{P}(G^\phi)$;
(2) if $A \in \mathscr{P}G$ and $A \leqq F \leqq G$, then $A \in \mathscr{P}F$.

Thus for example if $\mathscr{P}_1 G = \{A \mid A \text{ sn } \langle A, A^g \rangle \text{ for all } g \in G\}$, then $\mathscr{P}_1$ is a Sylow property. Denote the set of subnormal subgroups of G by sn G. Then clearly Theorem 7.4.1 is tantamount to showing that $\mathscr{P}_1 G \subseteq \text{sn}\, G$ for groups G with Max. The proof of the latter is facilitated by the following reduction to a limiting situation.

Theorem 7.4.2. *Let* $\mathscr{P}$ *be a Sylow property. Suppose that there exists a group* Γ *with* Max *such that* $\mathscr{P}\Gamma \nsubseteq \text{sn}\,\Gamma$. *Then there exists G, satisfying* Max, *in fact a section of* Γ, *such that* $\mathscr{P}G \nsubseteq \text{sn}\, G$ *and we have*

(i) *if* $1 < N \lhd F \leqq G$, *then* $\mathscr{P}(F/N) \subseteq \text{sn}\,(F/N)$;
(ii) *if* $1 < N$ *sn* $F \leqq G$ *and* $A \in \mathscr{P}F$, *then* $\langle A, N \rangle$ *sn* F.

Proof. We consider the set of all F/N such that $1 \leqq N \lhd F \leqq \Gamma$ and $\mathscr{P}(F/N) \nsubseteq \text{sn}\,(F/N)$ and we choose $G = F/N$ with N maximal for this to hold. Then (i) is clear.

In order to prove (ii) we first of all consider the case $N \lhd F$. Set $\bar{F} = F/N$ and suppose that $A \in \mathscr{P}F$. Then $\bar{A} \in \mathscr{P}\bar{F}$ since $\mathscr{P}$ is a Sylow property and hence $\bar{A}$ *sn* $\bar{F}$ by (i). It follows at once that $\langle A, N \rangle = AN$ *sn* F. We now assume only that N *sn* F and that (ii) is false. We choose N maximal subject to these conditions. Then $A \leqq N_G(N)$. For, if not, $N < N^A$ so that N^A *sn* F which implies that $\langle A, N \rangle = AN^A$ *sn* F, a contradiction. Since $N \ntriangleleft F$ there exists a series $N \lhd M \lhd \cdots \lhd F$ with $M \neq N$. By choice of N, $\langle M, A \rangle$ *sn* F. However, $N \lhd \langle M, A \rangle$ and by the normal case it then follows that $\langle A, N \rangle$ *sn* $\langle A, M \rangle$ *sn* F so that $\langle A, N \rangle$ *sn* F, a contradiction. □

Condition (ii) in Theorem 7.4.2 leads to the formulation of the following definition: we say that a subgroup A of a group G *preserves subnormality* in G, and write A *psn* G, if and only if we have $\langle A, N \rangle$ *sn* F whenever $A \leqq F$ and $1 < N$ *sn* $F \leqq G$. Denote the set of such A by psn G. Theorem 7.4.2(ii) says that $A \in \mathscr{P}G$ implies A *psn* G. We note that if G satisfies Max, then $\text{sn}\, G \subseteq \text{psn}\, G$. However, consideration of the symmetric group of degree 3 shows that the reverse inclusion does not hold.

Theorem 7.4.3. *Suppose that G is a group with* Max. *Then*

(1) psn G *is closed with respect to forming joins;*

(2) *if A psn F where $F \leqq G$, then there exists a unique largest subgroup $S_F(A)$ of F such that A sn $S_F(A)$;*

(3) *if A_1, A_2 psn F, $F \leqq G$, $A_1 \neq 1 \neq A_2$, then $S_F(A_1) = S_F(A_2)$ if and only if A_1, A_2 sn $\langle A_1, A_2 \rangle$.*

Proof. (1) is clear since groups with Max have the join property (Theorem 1.3.10).

(2) We need to show that if $A \in \text{psn}\, F$ and A *sn* F_i, where $F_i \leqq F$, $i = 1, 2$, then A *sn* $\langle F_1, F_2 \rangle$. Suppose that this is false and choose a counterexample with A maximal. If A is not normal in either F_1 or F_2, we may choose $f_i \in F_i$ such that $A \ntrianglelefteq A_i$, $A_i = \langle A, A^{f_i} \rangle$ *sn* F_i, $i = 1, 2$. Set $B = \langle A_1, A_2 \rangle$. Then B *psn* F, by (1). Now A_1, A_2 are subnormal in B since $A \lhd B$ and $A^{f_i} \in \text{psn}\, F$. It follows that A_i *sn* $\langle B, F_i \rangle$, $i = 1, 2$, by choice of A. By definition of *psn* we have $B = \langle B, A_i \rangle$ *sn* $\langle B, F_i \rangle$, $i = 1, 2$, and therefore B *sn* $\langle F_1, F_2 \rangle$, again by choice of A. Thus A *sn* $\langle F_1, F_2 \rangle$, a contradiction.

If A is normal in F_1, say, then we may assume that A is not normal in F_2 and an argument similar to, but easier than, the above leads to the desired conclusion.

(3) Suppose that A_i *sn* $\langle A_1, A_2 \rangle$, $A_i \in \text{psn}\, F$, $i = 1, 2$. Then if $S_F(A_i) = S_i$, we have $\langle A_1, A_2 \rangle \leqq S_1 \cap S_2$. Now $\langle A_1, A_2 \rangle \in \text{psn}\, F$ by (1) and it follows that $\langle A_1, A_2 \rangle$ *sn* S_1. Hence A_2 *sn* S_1 so that $S_1 \leqq S_2$ by uniqueness. Similarly $S_2 \leqq S_1$ and $S_1 = S_2$, as required. The converse is trivial. □

Combining Theorems 7.4.2 and 7.4.3 we obtain

Theorem 7.4.4. *Let $\mathscr{P}$ be a Sylow property and G be a group with* Max. *In trying to prove that $\mathscr{P}G \subseteq \text{sn}\, G$, we may without loss of generality assume that for all $F \leqq G$* (i) *$\mathscr{P}F \subseteq \text{psn}\, F$, and* (ii) *whenever $E \leqq F$, $E = \langle A_1, \ldots, A_r \rangle$ where A_i psn F, then $S_F(E)$ exists.*

We can now prove Theorem 7.4.1. Suppose that

$$\mathscr{P}_1 G = \{A \leqq G \mid A \textit{ sn } \langle A, A^g \rangle \text{ for all } g \in G\}.$$

Then in order to show $\mathscr{P}_1 G \subseteq \text{sn}\, G$ for G satisfying Max, we may assume by Theorem 7.4.4 that $S_G(A)$ exists and it therefore contains A^g for all $g \in G$. Hence

$$A \textit{ sn } A^G \lhd G$$

so that A *sn* G, as required. □

Remarks

1. The property $\mathscr{P}_2$ defined by

$$\mathscr{P}_2 G = \{A \leqq G \mid A^g A = A A^g \text{ for all } g \in G\}$$

is a Sylow property. However, the above techniques are not enough to prove that $\mathscr{P}_2 G \subseteq \text{sn}\, G$ for groups G with Max.

2. It is not difficult to deduce from Theorem 7.4.4 the following well-known result of Baer [7].

Theorem 7.4.5. *If G is a group with* Max, *then its Fitting subgroup is equal to the set of all left Engel elements of G.*

§7.5 Polycyclic-by-finite groups

Suppose that G is a group and ϕ a homomorphism of G. We say that ϕ is a *finite* homomorphism if G^{ϕ} is finite. One way of studying the subnormal structure of an infinite group G is to look at the subnormal structure of its finite images and then try to deduce something about G itself. Of course this method has chance of success only in cases where G has plenty of finite images, for example, if G is residually finite. The first question that arises in connection with subnormality is: given $H \leqq G$ and $H^{\phi}\, sn\, G^{\phi}$ for all finite homomorphisms ϕ of G, for what classes of groups G can we deduce $H\, sn\, G$? Our first result shows that the class of polycyclic-by-finite groups qualifies. This is well known to be the **P**-closure of the class consisting of finite or cyclic groups.

Theorem 7.5.1 (Kegel [70]). *Suppose that G is a polycyclic-by-finite group and $H \leqq G$ such that $H^{\phi}\, sn\, G^{\phi}$ for all finite homomorphisms ϕ of G. Then $H\, sn\, G$.*

We need a simple lemma on finitely generated torsion-free abelian groups.

Lemma 7.5.2. *Suppose that A is a finitely generated torsion-free abelian group and $B \leqq A$. Then there exists a positive integer m such that*

$$B = BA^{m} \cap \left(\bigcap_{p \in \pi} BA^{p} \right)$$

where π is any infinite set of primes.

Proof. By the basis theorem, A and B have bases $a_1, \ldots, a_n$ and $b_1, \ldots, b_t$, respectively, where $b_i = a_i^{k_i}$ and k_{i-1} divides k_i for $i = 2, \ldots, t$. Let $m = k_t$. Then $A^m \leqq B \times D$, where $D = \langle a_{t+1}, \ldots, a_n \rangle$ and it is easy to check that

$$BA^{m} \cap \left(\bigcap_{p \in \pi} BA^{p} \right) \leqq B\left(\bigcap_{p \in \pi} D^{p} \right) = B,$$

since π is infinite. Equality follows. □

Proof of Theorem 7.5.1. The *Hirsch length* $h(G)$ of G is defined to be the number of infinite cyclic factors in any series of G for which the factors are cyclic or finite. By Lemma 2.3.1, $h(G)$ is an invariant of G. We proceed by induction on $h(G)$, the result being trivial if $h(G) = 0$. Thus we may assume that G is infinite and so G has a torsion-free abelian normal subgroup $A \neq 1$. Set $B = A \cap H$ and let m be the integer produced by Lemma 7.5.2. Then A^m is infinite and so by the induction hypothesis $A^m H \, sn \, G$. Hence $[A, {}_n H] \leqq A^m B$, for some n. Moreover $A^p H \, sn \, G$ and since $|A/A^p| = p^r$, where $r = \text{rank } A$, we have $A^p H \lhd^r AH$. Hence we have

$$[A, {}_{n+r} H] \leqq A^m B \cap \left(\bigcap_{p \in \pi} A^p B \right),$$

where π is any set of primes. By Lemma 7.5.2, $[A, {}_{n+r} H] \leqq B \leqq H$ so that $H \, sn \, AH$. But $AH \, sn \, G$ by the induction hypothesis, so we have $H \, sn \, G$, as required. □

Theorem 7.5.1 implies the solubility of a decision problem for polycyclic-by-finite groups.

Corollary 7.5.3. *There is an algorithm for deciding whether or not a given subgroup of a polycyclic-by-finite group is subnormal.*

In [75] Lennox shows that there is no analogous result for finitely generated metabelian minimax groups. If we take G to be the group with presentation

$$G = \langle a, x, y \mid a^x = a^{-1}, a^y = a^2, [x, y] = 1 \rangle,$$

then it is easy to check that with $H = \langle a, x \rangle$, $H^\phi \lhd G^\phi$ for any finite homomorphism ϕ of G, but H, though ascendant, is not subnormal in G.

Mal'cev in [99] proved that each subgroup H of a polycyclic-by-finite group G is closed in the profinite topology on G, that is H is equal to the intersection of all subgroups of finite index in G which contain H. (For a short proof see Wilson [169].)

However, in the above example the subgroup H is not closed in the profinite topology on G. In [16] Brewster and Lennox prove, by a detailed analysis of the profinite and related topologies, that *if G is a soluble minimax group and H is a closed subgroup in G such that H^ϕ sn G^ϕ for all finite homomorphisms ϕ of G, then H sn G.*

They also give an example to show that the result fails to hold if we omit the requirement that G be minimax, even for closed subgroups of a finitely generated metabelian group.

Kegel's theorem can be used to extend many of the foregoing subnormality criteria from the class of finite groups to that of polycyclic-by-finite groups. Sometimes the extension is immediate. As an example of a case where it is not immediate, but easy, we give the extension of Wielandt's theorem (Theorem 7.2.5).

Theorem 7.5.4. *Suppose that G is a polycyclic-by-finite group, $A, B \leqq G$ and $AB^x = B^x A$ for all $x \in G$. Then we have*

(a) *if $G = AB^G = BA^G$, then $G = AB$;*

(b) $A^B \cap B^A \, sn \, G$;

(c) *if* $AB \leqq H \leqq G$, *then* $A^H \cap B^H$ *sn* G;
(d) *if* X, Y *are arbitrary subsets of* G, *then* $[A^X, B^Y]$ *sn* G.

Proof. (a) By Kegel's theorem we have at once that $G = ABN$ for all normal subgroups N of finite index in G. But AB is a subgroup of G by hypothesis and so, by Mal'cev's theorem mentioned above, $G = AB$.

(b) This is not an immediate consequence of Kegel's result, since if we set $C = A^B \cap B^A$ and take ϕ to be a finite homomorphism of G, then denoting images under ϕ by bars, we do not necessarily have $\bar{C} = \bar{A}^{\bar{B}} \cap \bar{B}^{\bar{A}}$. However, once we have (b), (c) and (d) are deduced as in the finite case. We prove (b) as follows. Let N be the kernel of ϕ. We work in $G/N \cong G^\phi$. By Theorem 7.2.5(b), $\bar{A}^{\bar{B}} \cap \bar{B}^{\bar{A}}$ *sn* $\bar{G}$ which implies that $D = (A^B N) \cap (B^A N)$ *sn* G. Clearly $C \leqq D$ so that for sufficiently large n, $[G, {}_nC] \leqq D \leqq ABN$. Therefore on commutating once more with C we obtain

$$[G, {}_{n+1}C] \leqq [ABN, C] \leqq CN,$$

since $N \lhd G$ and $C \lhd AB$. Thus we do in fact have CN *sn* G, that is C^ϕ *sn* G^ϕ. By Theorem 7.5.1, C *sn* G as required. □

§7.6 Join-subnormality

Suppose that H and K are subgroups of a group G. Following Wielandt [163], we shall say that H and K are *join-subnormal* if both H and K are subnormal in their join $\langle H, K \rangle$. Using this concept we can rephrase Theorem 7.3.3(ii)—*if* G *is a finite group and* $H \leqq G$ *is such that any two conjugates of* H *are join-subnormal, then* H *is subnormal in* G. Thus in this situation H^G is generated by pairwise join-subnormal subgroups and if $1 \neq H \neq G$, we conclude that $H^G < G$ and so G is not simple. Wielandt considers the case where a finite group G is generated by pairwise join-subnormal subgroups $G_1, \ldots, G_n$ and raises the question of the influence of this fact on the normal structure of G—must the G_i be subnormal in G, or can one at least say that G is non-simple in non-trivial cases? Wielandt points out that both of these questions have a negative answer. The alternating group G on 9 symbols is generated by the three permutations $g_1 = (123)(456)(789)$, $g_2 = (147)$ and $g_3 = (259)$. If we set $G_i = \langle g_i \rangle, i = 1, 2, 3$, then G_i and G_k are join-subnormal since they generate a 3-group—indeed for $i < k$ the G_i-conjugates of g_k generate an abelian group. However, if additional conditions are imposed on the G_i, we obtain useful positive answers.

Theorem 7.6.1 (Wielandt [163]). *Suppose that the finite group* G *is generated by pairwise permutable subgroups* $G_1, \ldots, G_n$ *which are pairwise join-subnormal. Then each* G_i *is subnormal in* G.

This result will be superseded by Theorem 7.7.1. The proof depends on

Wiedlandt's theory of operators developed in Chapter 4. We need one further operator not mentioned in that chapter.

Suppose $\mathfrak{X}$ is a class of groups. For a group H we use $H_{\mathfrak{X}}$ to denote the join of all subnormal subgroups of H which belong to $\mathfrak{X}$. We then have

Proposition 7.6.2. *Suppose that $\mathfrak{X}$ is a class of perfect simple groups and H is a subnormal subgroup of a group G. Then $H_{\mathfrak{X}}$ normalizes every subnormal subgroup of G and $H \mapsto H_{\mathfrak{X}}$ is an operator.*

The first assertion follows at once from Theorem 4.6.1 and the second from

Lemma 7.6.3. *Suppose H and K are join-subnormal and S is a non-abelian simple normal subgroup of $\langle H, K \rangle$. Then either $S \leqq H$ or $S \leqq K$.*

Proof. Suppose that $S \nleqq H$. Then $S \cap H = 1$ and by Remark 3 on p. 54 we have $[H, S] = 1$. If in addition $S \nleqq K$, then $[K, S] = 1$ so that S is in the centre of $\langle H, K \rangle$. However, S is non-abelian, a contradiction. □

Proof of Theorem 7.6.1. We prove $G_1 \, sn \, G$. Suppose the result to be false. We choose a minimal counterexample $(G; G_1, \ldots, G_n)$ in the sense that $d = |G| + |G_1| + \cdots + |G_n|$ is least possible with G_1 not subnormal in G. Clearly $n \geqq 3$ and $G_n \neq 1$. We select a maximal normal subgroup M_n of G_n and let $\mathfrak{X}$ be the P-closure of the class (G_n/M_n). Then by Theorem 4.1.3 the mapping $H \mapsto H^{\mathfrak{X}}$ is an operator which we apply to the groups $G_2, \ldots, G_n$, but not to G_1. According to (4) and (5) on p. 134, the pairwise join-subnormal subgroups $G_1, G_2^{\mathfrak{X}}, \ldots, G_n^{\mathfrak{X}}$ are pairwise permutable. Since $|G_n^{\mathfrak{X}}| < |G_n|$, the minimality hypothesis leads at once to the conclusion that

$$G_1 \, sn \, \langle G_1, G_2^{\mathfrak{X}}, \ldots, G_n^{\mathfrak{X}} \rangle = G_1 N$$

where

$$N = \langle G_1^{\mathfrak{X}}, \ldots, G_n^{\mathfrak{X}} \rangle \lhd G.$$

The normality follows easily from the fact that

$$\langle G_i, G_k \rangle^{\mathfrak{X}} = \langle G_i^{\mathfrak{X}}, G_k^{\mathfrak{X}} \rangle \lhd \langle G_i, G_k \rangle, \qquad \text{for all } i, k.$$

If $N \neq 1$, then $|G/N| < |G|$ and we conclude that $G_1 N \, sn \, G$ from which we get the contradiction $G_1 \, sn \, G$. Hence $N = 1$, from which it follows that all composition factors of each G_i are isomorphic to $G_n/M_n = S$, say.

Suppose that S is perfect. We then apply the same argument as above with the operator $H \mapsto H_{\mathfrak{X}}$ instead of $H \mapsto H^{\mathfrak{X}}$. Corresponding to N we obtain the normal subgroup $L = \langle (G_1)_{\mathfrak{X}}, \ldots, (G_n)_{\mathfrak{X}} \rangle$, which this time cannot be 1 since it contains the minimal subnormal subgroups of G_n. Hence $G_1 L \, sn \, G$. However, L normalizes G_1 by Proposition 7.6.2 and again we obtain the contradiction $G_1 \, sn \, G$.

Hence S is a cyclic group of order p, a prime, and therefore each G_i is a p-group. But then so is their product $G = G_1 \cdots G_n$ and hence again we have, as our final contradiction, that $G_1 \; sn \; G$. □

Wielandt's main result on join-subnormality is

Theorem 7.6.4. *Suppose that G is a finite group generated by pairwise join-subnormal subgroups $G_1, \ldots, G_n$. Then*

(a) *G_j is subnormal in G, for fixed j, if $G_j^\phi \; sn \; G^\phi$ for every homomorphism ϕ of G which maps all the G_i onto p-groups (for the same prime p);*

(b) *G_i is subnormal in G, for all $i = 1, 2, \ldots, n$, if G^ϕ is a p-group for each of the homomorphisms ϕ mentioned in* (a).

That these sufficient conditions are also necessary is clear.

An interesting criterion for the non-simplicity of a group can be deduced from Theorem 7.6.4.

Theorem 7.6.5. *Suppose G is a finite group which is generated by pairwise join-subnormal proper subgroups $G_1, \ldots, G_n$ which are not all p-groups for the same prime p. Then G is not simple.*

Proof. Suppose G is simple. Clearly we may assume that $G_1 \neq 1$. Hence G_1 is not subnormal in G since $G_1 < G$. By Theorem 7.6.4(a) there exists a homomorphism ϕ of G such that all G_i^ϕ are p-groups and G_1^ϕ is not subnormal in G^ϕ. Hence $G_1^\phi \neq 1$ and so ϕ is an isomorphism since G is simple. Hence all G_i are p-groups, a contradiction. □

The example given in the introduction to this section shows that the alternating group on 9 symbols is generated by 3 pairwise join-subnormal subgroups of order 3.

For the proof of Theorem 7.6.4 we need two auxiliary results, the first of which shows that if G is generated by pairwise join-subnormal subgroups, then their nilpotent residuals are subnormal in G.

Proposition 7.6.6. *Suppose the finite group G is generated by pairwise join-subnormal subgroups $G_1, \ldots, G_n$. Then if $G_i^{\mathfrak{N}}$ is the nilpotent residual of G_i and $N = \langle G_1^{\mathfrak{N}}, \ldots, G_n^{\mathfrak{N}} \rangle$, we have* (a) *$G_i^{\mathfrak{N}} \; sn \; G$ and* (b) *$G_i \; sn \; G_i N$, for all i.*

Proof. (a) By Theorem 4.4.1 the $G_i^{\mathfrak{N}}$ satisfy the hypotheses of Theorem 7.6.1 and therefore $G_i^{\mathfrak{N}} \; sn \; N$. But $N \lhd G$ (see the argument in the proof of Theorem 7.6.1) and so $G_i^{\mathfrak{N}} \; sn \; G$.

(b) Again by Theorem 4.4.1 the subgroups $G_1, G_2^{\mathfrak{N}}, \ldots, G_n^{\mathfrak{N}}$ satisfy the hypotheses of Theorem 7.6.1 and hence G_1 is subnormal in their join $G_1 N$. □

We now consider the group $\bar{G} = G/N$, where G and N are as in the statement of Proposition 7.6.6. Clearly $\bar{G}$ is generated by the subgroups $\bar{G}_i = G_i N/N$, which are all nilpotent since $G_i^{\mathfrak{N}} \leqq N$. Of course the G_i are pairwise join-subnormal since the G_i are. In the next theorem Wielandt describes the structure of such a join of pairwise join-subnormal nilpotent subgroups showing that it is essentially a central product of the joins of corresponding Sylow subgroups.

Theorem 7.6.7. *Suppose that the finite group G is generated by pairwise join-subnormal nilpotent subgroups $G_1, \ldots, G_n$.*

(a) *Let p be a prime and P_i, C_i be the Sylow p-subgroup and p-complement, respectively, of G_i. Let P be the join of the P_i, C that of the C_i. Then*

$$P \lhd G, \quad C \lhd G, \quad PC = G, \quad [P, C] = 1.$$

(b) *Suppose further that $p, q, \ldots, r$ are the distinct prime divisors of $\prod_{i=1}^{n} |G_i|$ and define $Q, \ldots, R$ in analogous fashion to P. Then $G = PQ \cdots R$ and the factors centralize each other.*

We note that the example of A_9 cited earlier shows that P need not be a p-group.

Proof. (a) Since the G_i are nilpotent, it follows from Theorem 4.4.1 that $\langle G_i, G_k \rangle$ is nilpotent. Thus P_i, C_k are subgroups of coprime order in a nilpotent group and hence commute elementwise. Therefore $[P, C] = 1$. Moreover

$$PC = \langle P, C \rangle = \langle P_1, \ldots, P_n, C_1, \ldots, C_n \rangle = \langle G_1, \ldots, G_n \rangle = G.$$

(b) The subgroups C, $C_1, \ldots, C_n$ satsify the same hypotheses as G, $G_1, \ldots, G_n$ and since $\prod_{i=1}^{n} |C_i|$ has only the prime divisors $q, \ldots, r$, we obtain the desired result by induction on the number of prime divisors. □

We are now in a position to prove the main result, Theorem 7.6.4. So suppose that G is a finite group generated by pairwise join-subnormal subgroups $G_1, \ldots, G_n$. We assume that G_j^{ϕ} *sn* G^{ϕ} for every homomorphism ϕ of G which maps all the G_i onto p-groups for the same prime p. We are required to prove that G_j *sn* G.

We suppose that this is false and choose a counterexample with $d = |G| + |G_1| + \cdots + |G_n|$ minimal. Thus we have at once that if $1 \neq N \lhd G$, then $G_j N$ *sn* G. We claim that all the G_i are nilpotent. If not, then $1 \neq \langle G_1^{\mathfrak{N}}, \ldots, G_n^{\mathfrak{N}} \rangle = N \lhd G$ and so $G_j N$ *sn* G. However, G_j *sn* $G_j N$ by Proposition 7.6.6. Thus G_j *sn* G, a contradiction.

However, the G_i are not all p-groups for the same prime p, for in that case G_j *sn* G by hypothesis, taking ϕ to be the identity automorphism. Thus $\Pi\, |G_i|$ has at least 2 distinct prime divisors. Taking one of them to be p, we must have at least one $C_i \neq 1$, in the notation of Theorem 7.6.7. By that result $[P_j, C] = 1$ and $C \lhd G$. Hence G_jC *sn* G and we obtain

$$P_j \lhd P_jC = P_jC_jC = G_jC \, sn \, G.$$

Thus the Sylow p-subgroup of G_j is subnormal in G. Since this holds for all Sylow subgroups, we have that G_j *sn* G, a contradiction. We have proved (a). Part (b) follows at once since p-subgroups are subnormal in their join if and only if this is a p-group. □

We finally record the suggestion made by Wielandt ([163] and [164]) that progress with join-subnormality might well be made by considering the problems from a graph-theoretic point of view (vertices are subgroups and an edge between two vertices denotes join-subnormality).

Wielandt also asks the question: suppose that A and B are subgroups of a finite group such that for each $a \in A$, $b \in B$ there exists a positive integer n such that $[b, {}_na] \in A$ and $[a, {}_nb] \in B$. Are A and B join-subnormal?

§7.7 The subnormalizer of a subgroup

Suppose that H is a subgroup of a group G. Then the normalizer $N_G(H)$ of H can of course be defined by

$$N_G(H) = \{g \mid H \lhd \langle H, g \rangle\}$$

and H is a normal subgroup of G if and only if $N_G(H) = G$. Following Wielandt we may in analogous fashion form the set

$$S_G(H) = \{g \mid H \, sn \, \langle H, g \rangle\}$$

and call it the *subnormalizer* of H in G. (The notation was anticipated in Theorem 7.4.3.) It is an immediate corollary of Theorem 7.3.3(i) that for finite groups G, H is subnormal in G if and only if $S_G(H) = G$. However, whereas $N_G(H)$ is always a subgroup, it is not in general true, even for finite groups, that $S_G(H)$ is a subgroup. As an illustrative example the symmetric group G of degree 5 can be generated by two subgroups A, B each of order 8 whose intersection H has order 2. Thus H *sn* A and H *sn* B, so that both A and B are contained in $S_G(H)$. If $S_G(H)$ were a subgroup, then we would have $S_G(H) = G$ and so H *sn* G, which is not the case. Another example is of course the alternating group of degree 9; see p. 234.

One situation, however, where the subnormalizer is always a subgroup is given to us by Wielandt's maximizer lemma (Lemma 7.3.1)—if G is a finite group and H is a subgroup of G such that H is not subnormal in G but is subnormal in every proper subgroup K of G with $H \leqq K$, then $S_G(H)$ is a (proper) subgroup of G. In fact $S_G(H)$ is the Wielandt maximizer of H here. Also for any finite group G and subgroup H, it is easy to see that $S_G(H)$ is a subgroup if and only if for all subgroups A, B of G with H *sn* A, H *sn* B, it follows that H *sn* $\langle A, B\rangle$. This latter condition therefore takes on interest in its own right.

In the example of the symmetric group G of degree 5 mentioned above it is clear that the subgroups A and B do not permute, for otherwise $G = AB$ would be a 2-group. That this is no coincidence is shown by the following theorem, first proved by Maier in [94] for soluble subgroups and then for arbitrary subgroups by Wielandt [165] (see also Maier [95]).

Theorem 7.7.1. *Suppose that $G = AB$ is a finite group, with A and B subgroups and H sn A, H sn B. Then H sn G.*

Proof. Suppose that the result is false. We choose a counterexample G of minimal order for which $|G:A|+|H|$ is also minimal and proceed to derive a contradiction. We deduce

(i) A is a maximal subgroup of G. For if $A < U < G$, we have $G = UB$ and $U = A(B \cap U)$. Then by the minimality of $|G|$ we have H *sn* U and since $|G:A|+|H| > |G:U|+|H|$, we have H *sn* $UB = G$, a contradiction.

(ii) $A_G = \mathrm{core}_G(A) = 1$. For if not and $N = A_G$, then we have HN *sn* G and since $HN \leqq A$, we have H *sn* HN. Hence H *sn* G, a contradiction.

(iii) There are no non-trivial subnormal subgroups of G contained in A. This follows from (i), (ii) and Lemma 7.3.16.

(iv) H is simple. For, if $1 < K \underset{\neq}{\lhd} H$, then K *sn* A and K *sn* B so that it follows from $|G:A|+|H| > |G:A|+|K|$ that K *sn* G, which is not so by (iii).

We must now distinguish the cases H abelian and non-abelian.

Case 1. $|H| = p$, p a prime. Since H is subnormal in both A and B it follows that H^A and H^B are p-groups. If S is a Sylow p-subgroup of A, then $H^A \leqq S$. Hence H is contained in all Sylow p-subgroups of both A and B. According to a theorem of Wielandt (see [63], VI 4.7) there are Sylow p-subgroups A_p and B_p of A and B, respectively, such that A_pB_p is a Sylow p-subgroup of G. Suppose $g \in G$. Then $g = ab$, $a \in A$, $b \in B$, and it follows from the above remarks that $\langle H^a, H^{b^{-1}}\rangle$ is contained in A_pB_p and is therefore a p-group. Transforming by b shows that $\langle H, H^g\rangle$ is a p-group so that H *sn* $\langle H, H^g\rangle$. It now follows from Theorem 7.3.3 that H *sn* G, a contradiction. Hence H is not abelian.

Case 2. H non-abelian simple. We define

$$\mathscr{S} = \{K \mid H \leqq K \leqq A \cap B,\ K\ sn\ A,\ K\ sn\ B,\ K \text{ is semisimple}\}.$$

We recall here that a semisimple group is a direct product of non-abelian simple groups (see §2.2). Since $H \in \mathscr{S}$, we have $\mathscr{S} \neq \varnothing$ and so

$$C = \langle K \mid K \in \mathscr{S}\rangle \neq 1.$$

By Theorem 2.2.10 we have $H \lhd C$, $C \in \mathscr{S}$, but C is not subnormal in G, since H is not subnormal in G. By Theorem 7.3.4(i), there exists $g \in \langle C, C^g \rangle$, $g \notin C$. We may write $g = ab$, $a \in A$, $b \in B$. Thus $ab \in \langle C, C^{ab} \rangle$ so that $ba = (ab)^{b^{-1}} \in \langle C^{b^{-1}}, C^a \rangle$. Now C and C^a are semisimple subnormal subgroups of A and hence C^a normalizes C, again by Theorem 2.2.10. Similarly $C^{b^{-1}}$ normalizes C. Hence $ba \in N_G(C)$ and so $C^{ba} = C$. Therefore $K = C^b = C^{a^{-1}}$ is semisimple and subnormal in A and in B and hence the same is true of $\langle H, K \rangle$. Thus $\langle H, K \rangle \in \mathscr{S}$ and so $K \leqq C$, that is $K = C$. Then a, b and g all normalize C and therefore $g \in C$, a contradiction. □

It is easy to deduce from Theorem 7.7.1 an important subnormality criterion.

Theorem 7.7.2. *Suppose that $G = AB$ is a finite group, where A and B are subgroups, and $H \leqq G$. Then H is subnormal in G if and only if H sn H^A and H sn H^B.*

Proof. Suppose that H sn H^A and H sn H^B. Then $H^A \lhd \langle H, A \rangle = A^*$, say, and so H sn A^*. Similarly H sn $B^* = \langle H, B \rangle$. Clearly $G = A^*B^*$ and by Theorem 7.7.1 H sn G as required. The converse is trivial. □

As Wielandt also shows in [165], some of the results on join-subnormality (§7.6) can be used to deduce the following subnormality criteria from Theorem 7.7.2.

Theorem 7.7.3. *Suppose that H is a subgroup of the finite group $G = AB$, where A and B are subgroups.*

(a) *If H sn $HH^x = H^xH$ for all $x \in A \cup B$, then H sn G.*

(b) *Suppose that H sn $\langle H, H^x \rangle$ for all $x \in A \cup B$. Then $H^{\mathfrak{N}}$ sn G. Thus if $H = H'$, then H sn G.*

(c) *Suppose that $HH^x = H^xH$ for all $x \in A \cup B$. Let P be a normal Sylow subgroup of H. Then P sn G. If H is nilpotent, then H sn G.*

For the proof we need the following criteria for H sn $\langle H, X \rangle$.

Proposition 7.7.4. *Suppose that H, X are subgroups of the finite group G.*

(a) *If H sn $HH^x = H^xH$ for all $x \in X$, then H sn $\langle H, X \rangle$.*

(b) *If $HH^x = H^xH$ for all $x \in X$, then a normal Sylow p-subgroup P of H is subnormal in $\langle P, X \rangle$.*

Proof. (a) Since H sn HH^x for all elements $x \in X$, it follows that H^x sn H^xH^y for all $x, y \in X$ and so conjugates of H under X are pairwise join-subnormal. By Theorem 7.6.1, H sn $\langle H^x | x \in X \rangle$. But the latter group is normal in $\langle H, X \rangle$.

(b) By the result of Wielandt on Sylow subgroups of products mentioned in the proof of Case 1 of Theorem 7.7.1, we have that $PP^x = P^xP$ is a Sylow p-subgroup of HH^x for each $x \in X$. But then P sn PP^x and so P sn $\langle P, X \rangle$ by (a). □

We now prove Theorem 7.7.3. For the first part we assume that $H \leqq G = AB$ and that H *sn* $HH^x = H^xH$ for all $x \in A \cup B$. By Proposition 7.7.4(a) we obtain H *sn* $\langle H, A\rangle$ and H *sn* $\langle H, B\rangle$. That H *sn* G now follows from Theorem 7.7.2.

For the second part we assume that H *sn* $\langle H, H^x\rangle$ for all $x \in A \cup B$. By Proposition 7.6.6(a), $H^{\mathfrak{N}}$ *sn* $\langle H, A\rangle$ and we again apply Theorem 7.7.2.

Finally in the third part we have from Proposition 7.7.4(b) that P *sn* $\langle P, A\rangle$ and we apply Theorem 7.7.2 once more. If H is nilpotent, then all of its Sylow subgroups are subnormal in G and so also is H by Wielandt's join theorem. □

Wielandt gives examples to show that certain conditions rather surprisingly fail to enable one to deduce results of the type H *sn* A, H *sn* $B \Rightarrow H$ *sn* $\langle A, B\rangle$. For example not even $H \leqq \zeta_1(A)$, H *sn* B and solubility of $\langle A, B\rangle$ suffice to ensure H *sn* $\langle A, B\rangle$. For, let p be an odd prime and consider the subgroups $H = \langle h\rangle$, $A = \langle h, a\rangle$, $B = \langle h, b\rangle$ of $GL(3, p)$, where

$$h = \begin{pmatrix} 1 & 0 & 0 \\ 0 & 1 & 0 \\ 0 & 0 & -1 \end{pmatrix}, \qquad a = \begin{pmatrix} 1 & 1 & 0 \\ 0 & 1 & 0 \\ 0 & 0 & 1 \end{pmatrix}, \qquad b = \begin{pmatrix} 1 & 0 & 0 \\ 0 & 0 & 1 \\ 0 & 1 & 0 \end{pmatrix}.$$

It is clear that $H \leqq \zeta_1(A)$ and H *sn* B. Suppose, however, that H *sn* $\langle A, B\rangle$. Let $K = b^{-1}Hb$. Then $K = \langle k\rangle$, where

$$k = \begin{pmatrix} 1 & 0 & 0 \\ 0 & -1 & 0 \\ 0 & 0 & 1 \end{pmatrix},$$

and we deduce that K *sn* $\langle A, B\rangle$ and hence K *sn* $\langle k, a\rangle$. But this is not the case since a has order p and is inverted under conjugation by k. Therefore H is not subnormal in $\langle A, B\rangle$. Of course the example of A_9 (p. 234) has the same features except for solubility.

Conjectures (Wielandt [165]). *Suppose that $H \leqq G = AB$, a finite group, with A and B subgroups. Then*

(a) *H sn G if $HH^x = H^xH$ for all $x \in A \cup B$;*

(b) *H sn G if H sn $\langle H, H^x\rangle$ for all $x \in A \cup B$;*

(c) *H sn G if H sn $\langle H, x\rangle$ for all $x \in A \cup B$.*

The first two conjectures, if true, would strengthen Theorem 7.7.3 considerably and one might hope that they could be proved by using corresponding analogues for Proposition 7.7.4. Wielandt gives examples to show that these analogues are actually false so that other methods will have to be found.

The *subnormalizer* $S_G(H)$ defined at the beginning of this section is not the only possible candidate for that role. Indeed other subnormality criteria give rise to 'subnormalizers'. For example, in view of Theorem 7.3.3(ii) and (iii) and

Theorem 7.3.11(i), we might define

$$S_G^1(H) = \{g \mid H \text{ } sn \text{ } \langle H, H^g \rangle\},$$
$$S_G^2(H) = \{g \mid H \text{ } sn \text{ } \langle H, H^{h^g} \rangle \text{ for all } h \in H\},$$
$$S_G^3(H) = \{g \mid [g, {}_n h] \in H \text{ for all } h \in H, \, n = n(g, h)\}.$$

Note that

$$S_G(H) \subseteq S_G^1(H) \subseteq S_G^2(H) \subseteq S_G^3(H).$$

The problem is which, if any, of these is a good candidate for the elusive subnormalizer. Perhaps it is in this direction that progress will be made in the next 40 years.

BIBLIOGRAPHY

1. Abramovskiĭ, I. N. Locally generalized Hamiltonian groups, *Sibirsk. Mat. Ž.* **7**, 481–5 (1966) = *Siberian Math. J.* **7**, 391–3 (1966).
2. Abramovskiĭ, I. N. The structure of locally generalized Hamiltonian groups, *Leningrad Gos. Ped. Inst. Učen. Zap.* **302**, 43–9 (1967).
3. Alperin, J. L. and Lyons, R. On conjugacy classes of p-elements, *J. Algebra* **19**, 536–7 (1971).
4. Baer, R. Group elements of prime power index, *Trans. Amer. Math. Soc.* **75**, 20–47 (1953).
5. Baer, R. Nilgruppen, *Math. Z.* **62**, 402–37 (1955).
6. Baer, R. Lokal Noethersche Gruppen, *Math. Z.* **66**, 341–63 (1957).
7. Baer, R. Engelsche Elemente Noetherscher Gruppen, *Math. Ann.* **133**, 256–70 (1957).
8. Baer, R. Polyminimaxgruppen, *Math. Ann.* **175**, 1–43 (1968).
9. Baer, R. and Heineken, H. Radical groups of finite abelian subgroup rank, *Illinois J. Math.* **16**, 533–80 (1972).
10. Bartels, D. Subnormality and invariant relations on conjugacy classes in finite groups, *Math. Z.* **157**, 13–17 (1977).
11. Baumslag, G. Wreath products and p-groups, *Proc. Cambridge Philos. Soc.* **55**, 224–31 (1959).
12. Baumslag, G., Kovács, L. G. and Neumann, B. H. On products of normal subgroups, *Acta Sci. Math.* (Szeged) **26**, 145–7 (1965).
13. Best, E. and Taussky, O. A class of groups, *Proc. Roy. Irish Acad. Sect.* **A47**, 55–62 (1942).
14. Bradway, R. H., Gross, F. and Scott, W. R. The nilpotence class of core-free quasinormal subgroups, *Rocky Mountain J. Math.* **1**, 375–82 (1971).
15. Brewster, D. C. A criterion for the permutability of subnormal subgroups, *J. Algebra* **36**, 85–7 (1975).
16. Brewster, D. C. and Lennox, J. C. Subnormality and quasinormality in soluble minimax groups, *J. Algebra* **48**, 368–81 (1977).
17. Brookes, C. J. B. Groups with every subgroup subnormal, *Bull. London Math. Soc.* **15**, 235–8 (1983).
18. Busetto, G. Proprietà di immersione dei sottogruppi quasinormali periodici, *Sem. Mat. Univ. Catania* **34**, 243–9 (1979).
19. Camina, A. Hypernormalizing groups, *Math. Z.* **100**, 59–68 (1967).
20. Camina, A. Finite soluble hypernormalizing groups, *J. Algebra* **8**, 362–75 (1968).
21. Camina, A. The Wielandt length for finite groups, *J. Algebra* **15**, 142–8 (1970).
22. Cappit, D. On groups with every subgroup 2-subnormal, *J. London Math. Soc.* (2) **2**, 17–18 (1973).
23. Casolo, C. Gruppi finiti risolubili in cui tutti i sottogruppi subnormale hanno difetto al più 2, *Rend. Sem. Mat. Univ. Padova* **71**, 257–71 (1984).
24. Casolo, C. On groups with all subgroups subnormal, *Bull. London Math. Soc.* **17**, 397 (1985).
25. Černikov, S. N. Endlichkeitsbedingungen in der Gruppentheorie, Math. Forschungsberichte XX, Berlin: VEB Deutscher Verlag der Wissenschaften 1963; (German translation of Finiteness conditions in the general theory of

groups, *Uspehi Mat. Nauk.* **14**, 45–96 (1959) = *Amer. Math. Soc. Translations* (2) **84**, 1–67 (1969) (with a supplement by the author)).
26. Curtis, C. and Reiner, I. *Representation theory of finite groups and associative algebras*, New York: Interscience 1962.
27. Dark, R. S. A prime Baer group, *Math. Z.* **105**, 294–8 (1968).
28. Dedekind, R. Über Gruppen, deren sämtliche Teiler Normalteiler sind, *Math. Ann.* **48**, 548–61 (1897).
29. Deskins, W. E. On quasinormal subgroups of finite groups, *Math. Z.* **82**, 125–32 (1963).
30. Dietzmann, A. P. On p-groups, *Dokl. Akad. Nauk. SSSR* **15**, 71–6 (1937).
31. Dixon, J. D. *The structure of linear groups*, London: Van Nostrand 1971.
32. Drukker, M., Robinson, D. J. S. and Stewart, I. N. The subnormal coalescence of some classes of groups of finite rank. *J. Austral. Math. Soc.* **16**, 324–7 (1973).
33. Fitting, H. Beiträge zur Theorie der Gruppen endlicher Ordnung, *Jahresber. Deutsch. Math. Verein.* **48**, 77–141 (1938).
34. Fuchs, L. *Abelian groups*, Oxford: Pergamon 1960.
35. Fuchs, L. *Infinite abelian groups*, vol. 1, New York: Academic Press 1970.
36. Gaschütz, W. Gruppen, in denen das Normalteilersein transitiv ist, *J. reine angew. Math.* **198**, 87–92 (1957).
37. Gross, F. Subnormal, core-free, quasinormal subgroups are solvable, *Bull. London Math. Soc.* **7**, 93–5 (1975).
38. Gross, F. Infinite permutable subgroups, *Rocky Mountain J. Math.* **12**, 333–43 (1982).
39. Gruenberg, K. W. The upper central series in soluble groups, *Illinois J. Math.* **5**, 436–66 (1961).
40. Gruenberg, K. W. *Cohomological topics in group theory*, Lecture Notes in Mathematics 143, Berlin: Springer-Verlag (1970).
41. Hainzl, J. *Über die Vertauschbarkeit subnormalen Untergruppen*, Dissertation, Tübingen (1962).
42. Hall, M. A topology for free groups and some related groups, *Ann. of Math.* **52**, 127–39 (1950).
43. Hall, M. *The theory of groups*, New York: Macmillan 1959.
44. Hall, P. Finiteness conditions for soluble groups, *Proc. London Math. Soc.* (3) **4**, 419–36 (1954).
45. Hall, P. Finite-by-nilpotent groups, *Proc. Cambridge Philos. Soc.* **52**, 611–6 (1956).
46. Hall, P. Some sufficient conditions for a group to be nilpotent, *Illinois J. Math.* **2**, 787–801 (1958).
47. Hall, P. On the finiteness of certain soluble groups, *Proc. London Math. Soc.* (3) **9**, 595–622 (1959).
48. Hall, P. The Frattini subgroups of finitely generated groups, *Proc. London Math. Soc.* (3) **11**, 327–52 (1961).
49. Hall, P. On non-strictly simple groups, *Proc. Cambridge Philos. Soc.* **59**, 531–53 (1963).
50. Hall, P. and Hartley, B. The stability group of a series of subgroups, *Proc. London Math. Soc.* (3) **16**, 1–39 (1966).
51. Hartley, B. A note on the normalizer condition, *Proc. Cambridge Philos. Soc.* **74**, 11–15 (1973).

52. Hartley, B. On the normalizer condition and barely transitive groups, *Algebra i Logika* **13**, 589–602 (1975).
53. Hartley, B. and Peng, T. A. Subnormality, ascendancy, and the minimal condition on subgroups, *J. Algebra* **41**, 58–78 (1976).
54. Hawkes, T. O. Groups whose subnormal subgroups have bounded defect, *Arch. Math.* (Basel) **43**, 289–94 (1984).
55. Heineken, H. A class of three-Engel groups, *J. Algebra* **17**, 341–5 (1971).
56. Heineken, H. and Mohamed, I. J. A group with trivial centre satisfying the normalizer condition, *J. Algebra* **10**, 368–76 (1968).
57. Heineken, H. and Mohamed, I. J. Groups with normalizer condition, *Math. Ann.* **198**, 179–87 (1972).
58. Hickin, K. K. and Phillips, R. E. On classes of groups defined by systems of subgroups, *Arch. Math.* (Basel) **24**, 346–50 (1973).
59. Higman, G., Neumann, B. H. and Neumann, H. Embedding theorems for groups, *J. London Math. Soc.* **24**, 247–54 (1949).
60. Hirsch, K. A. Über lokal-nilpotente Gruppen, *Math. Z.* **63**, 290–4 (1955).
61. Huppert, B. Monomiale Darstellung endlicher Gruppen, *Nagoya Math. J.* **6**, 93–4 (1953).
62. Huppert, B. Über das Produkt von paarweise vertauschbaren zyklischen Gruppen, *Math. Z.* **58**, 243–64 (1953).
63. Huppert, B. *Endliche Gruppen I*, Berlin-Heidelberg-New York: Springer 1967.
64. Itô, N. and Szép, J. Über die Quasinormalteiler von endlichen Gruppen, *Acta Sci. Math.* (Szeged) **23**, 168–70 (1962).
65. Iwasawa, K. Einige Sätze über freie Gruppen, *Proc. Imp. Acad. Tokyo* **19**, 272–4 (1943).
66. Iwasawa, K. On the structure of infinite M-groups, *Japanese J. Math.* **18**, 709–28 (1943).
67. Kargapolov, M. I. Some problems in the theory of nilpotent and soluble groups, *Dokl. Akad. Nauk. SSSR* **127**, 1164–6 (1959).
68. Kegel, O. H. Produkte nilpotenter Gruppen, *Arch. Math.* (Basel) **12**, 90–3 (1961).
69. Kegel, O. H. Sylow-Gruppen und Subnormalteiler endlicher Gruppen, *Math. Z.* **78**, 205–21 (1962).
70. Kegel, O. H. Über den Normalisator von subnormalen und erreichbaren Untergruppen, *Math. Ann.* **163**, 248–58 (1966).
71. Koppe, H. Über Untergruppen, die mit ihren Konjugierten vertauschbar sind, *Math. Z.* **152**, 99 (1976).
72. Kuroš, A. G. *The theory of groups*, 2nd ed. (2 vols.), New York: Chelsea 1960.
73. Kuroš, A. G. and Černikov, S. N. Soluble and nilpotent groups, *Uspehi Mat. Nauk.* **2**, 18–59 (1947) = *Amer. Math. Soc. Translations* **80** (1953).
74. Lennox, J. C. A note on orthogonality and permutability of subnormal subgroups, *Proc. Cambridge Philos. Soc.* **72**, 351–5 (1972).
75. Lennox, J. C. Finitely generated metabelian groups are not subnormality separable, *Math. Z.* **149**, 201–2 (1976).
76. Lennox, J. C. On groups in which every subgroup is almost subnormal, *J. London Math. Soc.* (2) **15**, 221–31 (1977).
77. Lennox, J. C. A note on quasinormal subgroups of finitely generated groups, *J. London Math. Soc.* (2) **24**, 127–8 (1981).
78. Lennox, J. C. On quasinormal subgroups of certain finitely generated groups, *Proc. Edinburgh Math. Soc.* **26**, 25–8 (1983).

79. Lennox, J. C., Segal, D. and Stonehewer, S. E. The lower central series of a join of subnormal subgroups, *Math. Z.* **154**, 85–9 (1977).
80. Lennox, J. C. and Stonehewer, S. E. The join of two subnormal subgroups, *J. London Math. Soc.* (2) **22**, 460–66 (1980).
81. Lennox, J. C. and Wilson, J. S. A note on permutable subgroups, *Arch. Math.* (Basel) **28**, 113–6 (1977).
82. McCaughan, D. J. Subnormal structure in some classes of infinite groups, *Bull. Austral. Math. Soc.* **8**, 137–50 (1973).
83. McCaughan, D. J. Subnormality in soluble minimax groups, *J. Austral. Math. Soc.* **17**, 113–28 (1974).
84. McCaughan, D. J. and McDougall, D. The subnormal structure of metanilpotent groups, *Bull. Austral. Math. Soc.* **6**, 287–306 (1972).
85. McCaughan, D. J. and McDougall, D. Criteria for subnormality, *Arch. Math.* (Basel) **29**, 451–4 (1977).
86. McCaughan, D. J. and Stonehewer, S. E. Finite soluble groups whose subnormal subgroups have defect at most two, *Arch. Math.* (Basel) **35**, 56–60 (1980).
87. Macdonald, I. D. Some explicit bounds in groups with finite derived groups, *Proc. London Math. Soc.* (3) **11**, 23–56 (1961).
88. McDougall, D. Soluble minimax groups with the subnormal intersection property, *Math. Z.* **114**, 241–4 (1970).
89. McLain, D. H. A characteristically-simple group, *Proc. Cambridge Philos. Soc.* **50**, 641–2 (1954).
90. McLain, D. H. Remarks on the upper central series of a group, *Proc. Glasgow Math. Assoc.* **3**, 38–44 (1956).
91. McLain, D. H. Local theorems in universal algebras, *J. London Math. Soc.* **34**, 177–84 (1959).
92. Mahdavianary, S. K. A special class of three-Engel groups, *Arch. Math.* (Basel) **40**, 193–9 (1983).
93. Maier, R. Normality conditions for quasinormal subgroups in finite groups, *Math. Z.* **123**, 310–14 (1971).
94. Maier, R. Um problema da teoria dos subgrupos subnormais, *Bol. Soc. Brasil Mat.* **8**, 127–30 (1977).
95. Maier, R. Subnormalidade em produtos, *Trabalho de Matemática* No. 167, Universidade de Brasília (1980).
96. Maier, R. and Schmid, P. The embedding of quasinormal subgroups in finite groups, *Math. Z.* **131**, 269–72 (1973).
97. Mal'cev, A. I. On a general method for obtaining local theorems in group theory, *Ivanov. Gos. Ped. Inst. Učen. Zap.* **1**, 3–9 (1941).
98. Mal'cev, A. I. On certain classes of infinite soluble groups, *Mat. Sb.* **28**, 567–88 (1951) = *Amer. Math. Soc. Translations* (2) **2**, 1–21 (1956).
99. Mal'cev, A. I. On homomorphisms onto finite groups, *Ivanov. Gos. Ped. Inst. Učen. Zap.* **18**, 49–60 (1958).
100. Martin, J. E. and Pamphilon, J. A. Engel elements in groups with the minimal condition, *J. London Math. Soc.* (2) **6**, 281–6 (1973).
101. Meldrum, J. D. P. On the Heineken–Mohamed groups, *J. Algebra* **27**, 437–44 (1973).
102. Neumann, B. H. Groups with finite classes of conjugate subgroups, *Math. Z.* **63**, 76–96 (1955).
103. Nymann, D. S. Dedekind groups, *Pacific J. Math.* **21**, 153–9 (1967).

104. Ore, O. On the application of structure theory to groups, *Bull. Amer. Math. Soc.* **44**, 801–6 (1938).

105. Peng, T. A. Finite groups with pronormal subgroups, *Proc. Amer. Math. Soc.* **20**, 232–4 (1969).

106. Peng, T. A. A criterion for subnormality, *Arch. Math.* (Basel) **26**, 225–30 (1975).

107. Peng, T. A. A note on subnormality, *Bull. Austral. Math. Soc.* **15**, 59–64 (1976).

108. Phillips, R. E. Some generalizations of normal series in infinite groups, *J. Austral. Math. Soc.* **14**, 496–502 (1972).

109. Plotkin, B. I. On some criteria of locally nilpotent groups, *Uspehi Mat. Nauk.* **9**, 181–6 (1954) = *Amer. Math. Soc. Translations* (2) **17**, 1–7 (1961).

110. Robinson, D. J. S. Groups in which normality is a transitive relation, *Proc. Cambridge Philos. Soc.* **60**, 21–38 (1964).

111. Robinson, D. J. S. Joins of subnormal subgroups, *Illinois J. Math.* **9**, 144–68 (1965).

112. Robinson, D. J. S. On finitely generated soluble groups, *Proc. London Math. Soc.* (3) **15**, 508–16 (1965).

113. Robinson, D. J. S. On the theory of subnormal subgroups, *Math. Z.* **89**, 30–51 (1965).

114. Robinson, D. J. S. Wreath products and indices of subnormality, *Proc. London Math. Soc.* (3) **17**, 257–70 (1967).

115. Robinson, D. J. S. A note on finite groups in which normality is transitive, *Proc. Amer. Math. Soc.* **19**, 933–7 (1968).

116. Robinson, D. J. S. A property of the lower central series of a group, *Math. Z.* **107**, 225–31 (1968).

117. Robinson, D. J. S. Infinite soluble and nilpotent groups, London: Queen Mary College Mathematical Notes (1968).

118. Robinson, D. J. S. Groups which are minimal with respect to normality being intransitive, *Pacific J. Math.* **31**, 777–85 (1969).

119. Robinson, D. J. S. *Finiteness conditions and generalized soluble groups*, Part 1, New York–Heidelberg–Berlin: Springer 1972.

120. Robinson, D. J. S. *Finiteness conditions and generalized soluble groups*, Part 2, New York–Heidelberg–Berlin: Springer 1972.

121. Robinson, D. J. S. Groups whose homomorphic images have a transitive normality relation, *Trans. Amer. Math. Soc.* **176**, 181–213 (1973).

122. Robinson, D. J. S. *A course in the theory of groups*, New York–Heidelberg–Berlin: Springer 1980.

123. Roseblade, J. E. On certain subnormal coalition classes, *J. Algebra* **1**, 132–8 (1964).

124. Roseblade, J. E. The permutability of orthogonal subnormal subgroups, *Math. Z.* **90**, 365–72 (1965).

125. Roseblade, J. E. A note on subnormal coalition classes, *Math. Z.* **90**, 373–5 (1965).

126. Roseblade, J. E. On groups in which every subgroup is subnormal, *J. Algebra* **2**, 402–12 (1965).

127. Roseblade, J. E. A note on disjoint subnormal subgroups, *Bull. London Math. Soc.* **1**, 65–9 (1969).

128. Roseblade, J. E. The derived series of a join of subnormal subgroups, *Math. Z.* **117**, 57–69 (1970).

129. Roseblade, J. E. and Stonehewer, S. E. Subjunctive and locally coalescent classes of groups, *J. Algebra* **8**, 423–35 (1968).
130. Schenkman, E. *Group theory*, Princeton: Van Nostrand 1965.
131. Schmid, P. *Normalität bei Paaren subnormaler Untergruppen*, Dissertation, Tübingen (1970).
132. Schmid, P. Subnormale Einbettung von Paaren von Gruppen, *Math. Z.* **118**, 34–9 (1970).
133. Schmidt, O. J. Infinite soluble groups, *Mat. Sb.* **17**, 145–62 (1945).
134. Scott, W. R. *Group theory*, Englewood Cliffs: Prentice-Hall 1964.
135. Smith, H. Commutator subgroups of a join of subnormal subgroups, *Arch. Math.* (Basel) **41**, 193–8 (1983).
136. Smith, H. Hypercentral groups with all subgroups subnormal, *Bull. London Math. Soc.* **15**, 229–34 (1983).
137. Smith, H. Groups with the subnormal join property, *Canadian J. Math.* **37**, 1–16 (1985).
138. Stadelmann, M. Gruppen, deren Untergruppen subnormal vom Defekt zwei sind, *Arch. Math.* (Basel) **30**, 364–71 (1978).
139. Stonehewer, S. E. The join of finitely many subnormal subgroups, *Bull. London Math. Soc.* **2**, 77–82 (1970).
140. Stonehewer, S. E. Permutable subgroups of infinite groups, *Math. Z.* **125**, 1–16 (1972).
141. Stonehewer, S. E. Permutable subgroups of some finite p-groups, *J. Austral. Math. Soc.* **16**, 90–7 (1973).
142. Stonehewer, S. E. Nilpotent residuals of subnormal subgroups, *Math. Z.* **139**, 45–54 (1974).
143. Stonehewer, S. E. Permutable subgroups of some finite permutation groups, *Proc. London Math. Soc.* (3) **28**, 222–36 (1974).
144. Stonehewer, S. E. Subnormal composition factors of infinite groups, *Rend. Sem. Mat. Univ. Padova* **63**, 285–94 (1980).
145. Szép, J. Bemerkung zu einem Satz von O. Ore, *Publ. Math. Debrecen* **3**, 81–2 (1953).
146. Van Werkhooven, A. J. On the join of subnormal subgroups, *Proc. Amer. Math. Soc.* **42**, 1–7 (1974).
147. Vil'jams, N. N. Metadedekind and metahamiltonian groups, *Mat. Sb.* **76**, 634–54 (1968) = *Math. USSR-Sb.* **5**, 599–616 (1968).
148. Wehrfritz, B. A. F. Wielandt's subnormality criteria and linear groups, *J. Algebra* **67**, 491–503 (1980).
149. Wendt, E. Über eine Spezielle Klasse von Gruppen, *Math. Ann.* **55**, 479–82 (1901–1902).
150. Whitehead, J. On certain properties of subnormal subgroups, *Canadian J. Math.* **30**, 573–82 (1978).
151. Whitehead, J. Subnormality and ascendancy in soluble groups of finite rank, *Arch. Math.* (Basel) **34**, 10–14 (1980).
152. Wielandt, H. Eine Verallgemeinerung der invarianten Untergruppen, *Math. Z.* **45**, 209–44 (1939).
153. Wielandt, H. Vertauschbare nachinvariante Untergruppen, *Abh. Math. Sem. Univ. Hamburg* **21**, 55–62 (1957).
154. Wielandt, H. Über den Normalisator der subnormalen Untergruppen, *Math. Z.* **69**, 463–5 (1958).

155. Wielandt, H. Sylowtürme in subnormalen Untergruppen, *Math. Z.* **73**, 386–92 (1960).
156. Wielandt, H. Subnormale Hüllen in Permutationsgruppen, *Math. Z.* **79**, 381–8 (1962).
157. Wielandt, H. On the structure of composite groups, *Proc. Internat. Conf. Theory of Groups*, Austral. Nat. Univ. Canberra, August 1965, Gordon and Breach Science Publishers, 379–88 (1967).
158. Wielandt, H. Topics in the theory of composite groups, *Lecture Notes by Darrell J. Howarth*, Wisconsin (1967).
159. Wielandt, H. Subnormal subgroups and permutation groups, *Lecture Notes by F. Demana, W. McWorter and S. Sehgal*, Ohio (1971).
160. Wielandt, H. Subnormale Untergruppen endlicher Gruppen, *Lecture Notes by Max Selinka*, Tübingen (1971).
161. Wielandt, H. Vertauschbarkeit von Untergruppen und Subnormalität, *Math. Z.* **133**, 275–6 (1973).
162. Wielandt, H. Kriterien für Subnormalität in endlichen Gruppen, *Math. Z.* **138**, 199–203 (1974).
163. Wielandt, H. Über das Erzeugnis paarweise kosubnormaler Untergruppen, *Arch. Math.* (Basel) **35**, 1–7 (1980).
164. Wielandt, H. Zusammengesetzte Gruppen: Hölders Programm Heute, *Proc. Symp. Pure Math.* **37**, 161–73 (1980).
165. Wielandt, H. Subnormalität in faktorisierten endlichen Gruppen, *J. Algebra* **69**, 305–11 (1981).
166. Williams, J. P. The join of several subnormal subgroups, *Proc. Cambridge Philos. Soc.* **92**, 391–9 (1982).
167. Williams, J. P. Conditions for subnormality of a join of subnormal subgroups, *Proc. Cambridge Philos. Soc.* **92**, 401–17 (1982).
168. Wilson, J. S. The normal and subnormal structure of general linear groups, *Proc. Cambridge Philos. Soc.* **71**, 163–77 (1972).
169. Wilson, J. S. Polycyclic groups and topology, *Rend. Sem. Mat. Fis. Milano* **51**, 17–28 (1983).
170. Zacher, G. Caratterizzazione dei t-gruppi finiti risolubili, *Ricerche Mat.* **1**, 287–94 (1952).
171. Zassenhaus, H. *The theory of groups*, 2nd ed., New York: Chelsea 1958.

GENERALIZATIONS OF RESULTS

For ease of reference the following is an indication of how certain results (listed in column **1**) have been generalized or superseded (by those listed in column **2**).

1	2	1	2
1.3.1	3.4.1	4.1.1	4.3.1
1.3.2	3.4.1	4.1.2	4.3.1
1.3.3	3.4.1	4.2.4(i)	3.3.1
1.3.11	4.6.1	4.3.6	4.3.2
1.3.13	3.4.1	4.5.1	4.1.8
1.4.2	3.5.2	4.6.1(a)	4.6.3
1.4.6	3.5.2	6.1.2	6.3.1
1.6.2	3.4.1	6.2.1	6.2.2
1.6.4	3.1.1	6.3.2	6.3.6
1.6.6	4.3.1	6.4.14	6.4.16
1.7.1	3.4.1	7.1.2	7.1.6
1.7.2	3.4.1	7.1.3	7.1.12
1.7.4	3.4.1	7.1.4	7.1.10
1.7.7	3.1.1	7.2.1	7.2.7
1.7.13	3.4.1	7.2.3	7.2.6
2.1.1(iii)	2.2.4	7.2.5	7.5.4
2.1.1(iv)	2.2.5	7.3.3(i)	7.3.11(ii)
2.1.1(v)	2.2.8	7.3.3(ii)	7.3.6
2.1.5	3.1.1	7.3.3(iii)	7.3.11(iv)
2.3.3	2.3.7	7.3.12(ii)	7.3.14(i)
2.6.4	2.6.1	7.6.1	7.7.1
2.7.1	4.5.12		

INDEX